THE LEGO® MINDSTORMS® EV3 DISCOVERY BOOK

THE LEGO® MINDSTORMS® EV3 DISCOVERY BOOK

a beginner's guide to building and programming robots

laurens **valk**

no starch press

THE LEGO® MINDSTORMS® EV3 DISCOVERY BOOK. Copyright © 2014 by Laurens Valk.

Printed in USA
Third printing

19 18 17 16 15   3 4 5 6 7 8 9

ISBN-10: 1-59327-532-3
ISBN-13: 978-1-59327-532-7

Publisher: William Pollock
Production Editor: Serena Yang
Interior Design: Octopod Studios
Developmental Editor: Seph Kramer
Technical Reviewer: Claude Baumann
Copyeditor: Julianne Jigour
Compositors: Riley Hoffman and Alison Law
Proofreader: Paula L. Fleming
Indexer: BIM Indexing & Proofreading Services

For information on distribution, translations, or bulk sales, please contact No Starch Press, Inc. directly:
No Starch Press, Inc.
245 8th Street, San Francisco, CA 94103
phone: 415.863.9900; info@nostarch.com;
www.nostarch.com

The Library of Congress has cataloged the first edition as follows:
Valk, Laurens.
  The LEGO Mindstorms NXT 2.0 discovery book : a beginner's guide to building and programming robots / Laurens Valk.
     p. cm.
  Includes index.
  ISBN-13: 978-1-59327-211-1
  ISBN-10: 1-59327-211-1
  1. Robots--Design and construction--Popular works. 2. Robots--Programming--Popular works. 3. LEGO toys. I. Title.
  TJ211.15.V353 2010
  629.8'92--dc22
                              2010011157

about the author

Laurens Valk is a robotics engineer based in the Netherlands, where he earned a bachelor's degree in Mechanical Engineering from Delft University of Technology. He is a member of the MINDSTORMS Community Partners (MCP), a select group of MINDSTORMS enthusiasts who help test and develop new MINDSTORMS products. He started building robots with the EV3 system a year before its 2013 release, and one of his designs appears on the EV3 packaging as an official bonus robot.

Laurens enjoys designing robots and creating tutorials to build and program them, so that robot fans around the world can re-create the designs and learn more about robotics. He has worked on several LEGO robotics books, including the best-selling first edition of this book, *The LEGO MINDSTORMS NXT 2.0 Discovery Book* (No Starch Press, 2010). He blogs about robots at *http://robotsquare.com/*.

about the
technical reviewer

Claude Baumann has taught advanced LEGO MINDSTORMS robotics in after-school classes for 15 years. He created ULTIMATE ROBOLAB, a cross-compiler environment that allowed graphical programming of LEGO RCX firmware, and with it conceived the world's only self-replicating program for the LEGO RCX (some call it a virus). More recently, he participated as a MINDSTORMS Community Partner (MCP) during the development of the new EV3 Intelligent Brick. He has been the assessor of various high school robotics projects and is the author of *Eureka! Problem Solving with LEGO Robotics* (NTS Press, 2013), several articles, and conference presentations. His special interest is robotic sound localization. The head of a network of high-school boarding institutions in Luxembourg (EU), Claude is married with three children and three marvelous grandchildren.

acknowledgments

First and foremost, I'd like to thank the readers of the first edition. Your countless emails and comments from all over the world have been a true inspiration for writing this book, and many new topics were inspired by your feedback.

The philosophy and structure of this book is the same as that of the first edition, but because of the transition from LEGO MINDSTORMS NXT to EV3, it was essentially rewritten from scratch. This was only possible thanks to the help of many talented people.

Many thanks go to Claude Baumann for reviewing the book for technical accuracy and for suggesting improvements. Thanks also to Marc-André Bazergui, Martijn Boogaarts, Kenneth Madsen, and Xander Soldaat for testing prototypes of the robots featured in the book as early as 2012.

Further thanks to the people at No Starch Press for making the first edition a success and for working with me on this new edition. Thanks to my publisher William Pollock, to my editor Seph Kramer, to Serena Yang for keeping the project on schedule, to Riley Hoffman and Alison Law for laying out the raw text on colorful pages, and to Leigh Poehler for dealing with all my business-related questions over the past years.

Thanks to the LEGO Group for developing such an inspirational and educational robotics kit, and for involving the community early in the design process. Thanks to the LEGO MINDSTORMS EV3 team, including Camilla, David, Flemming, Henrik, Lars Joe, Lasse, Lee, Linda, Marie, Steven, and Willem.

Thanks to the LDraw community for developing the tools required to create the building instructions in this book. In particular, thanks to Philippe Hurbain for creating 3D LDraw models of the EV3 components, to Michael Lachmann for creating MLCad, to Travis Cobbs for creating LDView, and to Kevin Clague for developing LPub 4 and LSynth. Also, thanks to John Hansen for creating the EV3 screen capture tool.

Finally, thanks to my friends and family for your support throughout the lengthy process of writing this book. Most of all, thanks to Fabiënne for your endless encouragement to finish this project. Thank you—you're the best.

brief contents

contents in detail

PART I   GETTING STARTED

PART II   PROGRAMMING ROBOTS WITH SENSORS

6
understanding sensors... 61

7
using the color sensor 75

8
using the infrared sensor... 89

PART III   ROBOT-BUILDING TECHNIQUES

11
building with gears

PART IV   VEHICLE AND ANIMAL ROBOTS

PART V   CREATING ADVANCED PROGRAMS

Are you ready to discover the captivating world of robotics? As you're reading this book, I assume that you've selected the LEGO MINDSTORMS EV3 robotics set as your learning tool, and I think that's a great choice.

I first became involved with MINDSTORMS in 2005 when I was 13 years old, using the Robotics Invention System, the version available at the time. It started out as a hobby, but I found robots so fascinating that I decided to pursue an engineering degree. LEGO MINDSTORMS proved to be an excellent resource to get familiar with many robotics and engineering concepts, such as programming and working with motors and sensors.

The purpose of this book is to help you explore the many possibilities of MINDSTORMS in hopes that you'll have just as much fun with this robotics set as I have and that you'll learn a lot along the way!

why this book?

The LEGO MINDSTORMS EV3 robotics set includes numerous parts and instructions for five robots. I think you'll find that it's a lot of fun to build and program these robots, but exploring beyond these models can be a bit overwhelming when you're just getting started. The set provides you with the tools you need to make the robots work, but the user guide covers only a fraction of what you need to know in order to build and program your own robots.

This book is designed as a guidebook to help you discover the power of LEGO MINDSTORMS EV3 as you learn to invent, build, and program your very own robots.

is this book for you?

This book assumes no previous experience with either building or programming LEGO MINDSTORMS. As you read, you'll move from basic to advanced programming and learn to build increasingly sophisticated robots. New users should begin with Chapter 1 and then follow the step-by-step instructions in Chapter 2 to build and program a basic robot. More experienced MINDSTORMS users might simply start with a chapter they find challenging and move on from there. The advanced

programming chapters in Part V and the robot designs in Part VI will be especially interesting for more advanced readers.

how does this book work?

Although you could use it as such, this book isn't intended as a reference manual; it's more like a workbook. I've mixed together building, programming, and robotic challenges to avoid long, theoretical chapters that can be hard to wade through.

For example, you'll learn basic programming techniques at the same time that you learn to make your first robot move, and you'll learn about more advanced programming as you build new robots. This book takes a "learning by doing" approach, which I think is the best way to learn how to build and program MINDSTORMS robots.

the discoveries

To help you really understand the concepts presented in each chapter, I've included many so-called *Discoveries*, or challenges, throughout the book. The Discoveries will challenge you to expand the example programs or even come up with completely new programs. For example, once you've learned how to make the robot play sounds and display text on the screen, you'll be challenged to make a program that has the robot show subtitles on the screen while it speaks.

You'll also find *Design Discoveries* at the end of many chapters. These will give you ideas to modify or improve the robot you built in that chapter. For example, you'll be challenged to make a racing robot drive faster by adding gears between the motors and the wheels, and you'll even be challenged to design a new robot that turns your EV3 into an intruder alarm!

estimating difficulty and time

To help you choose which Discoveries you might want to tackle, I've provided an estimated difficulty level for each one. Easy Discoveries (⬚) can typically be solved by creating or expanding a program using techniques similar to the examples. Medium (⬚⬚) indicates that you'll have to look a bit further and perhaps combine the new theory with some techniques you learned earlier. Hard Discoveries (⬚⬚⬚) will put your creativity to the test by going beyond the presented examples.

In rating the difficulty, I'm assuming that you read the chapters in chronological order. That is, a challenge marked "hard" in Chapter 4 may actually be quite easy compared to a challenge marked "hard" in Chapter 19.

In addition, each Discovery provides an estimate for how much time is required to solve it, ranked as short (🕐), medium (🕐🕐), or long (🕐🕐🕐) challenges. Typically, short Discoveries can be solved by making just a few changes to the example program, while long Discoveries involve creating a completely new program.

The Design Discoveries generally take more time to solve because they involve both building and programming. They're rated by the estimated amount of required building (✳) and programming (▭).

finding solutions

Some of the Discoveries give you one or two hints to get started, but there are many ways to solve each challenge. It doesn't matter whether you follow the guidelines exactly; you might just find an innovative solution I didn't think of.

Similarly, the difficulty and time given for each Discovery are only estimates. Don't worry if you need a little longer to solve a particular problem. Just remember to have fun when giving the challenges a try!

You can find solutions to some of the Discoveries at *http://ev3.robotsquare.com/*. These solutions will get you started, but you'll need to be creative in order to solve the challenges that don't have a solution available for download.

what to expect in each chapter

Here's a brief overview of each of the six parts of the book. Some of the terms used here may be new to you, but you'll learn all about them as you read the book.

part I: getting started

Part I begins by taking you through the contents of the EV3 robotics set in Chapter 1. In Chapter 2, you'll build your first robot and learn about the EV3 brick. In Chapter 3, you'll meet the EV3 software, which you'll use to program robots. In Chapter 4, you'll learn to use this software to make your robot move as you create your first programs with basic programming blocks. Finally, in Chapter 5, you'll learn essential programming techniques, such as making your robot repeat actions and do more than one thing at the same time.

part II: programming robots with sensors

This part teaches you all about sensors, which are essential components of MINDSTORMS robots. In Chapter 6, you'll begin by adding the Touch Sensor to the robot you built earlier to learn the programming techniques required to use sensors.

Then, you'll explore the Color Sensor in Chapter 7, the Infrared Sensor and the infrared beacon in Chapter 8, and two types of built-in sensors in Chapter 9.

part III: robot building techniques

This part covers the LEGO Technic building elements that come with your EV3 set. You'll learn to use beams, axles, and connector blocks in Chapter 10, and you'll learn how to use gears in your robots in Chapter 11.

part IV: vehicle and animal robots

Once you've got a handle on using motors and sensors, you'll build two robots to put those skills to the test. You'll build the Formula EV3 Race Car in Chapter 12 and ANTY, a robotic ant, in Chapter 13.

part V: creating advanced programs

Part V is devoted to more advanced programming concepts. You'll learn about data wires (Chapter 14), how to process sensor values and do math on the EV3 (Chapter 15), and how to make the robot remember things with variables (Chapter 16). Finally, Chapter 17 will teach you how to combine all of these programming techniques to create a robot that lets you play an Etch-A-Sketch-like game on the EV3 screen.

part VI: machine and humanoid robots

Having learned about motors, sensors, and many advanced programming techniques, you'll create two complex robots in this part. In Chapter 18, you'll build and program the SNATCH3R, an autonomous robotic arm that can find, grab, lift, and move the infrared beacon autonomously.

Finally, in Chapter 19, you'll build LAVA R3X, the humanoid shown on the front cover that walks and talks. Its design was inspired by the famous Alpha Rex humanoid robot from the previous generation of LEGO MINDSTORMS.

the companion website

On the companion website (*http://ev3.robotsquare.com/*), you'll find links to other helpful websites, downloadable versions of all of the example programs in this book, and solutions to some of the Discoveries presented in this book.

conclusion

MINDSTORMS can inspire creativity and ingenuity in builders of all ages. So grab your EV3 robotics set, start reading Chapter 1, and enter the creative world of LEGO MINDSTORMS. I hope this book will help spark your imagination!

PART I

getting started

preparing your EV3 set

All of the robots in this book can be built with one *LEGO MINDSTORMS EV3* set (LEGO catalog #31313). If you have this set, shown in Figure 1-1, you're good to go. If you have the LEGO MINDSTORMS EV3 Education Core set (#45544), visit *http://ev3.robotsquare.com/* for a list of pieces you'll need to complete the projects in this book.

In this chapter, you'll learn about the EV3 brick and the other elements in the EV3 set. You'll also download and install the software that you need to program your robots.

what's in the box?

The LEGO MINDSTORMS EV3 set comes with a lot of Technic building pieces as well as electronic components, including motors, sensors, the EV3 brick, a remote control, and cables (see Figure 1-2). You'll learn to use each of these components as you read this book. In addition, the inside back cover contains a complete parts list.

EV3 robots use large- or medium-sized *motors* to power their wheels, arms, or other moving parts. They use *sensors* to take input from their surroundings, such as the color of a surface or the approximate distance to an object. *Cables* connect the motors and sensors to the EV3 brick. The *infrared remote control*—or simply, the *remote*—can be used to move and steer a robot remotely.

the EV3 brick

The *EV3 brick*, or simply the EV3, is a small computer that controls a robot's motors and sensors, enabling the robot to move around by itself. For example, you'll soon build a robot that automatically moves away from an object in its path. When a sensor tells the EV3 that an object is nearby, the EV3 triggers the motors to drive the robot away.

Figure 1-1: The LEGO MINDSTORMS EV3 set (#31313) contains all of the elements necessary to build the robots in this book.

Your robot is able to perform these actions due to the use of a *program*—a list of actions that the robot will perform, usually one at a time. You'll create programs on a computer with the LEGO MINDSTORMS EV3 programming software. Once you've finished creating a program, you'll send it to your EV3 brick with the USB cable that comes with the set, and your robot should be ready to do what it's now programmed to do.

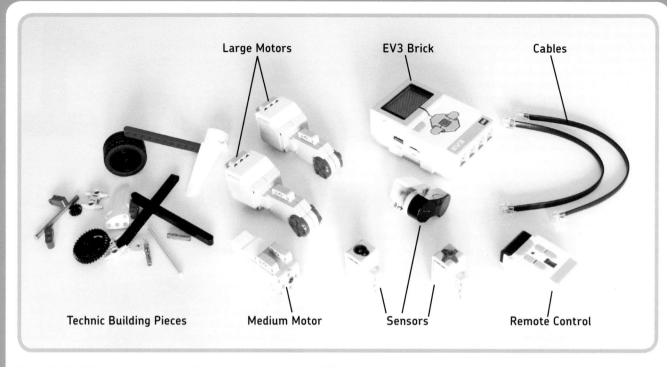

Figure 1-2: The EV3 set contains Technic building pieces, motors, sensors, the EV3 brick, a remote control, and cables.

To power your *EV3*, either insert six AA batteries (as shown in Figure 1-3) or use the LEGO EV3 rechargeable battery (#45501) and charger (#8887). The shape of the rechargeable battery pack makes your EV3 a bit bigger. While the robots in this book will work with this battery, the rechargeable battery won't fit in the TRACK3R robot featured on the LEGO MINDSTORMS EV3 packaging, unless you modify it slightly to make space.

To power the infrared remote control, insert two AAA batteries.

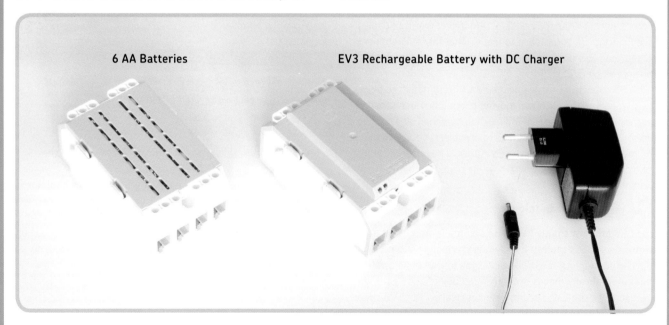

Figure 1-3: You can power the EV3 brick with six AA batteries or with the EV3 rechargeable battery.

sorting Technic elements

To save time when searching for LEGO Technic elements, consider sorting and storing your elements in an organizer like the one shown in Figure 1-4. Doing so will make it easier to build the models in this book and, later on, to design your own robots. You'll be able to see at a glance when a specific element runs out, so you won't have to waste time looking for pieces you don't have.

When sorting elements, it's best to do so by function. For example, separate beams from gears, axles, and so on. If you don't have enough bins for each type of element, store easily distinguishable items together. For example, store short grey axles with short red axles instead of mixing grey axles of various lengths.

The EV3 set comes with a set of stickers, one for each white "panel" element. Add all the stickers to the panels now, as shown in Figure 1-5; the sticker patterns will help you determine which sort of panel (large or small) to use later in this book.

the mission pad

The EV3 packaging includes a *mission pad* on the inside of the sleeve around the box, as shown in Figure 1-6. You can program your robots to interact with this pad by, for example,

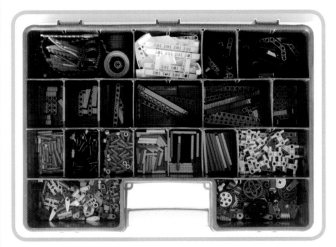

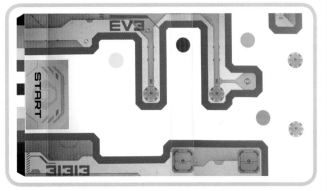

Figure 1-4: Example of an organizer containing the Technic elements in one EV3 set. The dimensions of this container are 45 cm x 33 cm x 9 cm (18 inches x 13 inches x 3.5 inches).

Figure 1-6: The mission pad. You'll find the pad by carefully cutting the sleeve around the EV3 packaging (look for a dashed line and a symbol of a pair of scissors).

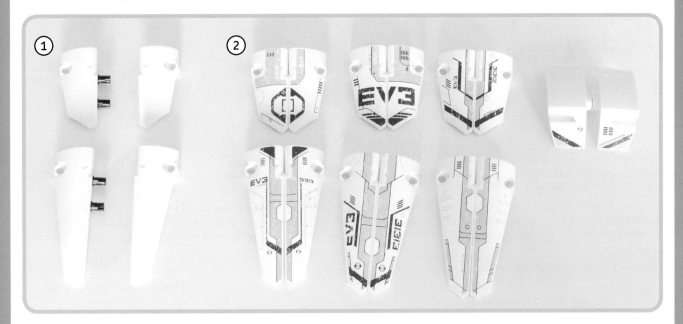

Figure 1-5: To add the stickers to the panels neatly, first connect complementary panel elements using two black pins, making it easier to align the stickers, and then remove the pins when you're done.

following the thick red line (see Chapter 7). For the activities in this book, you can go to *http://ev3.robotsquare.com/* to download and print your own track for the robot to follow.

controlling your robot

The EV3 set allows you to control your robot in many different ways, as shown in Figure 1-7. In this book, you'll learn how to write programs that make your robot do certain things automatically using the EV3 programming software, but you'll also learn to control your robots with a remote. You can steer and send commands to your robots with the infrared remote that comes with the set or download apps that let you use a smartphone or tablet as a remote control. These apps will allow you to control the motors and sensors on your robot and even create a customized remote. (See *http://ev3.robotsquare.com/* for an updated list of apps.)

downloading and installing the EV3 programming software

Before you can start writing programs for your robots, you need to download and install the EV3 programming software. An Internet connection is required in order to complete the following steps.

(If the computer you'll use for programming is not connected to the Internet, follow steps 1 and 2 on a computer with Internet access and download the installation file to a USB stick that is at least 1GB in size. Next, copy the file from the stick to your other computer and proceed with step 3.)

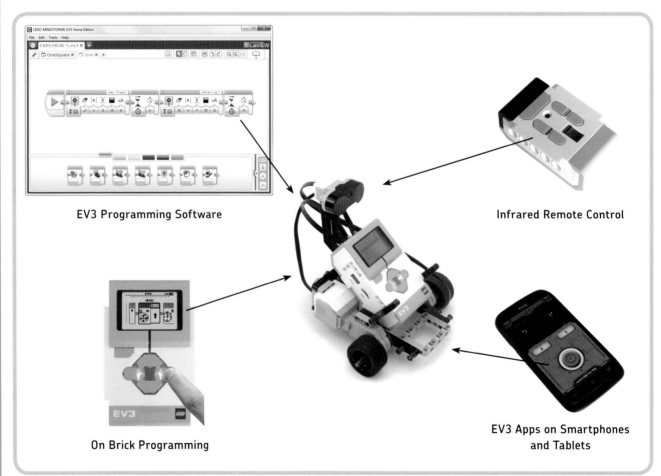

EV3 Programming Software

Infrared Remote Control

On Brick Programming

EV3 Apps on Smartphones and Tablets

Figure 1-7: You can create programs to make your robot move around automatically, or you can control it with a remote.

1. Go to *http://LEGO.com/ MINDSTORMS/*, Select your preferred language, click **Downloads**, and click **Download Software (PC/ MAC)**, as shown in Figure 1-8. Note that you can select any language you like, but this book will use the US English version.

2. On the next page, select your operating system (see Figure 1-9). For Windows XP, Windows Vista, Windows 7, Windows 8, and Windows 10, click **Download for Windows**, and save the installation file to your computer when asked. For Mac OS 10.6 and higher, choose **Download for OS X**.

NOTE If your download is taking a long time (the file is about 600MB), you can skip to Chapter 2 and start building! Just return here when the download is complete.

3. On Windows, double-click the file you've just downloaded and install the software according to the instructions shown on the screen (see Figure 1-10).
On a Mac, double-click the *.dmg* file you've downloaded and then double-click the package that appears. Follow the onscreen instructions to install the software.

4. Once the installation is complete (and you've restarted your computer when prompted), you should find a shortcut labeled *LEGO MINDSTORMS EV3 Home Edition* on your desktop. Double-click it to launch the software (no Internet access is required from now on).

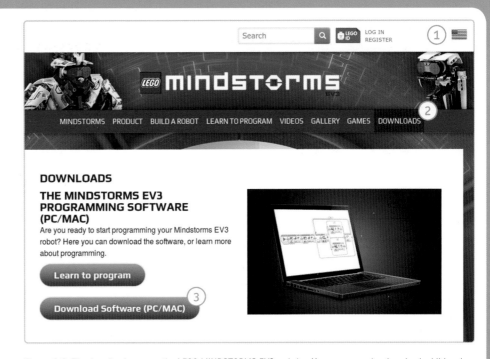

Figure 1-8: The download page on the LEGO MINDSTORMS EV3 website. Here, you can also download additional content, such as a user guide or additional programming blocks.

Figure 1-9: Select your operating system. When you download the software, choose the newest version available.

Figure 1-10: Installation in progress. Launch the installer by double-clicking the installation file that you've downloaded.

NOTE To update the software to a newer version, simply download the latest software and install it using the same steps. You do *not* need to uninstall the current version first.

conclusion

Now that you have everything you need to build and program a working robot, it's time to start building one. In Chapter 2, you'll learn more about the EV3 brick, motors, and the remote control as you build your first robot.

2

building your first robot

In Chapter 1, you learned that EV3 robots consist of motors, sensors, and the EV3 brick. To make it easy for you to understand how each of these work, you'll begin with only some of them.

In this chapter, you'll use the EV3 brick and two Large Motors to build a wheeled vehicle called the *EXPLOR3R*, as shown in Figure 2-1. You'll also add a receiver for the remote control. After you finish building the robot, you'll learn how to navigate the EV3 by using its buttons and how to control your robot remotely.

using the building instructions

The LEGO MINDSTORMS EV3 set contains many beams and axles, which come in a variety of lengths. To help you find the correct element, their lengths are indicated in the building instruction steps, as shown in Figure 2-2.

To find the length of a beam, simply count the number of holes (in the figure, the beam's length is denoted by the box with the "11"). To find the length of an axle, put it next to a beam and count the number of holes it covers (in the figure, the axle's length is indicated by the circle with "3").

When connecting beams or other elements using pins, be sure to select the correct pin based on its color, as shown in Figure 2-3. This is important because nonfriction pins spin freely (and are useful for smooth hinges) while friction pins resist rotation (and are more useful for rigid constructions).

Figure 2-1: The EXPLOR3R moves around on two front wheels and a support wheel in the back.

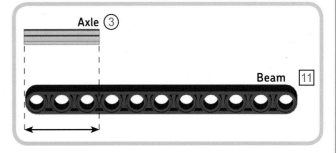

Figure 2-2: Beams and axles come in different lengths, so be sure to pick the correct ones while building. Determine the length as shown here or use the diagram on the front inside cover.

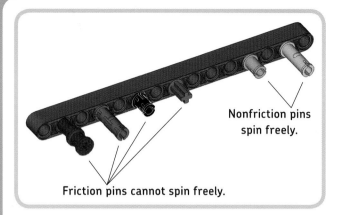

Nonfriction pins
spin freely.

Friction pins cannot spin freely.

Figure 2-3: The EV3 set contains friction pins and nonfriction pins. While
building with the instructions in this book, select the correct one based on
its color.

building the
EXPLOR3R

To begin building, select the pieces you'll need by referencing
the bill of materials, shown in Figure 2-4. Next, assemble the
robot as shown in the steps on the pages that follow.

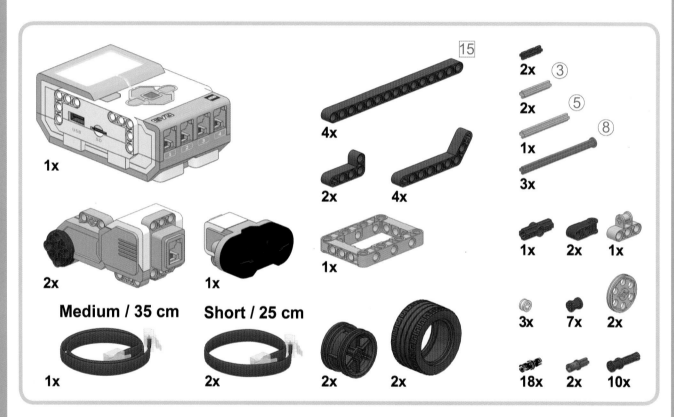

Figure 2-4: Bill of materials for the EXPLOR3R

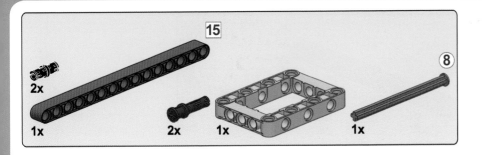

1

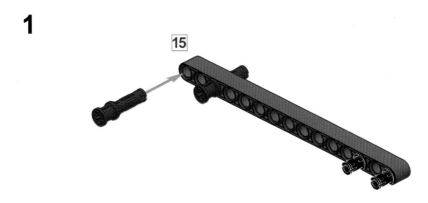

2

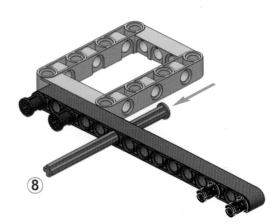

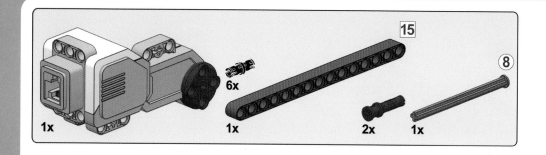

3

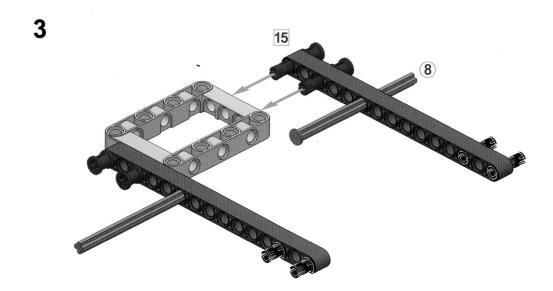

4

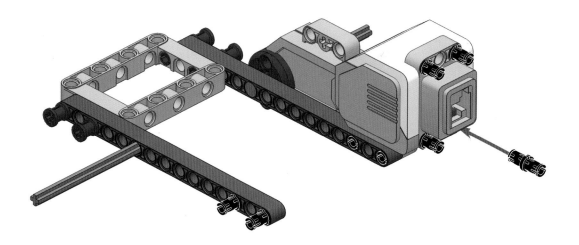

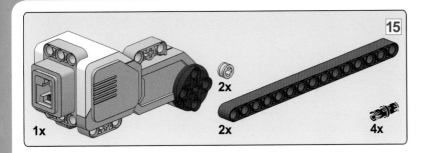

5

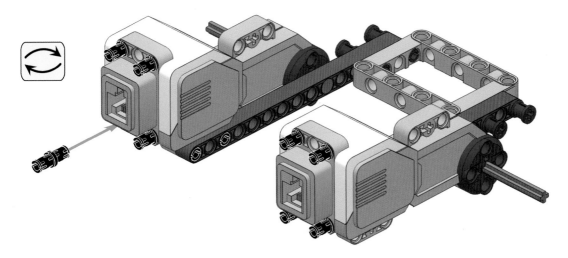

6

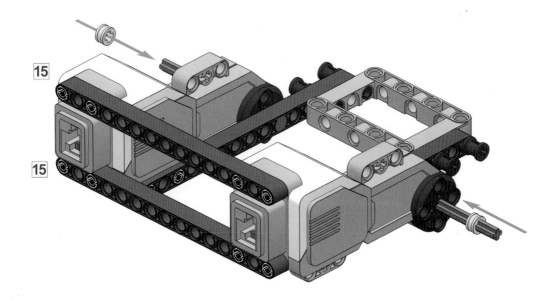

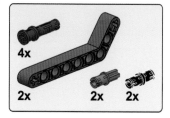

7

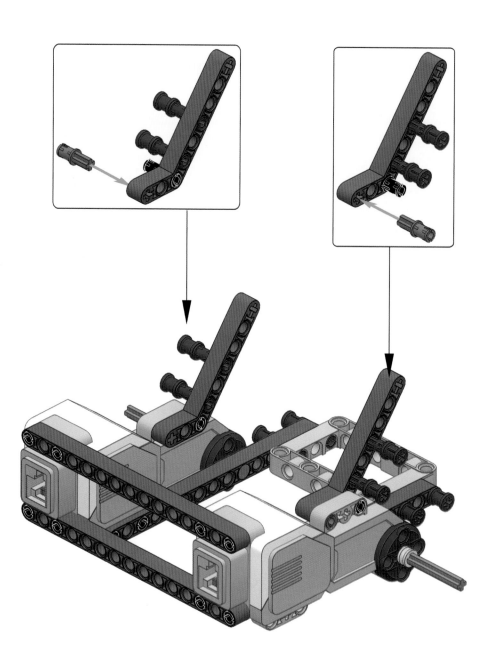

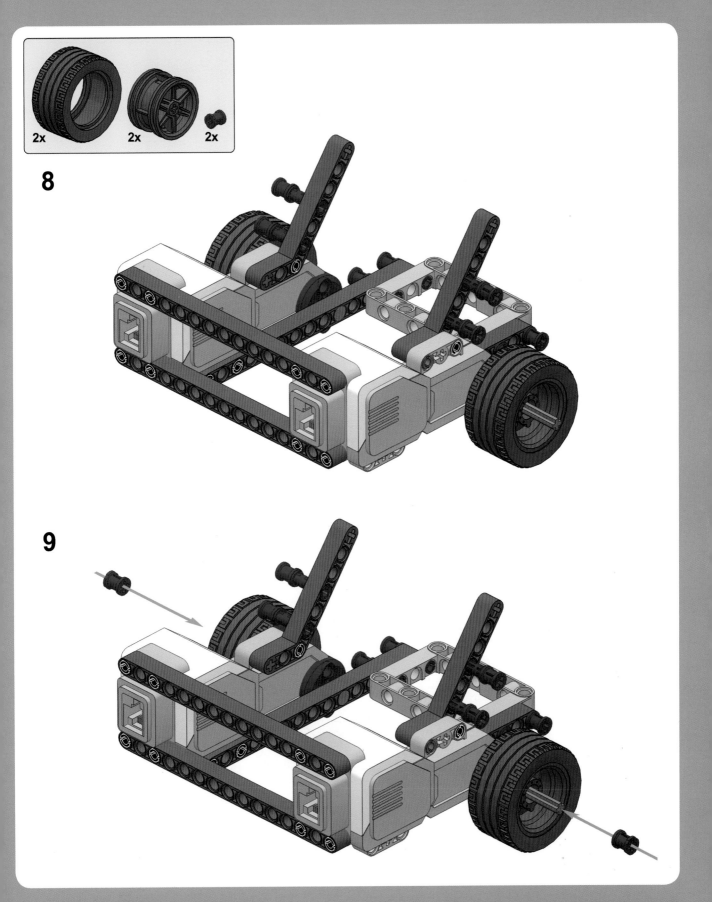

8

9

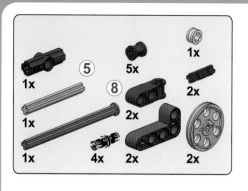

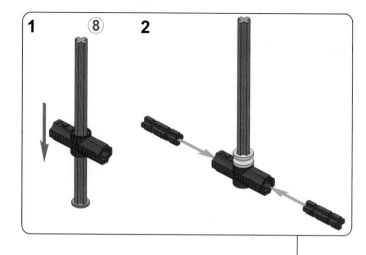

1

2

3

4

5

6

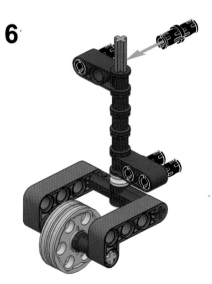

10

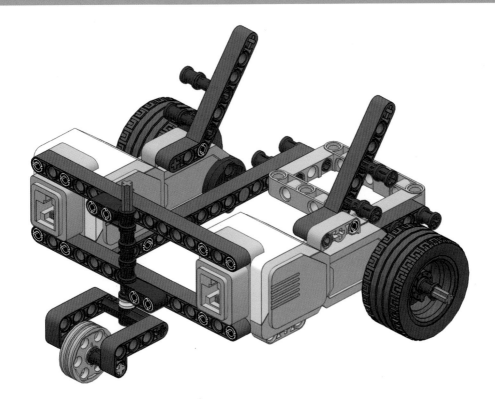

11

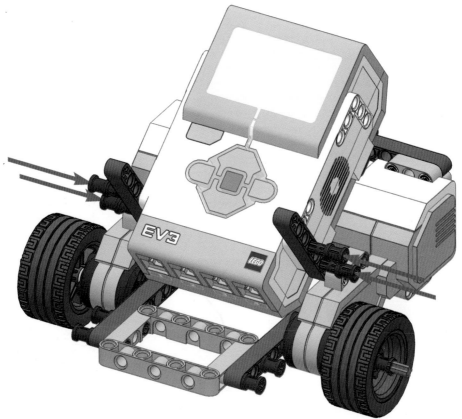

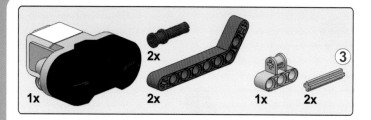

12

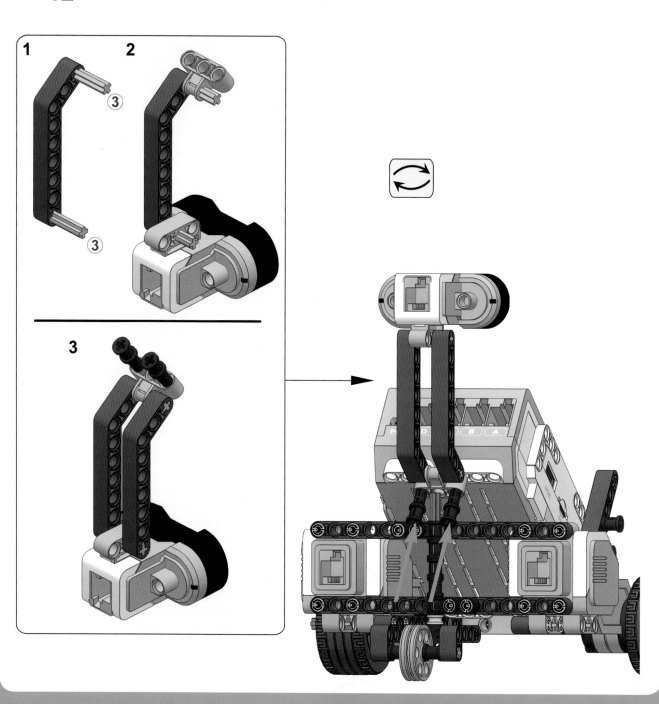

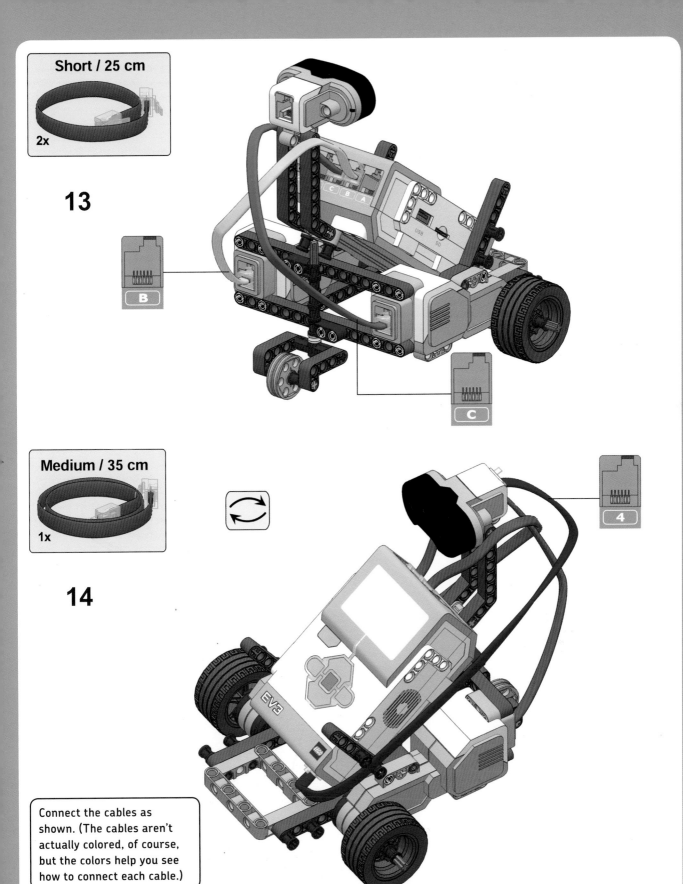

Short / 25 cm

2x

13

B

C

Medium / 35 cm

1x

14

4

Connect the cables as shown. (The cables aren't actually colored, of course, but the colors help you see how to connect each cable.)

output ports, input ports, and cables

Congratulations—you've finished building the EXPLOR3R!

Now let's talk about the wires you just connected to the EV3 brick. You connected the two Large Motors to *output ports* on the EV3, labeled B and C. Both Large and Medium Motors should always be connected to output ports—to port A, B, C, or D—as shown in Figure 2-5. Sensors should be connected to the *input ports* 1, 2, 3, or 4. (I'll discuss sensors in detail in Part II of this book.)

Your EV3 set has three types of cables: four short cables (25 cm, or 10 inches), two medium-sized ones (35 cm, or 14 inches), and one long cable (50 cm, or 20 inches). Always string the cables around your robot so as not to interfere with any rotating elements (like wheels), and make sure they don't drag on the floor when the robot moves around.

The EV3 brick has two USB connections. The one labeled *PC* at the top of the EV3 (see Figure 2-5) is used to transfer programs from your computer to your robot. The USB host on the side of the EV3 is used to connect external devices, such as a Wi-Fi Dongle, to the EV3. A microSD card slot next to the USB host port allows you to add to the 4MB of free space on the EV3's built-in storage. (The built-in storage will be enough for everything we do in this book.)

navigating the EV3 brick

Before you move on to programming in Chapter 3, let's try using the buttons on the EV3 brick (shown in Figure 2-6) to navigate around the menus and to run stored programs.

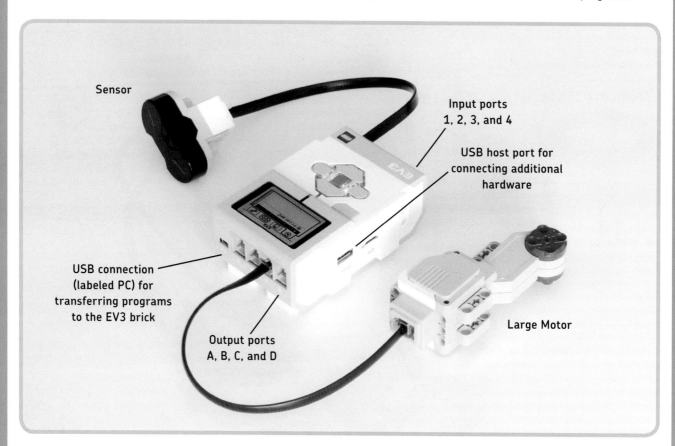

Sensor

Input ports 1, 2, 3, and 4

USB host port for connecting additional hardware

USB connection (labeled PC) for transferring programs to the EV3 brick

Output ports A, B, C, and D

Large Motor

Figure 2-5: You connect motors to output ports and sensors to input ports. The USB connection labeled PC *is used to transfer programs to the EV3.*

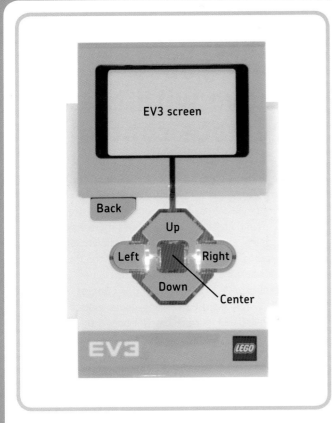

Figure 2-6: The EV3 screen, the EV3 buttons, and the brick status light around the buttons

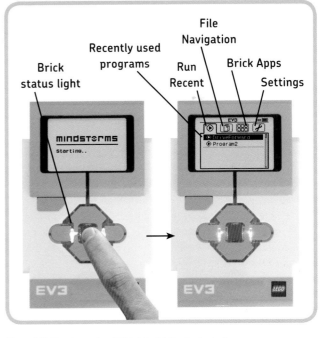

Figure 2-7: Turning on the EV3 brick with the Center button opens up a menu with four tabs. The Run Recent tab shown on the right contains recently used programs.

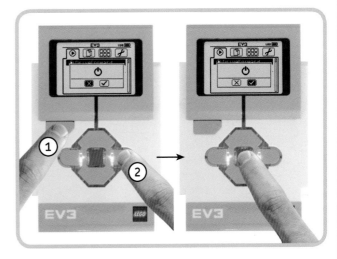

Figure 2-8: Turning off the EV3 brick

turning the EV3 on and off

To turn on the EV3, press the **Center** button, as shown in Figure 2-7. The *brick status light* should turn red while the EV3 starts up. Once startup is complete (in about 30 seconds), the status light should turn green and you should see a menu with four tabs on the screen. You'll learn to use each of these menus in later chapters. Each tab contains a specific set of files or functions, listed here from left to right:

Run Recent: This tab contains recently run programs.
File Navigation: This tab contains a folder for each programming project you've transferred to the EV3. Inside each folder, you'll find programs and related files, such as sounds.
Brick Apps: This tab contains applications for viewing sensors and controlling motors manually or remotely.
Settings: This tab lets you set user preferences, like Bluetooth visibility and sound volume.

To turn off the EV3, return to the Run Recent menu and press the **Back** button. When you see the power off icon, either select the check mark to turn the EV3 off or select the X to cancel (see Figure 2-8). If you're turning off the EV3 in order to replace the batteries, be sure to wait until the red status light is off, or you'll lose all programs that you sent to the EV3 since you turned it on.

selecting and running programs

You can switch to any of the four tabs with the **Left** and **Right** buttons. Pressing the **Back** button takes you back to the Run Recent tab. You can select any item on a tab using the **Up** and **Down** buttons. To choose a selected item, press the **Center** button.

EV3 robots begin performing their actions when you select and run a program that has been transferred to the EV3. Although you haven't transferred a program to the EV3 yet, you can try running a sample program that is already on the EV3 brick, called *Demo*. To test your EXPLOR3R, run this program by navigating to the File Navigation tab and selecting the *Demo* program, as shown in Figure 2-9.

If you've built everything properly, your robot should make some sounds, move forward, turn left twice, and display a pair of eyes on the screen. The green status light should blink while the program runs. To abort the running program, press the **Back** button. (Now that you've run the program once, it should appear on the Run Recent tab.)

NOTE The *Demo* program is made using On Brick Programming, but you'll run programs that you make on your computer in just the same way.

making your robot move with the remote control

Once you've finished building a robot, it's important to test its mechanical functions before you start programming in order to identify problems like missing cables or gears.

You can manually control your robot's motors with the Motor Control and IR Control apps, as shown in Figure 2-10. Motor Control lets you run each of the motors using the EV3 buttons. IR (Infrared) Control allows you to control the robot with a remote. Select **IR Control** on the Brick Apps tab, and use the infrared remote control to make your robot move (see Figure 2-10). Not only is this a simple way to test your robot remotely, but it's also a lot of fun!

NOTE The Infrared Sensor acts as the receiver for the infrared remote. You can't use the remote without the sensor. The IR Control app requires the sensor to be connected to input port 4, which is how the EXPLOR3R is set up.

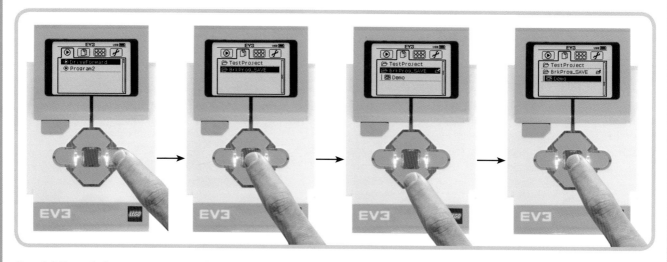

Figure 2-9: To run the Demo *program, go to the File Navigation tab, select the* BrkProg_Save *folder, open it with the Center button, select the* Demo *program with the Down button, and press the Center button. You'll find your own programs on the File Navigation tab as well. (In the figure, you can see a project I made called* TestProject.*)*

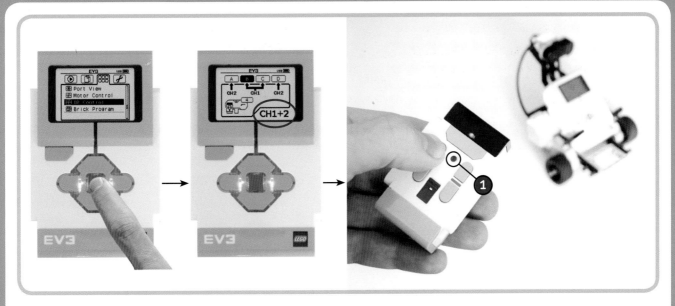

Figure 2-10: To activate Remote Control mode, navigate to the Brick Apps tab and select **IR Control**. If the screen doesn't show Ch1 + 2 in the lower-right corner, press the Center button again. In this configuration, you control the motors connected to ports B and C with the remote set to channel 1. (The red slider is all the way at the top.)

conclusion

In this chapter, you learned to work with two essential robot components: the EV3 brick and motors. When you ran the *Demo* program, the EV3 switched on the motors, which made the robot move. In Chapters 3 and 4, you'll learn how these programs work, as well as how to make your own programs with the EV3 software. The Infrared Sensor and infrared remote control will return in Part II of this book.

3

creating and modifying programs

After you build your robot, the next thing the robot needs is a program. For example, a program can make the EXPLOR3R drive forward and then steer left or right.

In this chapter, you'll learn how to create and edit programs using the EV3 software. While you can create programs for your robots without a computer using *On Brick Programming*, such programs are limited and won't let you access many of the features of the EV3 brick. (You'll find an introduction to On Brick Programming in Appendix B.)

a quick first program

First, you'll create and download a small program to your robot. To create the program, take the following steps:

1. Connect the robot to the computer using the USB cable that came with your set (see Figure 3-1) and make sure the EV3 brick is turned on. (You'll have to connect the robot to the computer each time you want to download a program to it.)

2. Launch the EV3 software by double-clicking the **LEGO MINDSTORMS EV3 Home Edition** shortcut on your desktop. Once the software loads, you should see the *lobby*, where you'll create new programs and open existing ones.

3. Open a new programming project by clicking the **+** symbol, as shown in Figure 3-2.

NOTE If you see a pop-up that says *Please update the programmable brick's firmware version*, follow the steps in "Updating the EV3 Firmware" on page 355.

Figure 3-1: The robot connected to the computer using the set's USB cable. Use the USB connection at the top of the EV3, as shown here.

4. Choose a Move Steering block and place it as shown in Figure 3-3. Remember that a program is basically a list of instructions for actions that the robot should perform. This block is an instruction that makes the robot move forward.

5. Now click the **Download and Run** button (see Figure 3-4). Your computer should download this simple program to your robot, and your robot should start moving forward. To download and run this program again, simply click this button again.

If your robot moves forward a short distance, you've created your first program. Congratulations!

NOTE If you can't download your program to the EV3 and the EV3 symbol on your computer screen is greyed out (EV3) rather than red (EV3), something may be wrong with the USB connection. Try removing the USB cable and plugging it back in again. If that doesn't help, turn your EV3 off and on again. You can find more help in Appendix A.

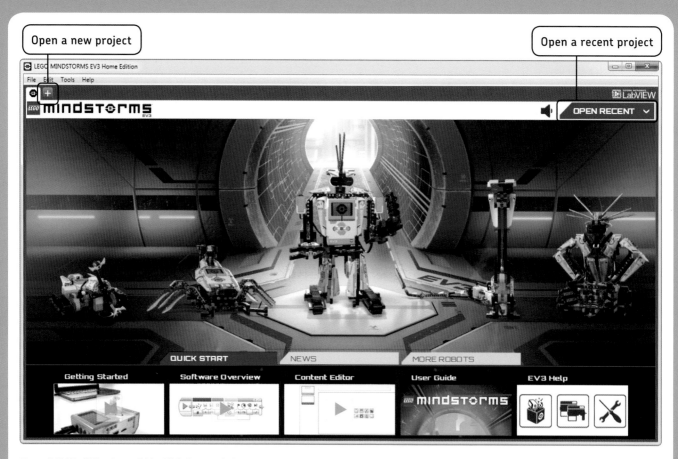

Open a new project

Open a recent project

Figure 3-2: The EV3 software lobby. Click the + symbol to start a new programming project.

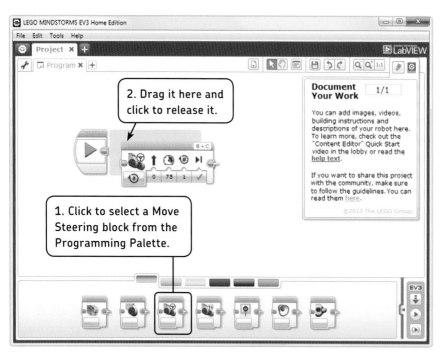

2. Drag it here and click to release it.

1. Click to select a Move Steering block from the Programming Palette.

Document Your Work   1/1

You can add images, videos, building instructions and descriptions of your robot here. To learn more, check out the "Content Editor" Quick Start video in the lobby or read the help text.

If you want to share this project with the community, make sure to follow the guidelines. You can read them here.

©2013 The LEGO Group.

Download and Run

Figure 3-4: Downloading a program to the robot and running it. The letters EV3 in red indicate that the robot is successfully connected to your computer.

Figure 3-3: Placing a block in a program. When you click to drop a block, it should snap to the orange Start block that's always there when you create a new program.

creating a basic program

Okay, your robot moved, but how did you do that? In the following sections, I'll explore various parts of the EV3 software to help you better understand how to create and edit basic programs before you move on to create more complex ones.

Your screen should look like Figure 3-5 after running your program. I'll discuss each of the marked sections in turn.

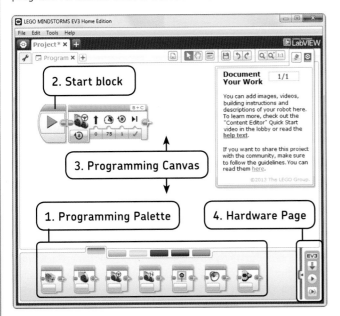

Figure 3-5: The EV3 software window has several sections. You used each of the labeled ones when you created your first program.

1. programming palette

Your EV3 programs consist of *programming blocks*. Each block instructs the robot to do something different, such as move forward or make a sound. You find blocks on the *Programming Palette* (see Figure 3-6).

There are several categories of blocks, each of which is found behind one of the colored tabs. You'll learn to use Action blocks (green), Flow blocks (orange), and My Blocks (light blue) in Chapters 4 and 5. You'll learn to use Flow blocks to control sensors in Part II of this book.

Sensor blocks (yellow) and Data Operations blocks (red) are discussed in Part V, and you'll meet some of the Advanced blocks (dark blue) throughout this book.

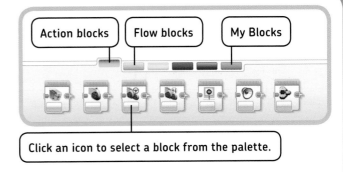

Figure 3-6: The Programming Palette

2. start block

Your programs always begin with the Start block. When you pick up the first block from the Programming Palette, you attach it to the Start block, as shown in Figure 3-7. The robot runs the blocks in your program one by one, from left to right, beginning with the block that's attached to the Start block.

If you accidentally delete the Start block, get a new copy from the orange tab of the Programming Palette.

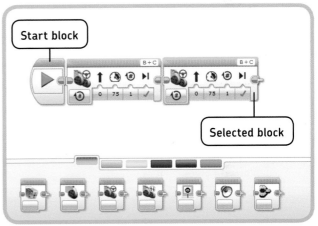

Figure 3-7: Once you've picked a block from the Programming Palette, place it on the Programming Canvas. If this is the first block in a program, place it right after the Start block.

3. programming canvas

You'll create your programs on the *Programming Canvas*. Once you've placed a block, you can move it with the left mouse button. (Click the block with the left mouse button, hold the button down, and drag the block.) If you drag a selection around multiple blocks, you can move all of the selected blocks with your mouse at once. To delete a block from the canvas, click to select it and then press the DEL key on your keyboard.

Normally, you'll place your programming blocks in a straight line, as shown in Figure 3-7, but sometimes it makes sense to arrange blocks differently to avoid clutter on the canvas. When you arrange blocks differently, you should connect the blocks with a *Sequence Wire*, as shown in Figure 3-8. A block that isn't snapped to another block or connected via a Sequence Wire will appear greyed out in your program, and it won't have any effect on your robot.

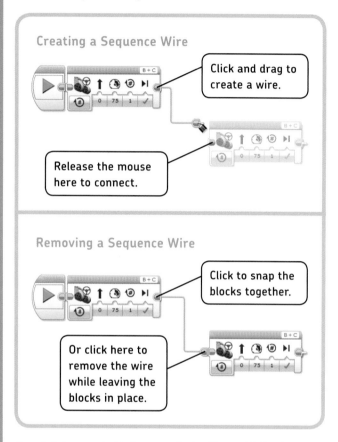

Figure 3-8: Programming blocks are usually placed in a straight line, but you can place them anywhere as long as you connect them with a Sequence Wire (top). You can remove a wire by clicking either end (bottom). Clicking the left end removes the wire and snaps the blocks together in a straight line; clicking the right end removes the wire and leaves the blocks in place.

4. hardware page

Use the Hardware Page to transfer your programs to the EV3 brick, to view the status of the EV3 and any connected devices, and to configure the connection between the EV3 and your computer. Click the toggle on the left to expand the Hardware Page, as shown in Figure 3-9. (I'll introduce you to many of its features throughout this book.)

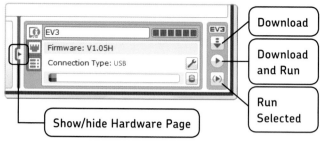

Figure 3-9: The Hardware Page

downloading and running a program

To transfer a program to the EV3 brick, make sure that the EV3 is connected to the computer and then click the **Download and Run** button on the Hardware Page. The robot should make a sound to indicate that the program has been transferred successfully, and the program should begin running automatically. The program will stop when it finishes running each of its blocks.

Once a program has been sent to the EV3 brick, the robot should be able to run that program even when you unplug the USB cable. Your programs remain in the EV3's memory when you turn off the EV3, allowing you to run them whenever you want.

manually running a program

When a program ends or you stop it by pressing the **Back** button on the EV3 brick, you should be able to restart it manually using the EV3's buttons, as discussed in Chapter 2. You'll find all of the programs that you've downloaded to your robot in the File Navigation tab of the EV3 brick. Programs you've just run should appear on the Run Recent tab.

NOTE Programs may disappear from the Run Recent tab, but you should still be able to find them on the File Navigation tab.

downloading a program without running it

It's not always a good idea to have a program run automatically once you've finished downloading it to your robot. For example, if your robot is sitting on your desk and you program it to move, it might just drive off the desk. To transfer a program to the EV3 without having it run automatically, click the **Download** button on the Hardware Page. Once the program finishes downloading (as indicated by the sound), disconnect the USB cable and then start the program using the buttons on the EV3 brick.

running selected blocks

Click **Run Selected** to run only the blocks you've selected. Doing so is useful for testing subsections of a large program. (To select multiple blocks, drag a selection around them or hold the SHIFT key while clicking the blocks you want to select. Click anywhere on the Canvas to deselect.)

NOTE In addition to using the USB cable, you can also transfer programs to the EV3 brick using Bluetooth or Wi-Fi. Appendix A shows how to use the Hardware Page to set up these wireless connections.

projects and programs

When you build a robot, you'll often want to create more than one program for it. While each program makes your robot behave differently, it makes sense to keep related programs together in a *project*. You'll now see how to manage a *project file* and the programs contained in it using the sections marked in Figure 3-10.

5. file organization

When you made your first program earlier, you actually created a new project file with one empty program in it, as shown in Figure 3-11. To add another empty program to the current project, click the **+** sign labeled *Add program*, as shown in the figure.

saving projects and programs

It's important to save your programs often while programming so you don't lose your hard work. To save all programs in your project at once, click **Save** on the toolbar or press CTRL-S. If you're saving a project for the first time, you should be prompted to choose a name for your project. In this case, enter **MyFirstProject** and then click **Save**.

To open a project you saved earlier, click **File ▸ Open Project** or use the **Open Recent** button in the lobby (see Figure 3-2), and locate your project file (usually in *Documents\LEGO Creations\MINDSTORMS EV3 Projects*).

To close a program or project, click the **x** on its tab, as shown in Figure 3-12. To switch to another open program in the current project, just click its tab. To reopen a closed program, use the Program List button on the toolbar (see "6. Toolbar" on page 30).

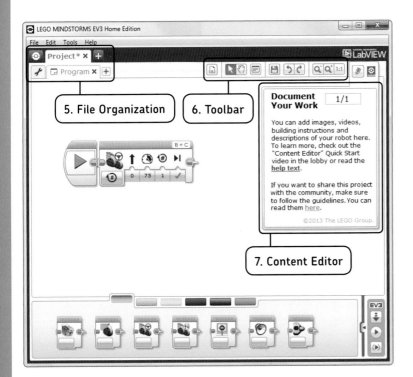

Figure 3-10: Tools for managing your project and the programs in it

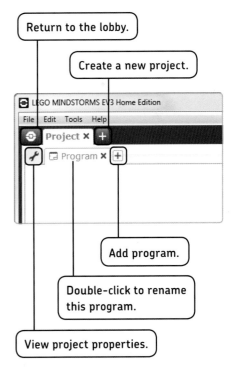

Figure 3-11: Opening new projects and programs

renaming projects and programs

To change the name of a *program*, double-click its tab and enter a new name. (For example, I renamed the first program in Figure 3-12 *DriveForward*.) To change the name of a *project*, you have to create a new project based on the current one by clicking **File ▸ Save Project As** and choosing a new name.

Be sure to choose descriptive names for your projects and programs so you can easily find them on the EV3 brick.

Figure 3-12: Tabs for multiple programs in a project. The * after MyFirstProject.ev3 indicates unsaved changes in your project.

finding projects and programs on the EV3 brick

When you click **Download** or **Download and Run**, the complete project should be transferred to the EV3 brick, as shown in Figure 3-13. In the File Navigation tab on the EV3 brick, you should find one folder for each project, containing all of its programs as well as any files used in the project, such as images or sounds. You start a program by selecting it and clicking the **Center** button.

(If you inserted a microSD card into the EV3 brick, you should see a folder called *SD_Card* on the File Navigation tab, and you should find the *MyFirstProject* folder inside it.)

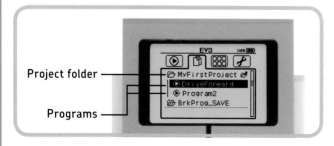

Figure 3-13: Programs are stored in project folders in the File Navigation tab. The folder name is the same as the project name.

NOTE It's usually a good idea to create one project for each of your robots, but downloading the project to the EV3 can take a long time if your project contains many programs, especially if your project includes sounds.

modifying project properties

Clicking the tool icon to the left of the program tabs in the EV3 software opens the *Project Properties* page (see Figure 3-14). Here you can add information about your project (give it a description, a representative photo, or even a video) and share it with others.

The Project Properties page lists all of the files in your project, including programs and sounds. To reopen a closed program in your project, just double-click its name. To delete a program from your project, select it and click **Delete**.

6. toolbar

Use the *toolbar* (see Figure 3-15) to open and save programs from your project, undo or redo changes to your program, and navigate around your program.

using select, pan, and zoom tools

When the *Select* tool button on the toolbar is blue, as shown in Figure 3-15, you should be able to use your mouse to place, move, and configure programming blocks on the Programming Canvas. You move around the Canvas using the arrow keys on your keyboard. You'll use the Select tool most of the time.

If you select the *Pan* tool, the mouse should move the Canvas. This is especially useful when you make large programs that don't fit on your computer screen. To navigate to a specific part of a program, select the Pan tool, click in the Programming Canvas, and drag the Canvas around by moving the mouse while holding the left mouse button. (Holding ALT while the Select tool is activated has the same effect.)

To get a better overview of a large program, *Zoom out* to fit more blocks on your screen. Click *Zoom in* or *Zoom reset* to return to normal view.

using the comment tool

Use the *Comment* tool to place comments on the Programming Canvas. These comments won't change the actions of your program, but they can help you remember what each part of a program is for. When you click **Comment** on the toolbar, a comment box should appear on the Canvas. Enlarge and move the box with your mouse, and double-click to enter a comment, as shown in Figure 3-16. To delete a comment, click the comment box and press DEL on your keyboard.

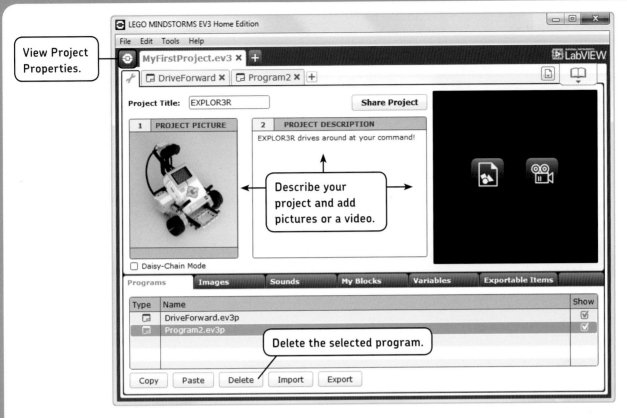

Figure 3-14: The Project Properties page. You can customize your project with pictures and a description or share it with other builders. Double-click a program to open it, or select it and click **Delete** to remove it from the project.

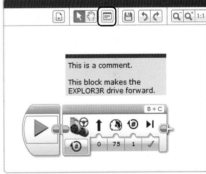

Figure 3-15: The toolbar

Figure 3-16: Comments in a program

duplicating one or more blocks

When programming, you'll sometimes want to duplicate a set of blocks rather than take new ones from the Palette one by one. To duplicate a set of selected blocks, drag them to a new position while holding CTRL, as shown in Figure 3-17. (You can do the same thing by clicking **Edit ▸ Copy** and **Edit ▸ Paste**, but you won't be able to control where the new blocks end up on the canvas.)

help documentation

For detailed overviews of all of the programming blocks (beyond the introductory information in this book), choose **Help ▸ Show EV3 Help**, as shown in Figure 3-18. For details about the Move Steering block, for example, open the help documentation and choose **Programming Blocks ▸ Action Blocks ▸ Move Steering**.

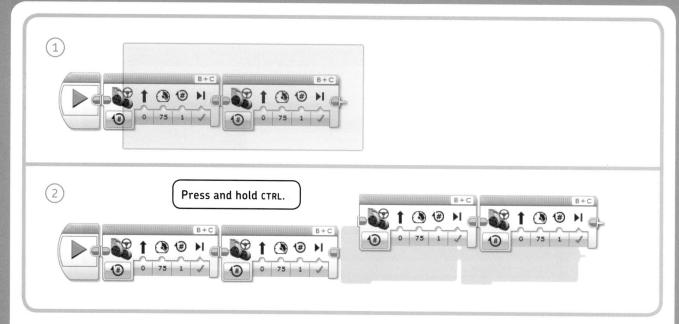

Figure 3-17: Duplicating a series of blocks: (1) Keep the left mouse button pressed while you drag a selection around the blocks you want to duplicate. (2) While holding down CTRL, drag the blocks next to the ones that were already there. (On a Mac, use the ⌘ key.)

If you activate *Context Help*, a small dialog will give you information about a selected block or button, as shown in Figure 3-18. Click **More Information** to view the related page in the help documentation.

You'll find an additional *user guide*, which describes the EV3 brick and other hardware in more detail, in the lobby.

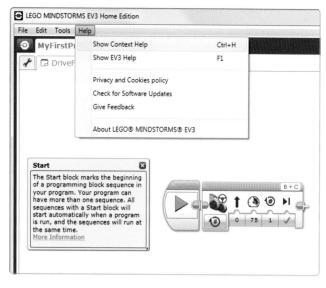

Figure 3-18: The Context Help window gives you information about the selected block (the Start block in this case).

7. content editor

The *Content Editor* on the right side of the screen (see Figure 3-19) is where you can add further information about your project, such as descriptions of how your program works or of how to build your robot, whether for your own use or to make it easy for others to rebuild a shared robot. You can add text and images to present your project to others much as you would in a slide presentation editor.

Usually, you'll want to hide the Content Editor to free up space on your screen.

building the official EV3 robots and bonus models

Once you've gained some experience in building and programming EV3 robots, you can build the 5 robots, such as EV3RSTORM, that you'll find in the lobby. You can also try some of the 12 bonus models found on the More Robots tab. I designed the RAC3 TRUCK, shown in Figure 3-20.

Toggle between Edit mode and View mode.

Click to show or hide the Content Editor.

Figure 3-19: Use the Content Editor to document your project (left). For example, you can add a picture that shows how a mechanism on the inside of your robot works (right).

Figure 3-20: Besides the 5 official EV3 projects, you can build 12 bonus robots, such as this racing truck.

conclusion

In this chapter, you've learned the basics of working with the EV3 programming software. You should now know how to create, edit, and save projects and programs, as well as how to transfer programs to the EV3 brick.

In Chapter 4, we'll take on some serious programming challenges!

4

working with programming blocks: action blocks

Chapter 3 taught you the basics of how to create a new program and transfer it to the EXPLOR3R robot. In this chapter, you'll learn how to use programming blocks to create working programs and how to make the EXPLOR3R move.

You'll also learn how to make your robot play sounds and display text or images on the EV3 screen, and you'll learn how to control the EV3 brick's colored light. After you've had a chance to practice a bit with the sample programs in this chapter, you'll be challenged to solve some programming puzzles by yourself!

how do programming blocks work?

EV3 programs consist of a series of programming blocks, each of which makes the robot do one particular thing, such as move forward for one second. Programs run the blocks one by one, beginning with the first block on the left. Once the first block finishes running, the program continues with the second block, and so on. Once the last block finishes running, the program ends.

Each programming block has one or more modes and several settings for each mode. You can modify a block's actions by changing its mode and settings. For example, in Figure 4-1, both blocks are Move Steering blocks, but they're in different modes. Because the blocks use different modes and different settings, they make the robot perform two different actions.

There are many types of programming blocks, each with its own name and unique icon so you can tell them apart. Different types of blocks are meant for different purposes, with similar blocks grouped together by color in the Programming Palette. In this chapter I'll focus on using *Action blocks* (the green blocks in the Programming Palette).

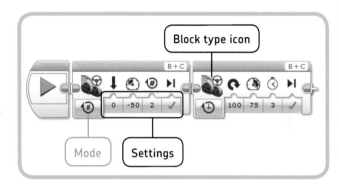

Figure 4-1: To configure a block's actions, change its mode and settings. For example, the first block makes the robot drive backward, while the second makes it steer right. (You'll learn to create this program in the next section.)

the move steering block

The *Move Steering block* controls the movement of a robot's motors. By using it in your program, you can make the EXPLOR3R move forward or backward and steer left or right. We used a Move Steering block in Chapter 3 to make the EXPLOR3R move forward for a short while.

the move steering block in action

Before we get into how the Move Steering block works, let's make a small program to see it in action. This program will make the EXPLOR3R drive backward until the wheels make two full rotations and then spin quickly to the right for three seconds. Because these are two different actions, you'll use two Move Steering blocks.

1. Create a new project named *EXPLOR3R-4*. You'll use this project for all of the programs you make in this chapter. Change the name of the first program to *Move*.

2. Pick two Move Steering blocks from the Programming Palette and place them on the Programming Canvas, as shown in Figure 4-2.

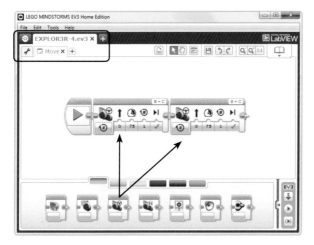

Figure 4-2: Creating the Move *program in the EXPLOR3R-4 project. Choose a Move Steering block from the Programming Palette and place it next to the Start block. Place the second block next to the first one.*

3. By default, the blocks you've just placed are configured to make the robot go forward for a little while, but we want to change the first Move Steering block so that it makes the robot drive backward for two wheel rotations. To accomplish this, change the settings on the first block, as shown in Figure 4-3.

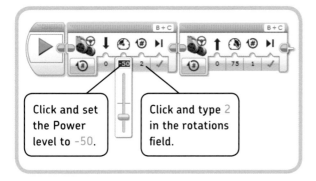

Click and set the Power level to -50.

Click and type 2 in the rotations field.

Figure 4-3: Configure the first block by changing the power setting to -50. To do this, drag the slider down to -50 or enter -50 manually. Negative values make the robot go backward. Next, enter 2 in the rotations field to make the robot stop moving when the wheels complete two rotations.

4. Now you want to modify the settings on the second block. This block will make the EXPLOR3R spin quickly to the right for three seconds. First, change its mode to **On for Seconds**, as shown in Figure 4-4.

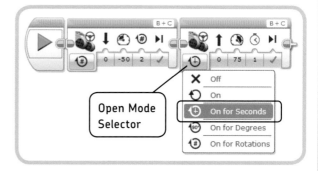

Open Mode Selector

× Off
↻ On
⏱ On for Seconds
↻90 On for Degrees
⏱ On for Rotations

Figure 4-4: Click the Mode Selector of the second block and select On for Seconds.

5. To make the robot spin quickly to the right for three seconds, change the settings on the second block, as shown in Figure 4-5.

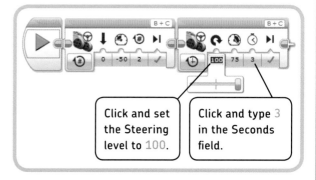

Click and set the Steering level to 100.

Click and type 3 in the Seconds field.

Figure 4-5: Configure the second block by dragging the steering slider all the way to the right and then entering 3 in the Seconds field.

6. Once you've configured both Move Steering blocks, you can download the program to your robot and run it. You'll know it's working if the EXPLOR3R goes backward until both wheels make two rotations and then turns around quickly for three seconds.

NOTE If your robot turns left rather than right, it's possible that you've connected the motors to the wrong ports on the EV3 brick. Review the steps on page 19 to check the cables on your robot.

understanding modes and settings

Let's look more closely at the settings on the blocks to better understand how that sample program really worked. Each block's action is determined by its mode and settings. Figure 4-6 shows several ways to configure the Move Steering block, besides the ones we've already used.

The Move Steering block has several *modes*, each of which can be selected by clicking the *Mode Selector* and each of which makes the block do something slightly different. For example, the first block in our program was in *On for Rotations* mode, which allows you to choose the number of motor rotations the robot will move for, while the second was in *On for Seconds* mode, which allows you to specify how long the motors will move in seconds.

The modes in the Move Steering block are as follows:

* *On*: Switch on motors.
* *Off*: Stop the motors.
* *On for Rotations*: Turn on motors for a specified number of complete rotations; then stop.
* *On for Seconds*: Turn on motors for a specified number of seconds; then stop.
* *On for Degrees*: Turn on motors for a specified number of degrees of rotation; then stop.

You'll learn to use the On and Off modes in "The On and Off Modes in Move Blocks" on page 45.

ports

In the *Ports* setting at the top right of the block, you can choose which output ports of the EV3 brick you've connected the driving motors to so that the program knows which motors it should switch on. The EXPLOR3R's motors are connected to ports B and C, so in our sample program, we left the block at its default setting of B+C.

steering

As you saw with the *Move* program, you can also make the robot steer. To adjust your robot's steering, click the *Steering* setting and drag the slider to the left (to make the robot steer to the left) or right.

How does the vehicle turn without a steering wheel? This block can control the robot's steering by controlling the two wheels independently. For the robot to move straight, both wheels will turn in the same direction at the same speed. To make a turn, one wheel will spin faster than the other, or the wheels can turn in opposite directions to make the robot turn in place. Figure 4-7 shows how different combinations of Power and Steering configurations make the EXPLOR3R turn.

power

The *Power* setting controls the speed of the motors. Zero power means that the wheels don't move at all, while 100 sets the motors to maximum speed. Negative values, such as –100 or –30, make the robot move backward, as you saw with the *Move* example program.

rotations, seconds, or degrees

Depending on which mode you select, the third setting on the Move Steering block lets you specify how long the motors should move. For instance, entering 3 in the Seconds setting in On for Seconds mode makes the EXPLOR3R move for three seconds.

On for Degrees mode allows you to specify the number of degrees the motors—and as a result, the wheels—should turn. Turning 360 degrees is the same as one full rotation of the wheels, while 180 degrees would be half a rotation. You can set the wheels to turn a number of full rotations when the block is in On for Rotations mode as you saw in the first block of the sample program.

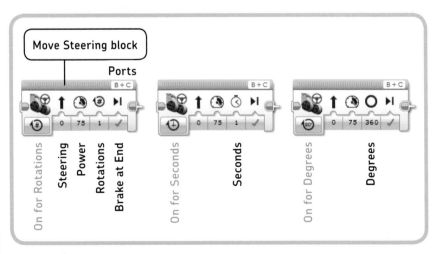

Figure 4-6: Modes (blue) and settings (black) on the Move Steering block. To select a different mode, click the Mode Selector and choose a mode from the drop-down menu. Most of the settings will be the same, regardless of which mode you choose.

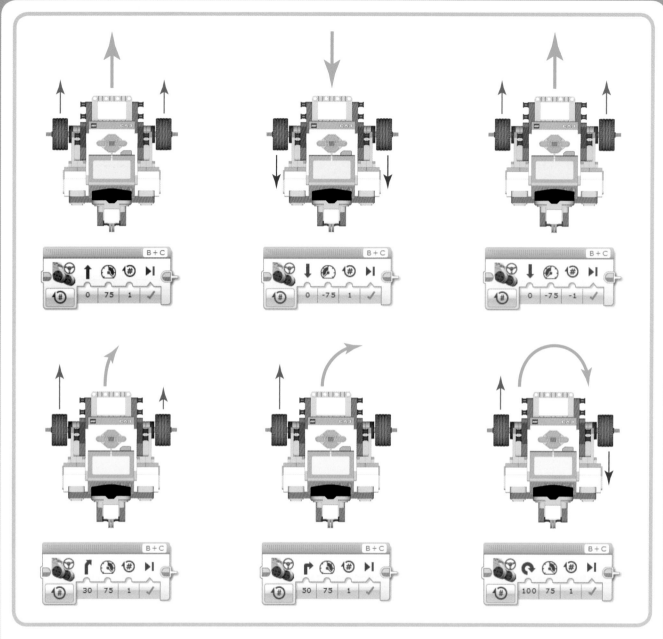

Figure 4-7: To make the robot turn, adjust the Steering setting; the block will control the speed and direction of both motors to make the robot turn. The red arrows indicate that a wheel rolls forward, and the blue ones mean that it rolls backward. The green arrow shows the direction in which the robot moves as a result.

NOTE Entering a negative power value, such as –75, and a positive rotation value, such as 1 rotation or 360 degrees, makes the robot go *backward*. The same holds for positive power and a negative rotation, such as 75 power and –1 rotation. However, if you enter a negative power value *and* a negative rotation, such as –75 power and –1 rotation, the robot goes *forward*! Refer to Figure 4-7 when you're unsure how a certain Power and Steering combination will affect your robot.

brake at end

The *Brake at End* option controls how the motors stop after they rotate the required number of degrees, rotations, or seconds. Selecting *true* (the check mark: ✓) makes the motors stop immediately, while *false* (✗) makes them stop gently.

making accurate turns

When you use a Move Steering block to make the robot turn 90 degrees, you might think you need to set the Degrees setting to 90, but this is not true. In fact, the Degrees setting specifies only how many degrees the motors—and, as a result, the wheels—turn. The actual number of degrees the motors should turn so the robot makes a 90-degree turn is different for every robot. Discovery #2 gets you started finding the appropriate number of degrees for your robot.

DISCOVERY #1: ACCELERATE!

Difficulty: ⬒ **Time:** 🕐

Now that you've learned some important information about the Move Steering block, you're ready to experiment with it. The goal in this Discovery is to create a program that makes the robot move slowly at first and then accelerate. To begin, place ten Move Steering blocks on the Programming Canvas and configure the first two as shown in Figure 4-8. Configure the third one in the same way, but set the motor's Power setting to 30, incrementing by 10 with each block until you reach the motor's maximum speed.

The blocks are now in On for Seconds mode. Once you've tested your program, change all 10 blocks to On for Rotations mode, with Rotations set to **1**, and run the program again. Which program takes longer to run? Can you explain the different behavior?

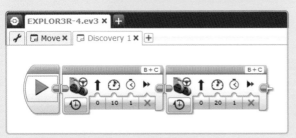

Figure 4-8: The first blocks of the program for Discovery #1. Remember to add a new program for each Discovery and save it when you're done so you can continue working on the programs later!

DISCOVERY #2: PRECISE TURNS!

Difficulty: ⬒ **Time:** 🕐

Can you get your robot to make a 90-degree turn in place? Create a new program with one Move Steering block configured in On for Degrees mode, as shown in Figure 4-9. Make sure that the Steering slider is all the way to the right, as in the *Move* program. How many degrees should the *wheels* turn for the robot to make an accurate 90-degree turn?

Begin by setting Degrees to 275. If that's not enough, try 280, 285, and so on, running the program each time to see whether the robot makes the required turn.

Once you find the right value to make a 90-degree turn, figure out which value you need to get the robot to make a 180-degree turn.

Figure 4-9: The program for Discovery #2. Which value makes your robot turn 90 degrees? Which value should you use for a 180-degree turn?

DISCOVERY #3: MOVE THAT BOT!

Difficulty: ⬒ **Time:** 🕐🕐

Create a program that uses three Move Steering blocks to make the EXPLOR3R move forward for three seconds at 50 percent power, turn 180 degrees, and then return to its starting position. When configuring the block that lets the robot turn around (the second block), use the Degrees value that you found in Discovery #2.

the sound block

It's fun to make programs that make the EXPLOR3R move, but things get even more fun when you program the EV3 to make sounds using the *Sound block*. Your robot can play two types of sounds: a simple tone, like a beep, or a prerecorded sound, such as applause or a spoken word like "hello." When you use a Sound block in your programs, the robot will seem more interactive and lifelike because it can "talk."

understanding the sound block settings

Even though every programming block allows the robot to do something different, all blocks are used in the same way. In other words, you can simply pick a Sound block from the Programming Palette and place it on the Programming Canvas as you did with the Move Steering block. Once the block is in place, select the mode and then adjust the settings to specify what sound you want to hear.

But before we create a program with Sound blocks, let's briefly explore the different modes and settings for this block, as shown in Figure 4-10. The four modes of the Sound block are as follows:

* *Play File*: Play a prerecorded sound, like "hello."
* *Play Tone*: Play a tone at a certain frequency (pitch) for a specified amount of time.
* *Play Note*: Play a note from a piano for a specified amount of time.
* *Stop*: Stop any sound that's currently playing.

file name

In Play File mode, you can choose a sound by clicking the *File Name* field and picking a file from a menu that appears. You can select a sound from categories like animals, colors, communication, and numbers. You can also record and add your own sounds using **Tools ▸ Sound Editor**. (See **Help ▸ Tools ▸ Sound Editor** for instructions.)

volume

Enter a number between 0 (soft) and 100 (loud) to set the volume of the sound you want to play.

play type

Use the Play Type setting to control what happens when the sound starts playing. Choose *Wait for Completion* (0) to pause the program until the sound stops playing. Choose *Play Once* (1) to let the program continue running the next block

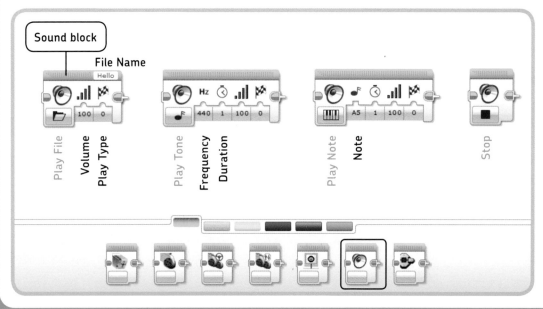

Figure 4-10: The four modes (in blue) of the Sound block and their settings (in black)

while the sound is still playing. If you select *Repeat* (2), the sound keeps repeating while the program runs the remaining blocks in your program. For most programs, you will select Wait for Completion (0).

note or tone

Depending on the mode you've selected, you can choose a *Note* from a piano keyboard or a *Tone* (a pitch) in hertz (Hz). A tone of 440 Hz (the default *Frequency* value) is clearly audible to the human ear, making it a useful sound for testing programs. For example, you could play the sound to indicate that a specific programming block has finished running.

duration

In the *Duration* box, enter the number of seconds you want the note or tone to play.

seeing the sound block in action

Now let's create a program called *SoundCheck* that makes the robot move and play sounds so you can see how the Sound block works.

To begin, create the new program shown in Figure 4-11. To create this program and the other examples throughout this book, follow these steps for each block in the program:

1. Locate the block on the Programming Palette using its color and icon and then place it on the canvas. For example, the Sound block is green, and its icon resembles a speaker.

2. Select the correct mode by looking at the icon shown in the Mode Selector. For example, the piano icon on the third block indicates that you should select Play Note.

3. Finally, enter the remaining settings in the block. For example, configure the Steering and Power settings in the Move Steering blocks. Your block should now look exactly the same as the one in the example program.

Once you've finished creating your program, download it to your robot and run it.

understanding the SoundCheck program

Now that you've run the program, let's go over how it works. The first Sound block makes the EXPLOR3R say "Goodbye." The Play Type is Wait for Completion, so the robot waits until the robot finishes saying "Goodbye" before moving to the next block. Next, a Move Steering block makes the robot drive forward for one rotation, and then another Sound block causes the EV3 to play a note. This block doesn't wait for the note to complete, so while the sound is playing, the second Move Steering block gets the robot to steer to the right for three seconds. Finally, the robot stops moving.

> ## DISCOVERY #5: WHICH DIRECTION DID YOU SAY?
>
> **Difficulty:** ▱▱ **Time:** ⏱
> Create a program like *SoundCheck* that has the robot announce its direction as it moves. While going forward, it should say "Forward," and while going backward, it should say "Backward." How do you configure the Play Type settings in the Sound blocks?

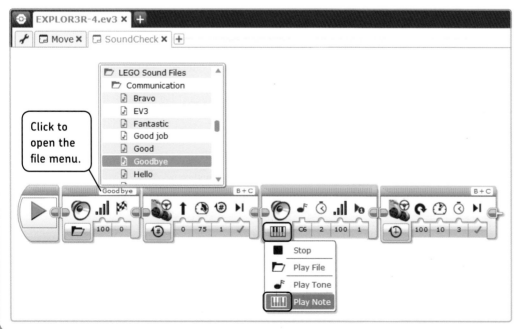

Figure 4-11: The SoundCheck program. To select a block's mode, open the Mode Selector and choose the icon from the list that matches the icon shown in the figure.

the display block

In addition to moving the robot around and playing sounds, an EV3 program can control the EV3's display. For example, you could create a program that makes the EV3 screen look like Figure 4-12. The screen is 178 pixels wide and 128 pixels tall. (*Pixels* are small dots that make up what you see on a screen.)

It's fun to play with the Display block, but perhaps more importantly, it's an effective way to test your programs. For example, you can display a sensor measurement on the screen to see whether the sensor is working properly, as you'll see in Part V of this book.

You can use the *Display block* to display an image (like a light bulb), text (like "Hello!"), or a shape (like a solid circle) on the EV3 screen. A single Display block can't put multiple images or text lines on the screen at once, so you would need to use a series of blocks to create the display shown in Figure 4-12.

Figure 4-12: Use Display blocks to show images, text, and shapes on the EV3 brick's screen.

understanding the display block settings

Once a Display block has put something on the EV3 screen, the program moves on to the next block—let's say a Move block. The EV3 screen keeps showing the image until another Display block is used to display something else. So, in this example, the image would remain on the screen while the robot moves.

When a program ends, the EV3 returns to the menu immediately, which means that if the last block in your program is a Display block, you won't have time to see what's on the display because the program will have finished. To see what's on the display, add a block—such as a Move block—to keep the program from ending instantly.

Figure 4-13 shows the four main modes of the Display block:

* *Image*: Display a selected image, such as a happy face, on the screen.
* *Shapes*: Display a line, circle, rectangle, or dot on the screen.
* *Text*: Show a text line on the screen.
* *Reset Screen*: Erase the screen and show the MINDSTORMS logo that you normally see when you run a program without Display blocks.

sub modes

Some modes have *sub modes*. When selecting Shapes mode in the Display block, for example, you'll have to choose among four sub modes—*Line*, *Circle*, *Rectangle*, and *Point*—as shown in Figure 4-13. Choosing *Shapes – Circle* mode makes the Display block show a circle shape on the EV3 screen, and you use the settings on the block to configure the circle's position, radius, fill, and color.

file name

When in Image mode, you can click the *File Name* field to choose an image from categories like eyes, expressions, objects, and LEGO. You can also create or load your own images by going to **Tools ▸ Image Editor**. (See **Help ▸ Tools ▸ Image Editor** for instructions.)

clear screen

The *Clear Screen* setting lets you choose whether to empty the screen before showing something new (when set to true) or to add something new to what's already on the screen (when set to false). You'll need a series of Display blocks to show more than one object on the screen. The first block should clear the screen before displaying something new, and the other blocks

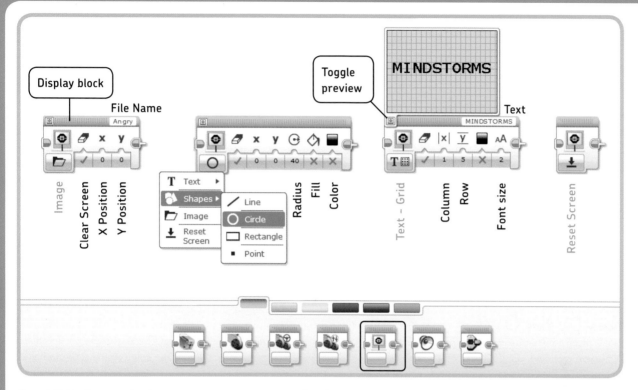

Figure 4-13: The four modes (blue) of the Display block and their settings (black). You can toggle the preview of the EV3 screen to see the result of your settings. When you're satisfied, close the preview.

should simply add something to the screen. To accomplish this, set the Clear Screen setting to true on the first block and set it to false on the blocks that follow. You'll see how this works in the *DisplayTest* example.

radius and fill

Some settings are specific to different modes of the Display block. For example, the *Radius* setting specifies the size of a circle, and *Fill* allows you to make a solid circle (true) or draw only the outline (false).

color

Usually, you'll want to set the *Color* setting to black (false), but if you've previously filled an area with, say, a black circle, you can still add text on top by setting its Color to white (true).

text and font size

In Text mode, enter a line of text that you want to display, such as *MINDSTORMS*, in the *Text* field, as shown in Figure 4-13. The text line may include numbers, and you can adjust the size of the text by setting the *Font Size* value to 0 (small), 1 (bold), or 2 (big).

x, y, column, and row

Whether you display an image, text (*Text – Pixels* mode), or shape, you can choose its position with the *X* (position relative to left end of the screen) and *Y* (position relative to the top) settings. If you preview the screen (see Figure 4-13), you should be able to easily adjust the values of X and Y.

In *Text – Grid* mode, you can specify a *Column* (0–20) and *Row* (0 –11) instead, making it easier to align multiple lines of text.

NOTE Remember that you can find further details about each block in the Help Documentation. For the Display block, go to *Help ▸ Show EV3 Help ▸ Programming Blocks ▸ Action Blocks ▸ Display.*

the display block in action

Let's test the functionality of the Display block by creating a program that puts things on the EV3 screen while the robot moves. Create a program called *DisplayTest*, and place three Display blocks and two Move Steering blocks on the Programming Canvas, as shown in Figure 4-14. Then, configure each block as shown. Once you've configured all the blocks, transfer the program to your robot and run it.

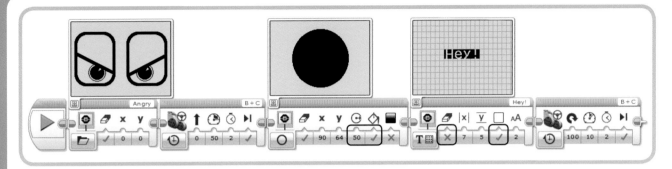

Figure 4-14: The DisplayTest *program makes your robot move while displaying things on the screen. The screen previews are shown here for convenience, but you can hide them if you like.*

DISCOVERY #7: SUBTITLES!

Difficulty: ▭ **Time:** ⏱

Create a program that uses four Sound blocks to say, "Hello. Good morning. Goodbye!" Use Display blocks to display what the robot says as subtitles on the EV3 screen and to clear the screen each time the robot starts saying something new. Do you place the Display blocks before or after the Sound blocks?

DISCOVERY #8: W8 FOR THE EXPLOR3R!

Difficulty: ▭▭ **Time:** ⏱⏱

Program the EXPLOR3R to drive in a figure eight, as shown in Figure 4-15. The robot should show faces on the screen as it moves. Choose different images from the Eyes category.

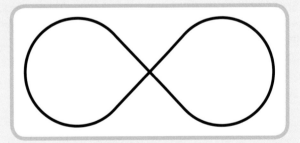

Figure 4-15: The drive path for Discovery #8. Try making the robot drive in a pattern that looks like this. It doesn't have to trace the exact line for now. (You'll learn how to do that in Chapter 7.)

understanding the DisplayTest program

In the *DisplayTest* program, the Display blocks put several things on screen, and the Move Steering blocks make the EXPLOR3R drive around. The first Display block clears the screen before it displays an image (the angry eyes). The robot then starts to move, and the image stays on the screen while the first Move Steering block is running. The next Display block (the solid circle) also clears the screen, removing the image with the angry eyes before displaying a circle.

The program moves on to another Display block, which puts white text in the circle. This block doesn't clear the screen in advance, so you'll see both the circle and the text on the screen. Finally, a Move block makes the robot turn right, and the program ends.

the brick status light block

The brick status light surrounding the EV3 buttons is normally green. It blinks while a program is running. With the *Brick Status Light block*, you can override this behavior and choose what this light does. Its three modes, shown in Figure 4-16, are as follows:

* *On*: Switch on the light and choose a *Color*: Green (0), Orange (1), or Red (2). The *Pulse* setting lets you choose whether the light should blink (true) or just stay on (false).
* *Off*: Switch off the light.
* *Reset*: Show the blinking green light that you normally see when a program runs.

Now make the *ButtonLight* program to test this functionality, as shown in Figure 4-17. The buttons should light up red when the robot says "Red" and green when it says "Green."

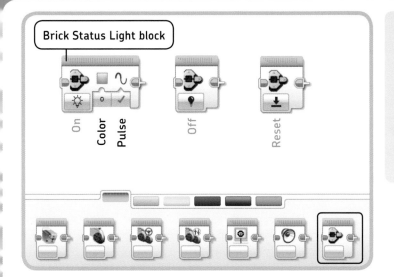

Figure 4-16: The three modes of the Brick Status Light block and their settings

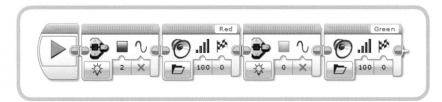

Figure 4-17: The ButtonLight *program*

DISCOVERY #9: TRAFFIC LIGHT!

Difficulty: 🔲 **Time:** ⏱

Modify the *ButtonLight* program to turn your robot into a traffic light. Make the robot say "Stop," "Activate," and "Go," while showing a red, orange, and green light, respectively.

the on and off modes in move blocks

Now that you have the hang of programming with some Action blocks, you're ready to examine the On and Off modes found in the Move Steering block and several other blocks. A block in *On* mode will switch on the motors and then instantly move to the next block. As the program continues, the motors keep running until the program reaches a block that tells them to stop or do something else. The *Off* mode turns the motors off.

To see why this is useful, create the *OnOff* program shown in Figure 4-18.

When you run the program, the robot should begin to move and say "LEGO MINDSTORMS" at the same time. When it's done speaking, it should stop moving and play a tone. The motors stop moving exactly when the robot finishes saying "LEGO MINDSTORMS," even though you didn't say beforehand how long they should stay on. This allows you to make the robot move *until* you stop it with the Off mode, regardless of what happens in between.

If you were to create a program with just one Move block in On mode, you might think that the robot would go forward indefinitely. In fact, this block switches on the motors, but then the program ends because it finishes running all of its blocks; when the program ends, the motors stop.

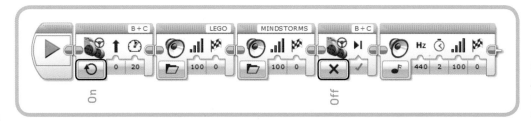

Figure 4-18: The OnOff *program. Carefully configure the modes in the Move Steering blocks.*

move tank, large motor, and medium motor blocks

Besides the Move Steering blocks, there are three more blocks that you can use to make motors move. The first is the *Move Tank block*, which you can use to control a vehicle with two tank treads or two wheels, such as the EXPLOR3R, much as you did before. But instead of choosing steering and power for the whole robot, you can now choose the power of each motor independently. By using different combinations of power for the left and right wheels, you can drive EXPLOR3R in different directions (see Figure 4-7).

The *Tank* program demonstrates how the Move Tank block works (see Figure 4-19). This block has essentially the same

functionality as the Move Steering block, but you control the settings differently. Use whichever block you prefer when making your programs for these types of vehicles.

If one of the motors is programmed to turn faster than the other (as in the *Tank* program), the block ends as soon as the fastest motor reaches the indicated number of rotations or degrees. (The same holds true for the Move Steering block.)

You use the *Large Motor block* to control individual Large Motors. This is useful for mechanisms that use only one motor, such as a claw that grabs an object. The modes and settings are the same as they are for the Move Tank block, but now they apply only to the selected motor.

The program in Figure 4-20 uses two Large Motor blocks to drive the left (B) and right (C) motors respectively, one at a time.

The *Medium Motor block* is the same as the Large Motor block, except you use it to control the single Medium Motor that comes with your EV3 set. (You'll see it in action in Chapter 12.)

further exploration

You've learned the basics of LEGO MINDSTORMS EV3 programming. Congratulations! You should now know how to program robots to make them move, play sounds, blink a light, and display text and images on the EV3 screen. Chapter 5 will teach you more about using programming blocks, including how to use blocks to pause a program and repeat a set of blocks.

But before you move on, try solving some of the Discoveries that follow to further fuel your programming skills.

NOTE Don't forget to save your programs after solving a Discovery. You may want to use them as a starting point for bigger programs later.

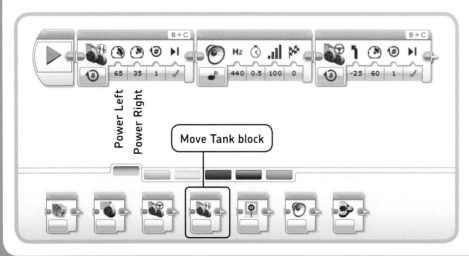

Figure 4-19: The Tank *program: The robot makes a gentle turn to the right with a Move Tank block. Then it stops, plays a sound, and moves again, now using a Move Steering block.*

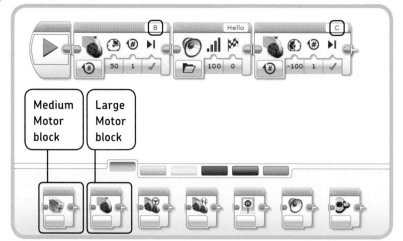

Figure 4-20: The LargeMotor program: The left motor turns forward, the robot says "Hello," and the right motor turns backward.

Medium Motor block

Large Motor block

DISCOVERY #11: CIRCLE TIME!

Difficulty: ◻️ **Time:** ⏱️⏱️

Can you make the EXPLOR3R drive in a circular pattern with a diameter of about 1 m (3 feet)? You'll need only one Move Steering block to accomplish this. How do you configure the Steering setting, and for how long should the motors run? How does the Steering setting affect the diameter of the circle? Does changing the motor speed have any effect on the diameter? When you're ready, try to accomplish the same effect with the Move Tank block.

DISCOVERY #12: NAVIGATOR!

Difficulty: ◻️◻️ **Time:** ⏱️⏱️⏱️

Create a program with Move Steering blocks that makes the EXPLOR3R drive in the pattern shown in Figure 4-21. While moving, the robot should display arrows on the EV3 screen that show the direction of its movement. When finished, it should display a stop sign. In addition to displaying its heading, the robot should speak the direction it's moving in. How do you configure the Play Type setting in the Sound blocks?

HINT You can find all the direction signs shown in Figure 4-21 in the list of images under the Display block's File Name setting.

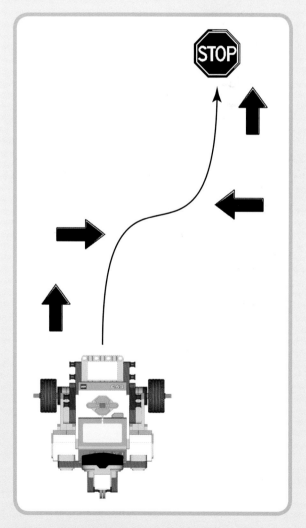

Figure 4-21: The driving pattern and the navigation signs for Discovery #12

DISCOVERY #13: ROBODANCER

Difficulty: ▭▭ **Time:** ⏱⏱⏱

Make the EXPLOR3R play musical beats and tones continuously (using Sound blocks) while it dances in zigzagging movements (using Move Steering blocks). After each movement, the robot should start playing a different sound.

HINT Try Repeat as the Play Type setting in some of the Sound blocks.

DESIGN DISCOVERY #1: ROBOCLEANER!

Building: ☀ **Programming:** ▭

Add some LEGO elements to your robot so that it can hold a dust cloth on the ground in front of it. Then, make a program with Move Steering blocks to make the robot move and clean your room. Instead of using a program, you can also remotely control your cleaning machine as described in Chapter 2. Cleaning has never been so much fun!

DESIGN DISCOVERY #2: ART WITH THE EXPLOR3R!

Building: ☀☀ **Programming:** ▭▭▭

In this Design Discovery, you're challenged to expand the EXPLOR3R robot design. Using LEGO pieces, create an attachment for your robot to hold a pen. As the robot drives over a big piece of paper, it will draw lines and figures with the pen. To get started, you can make it draw the figure-eight pattern in Discovery #8 on page 44.

Attaching a fixed pen to the robot is certainly fun for drawing simple figures, but with a fixed pen you're limited in what you can draw—or write—because the pen is constantly touching the paper. Use the Medium Motor in your EV3 set to lift up the pen; connect this motor to output port A with a cable. You can control this motor with the Medium Motor block. Can you make the robot write your name?

When you're done, photograph your design and add the photos to the Content Editor, as you learned in Chapter 3.

5

waiting, repeating, my blocks, and multitasking

The previous chapter taught you how to program your robot to perform a variety of actions, such as moving. In this chapter, you'll learn several programming techniques that give you more control over the order and flow of the blocks in your EV3 programs. You'll learn how to pause a program with Wait blocks, how to repeat a set of actions with Loop blocks, how to run multiple blocks simultaneously, and even how to make your own custom My Blocks.

the wait block

So far you've been using programming blocks that make the robot move, play sounds, or display something on the screen. Now you'll meet a block that does nothing more than pause the program for a set amount of time. This block, the *Wait block*, is shown in Figure 5-1.

using the wait block settings

You use the Wait block like any other programming block: Place it on the Programming Canvas and then configure its mode and settings. For this chapter, you'll use only the *Time* mode.

In Time mode, the Wait block simply pauses the program for a certain amount time, such as five seconds. Once the time has elapsed, the program continues with the next programming block. The number of seconds you enter in the *Seconds* box can be either an integer, such as 14, or a number with a decimal, such as 1.5. To pause the program for 50 milliseconds (0.05 seconds), for example, you would enter 0.05.

seeing the wait block in action

Why would you want to use a block that doesn't do anything but wait? Here's an example of the Wait block in action. Create a new project, save it as *EXPLOR3R-5*, and create a program called *WaitDisplay*, according to the instructions shown in Figure 5-1.

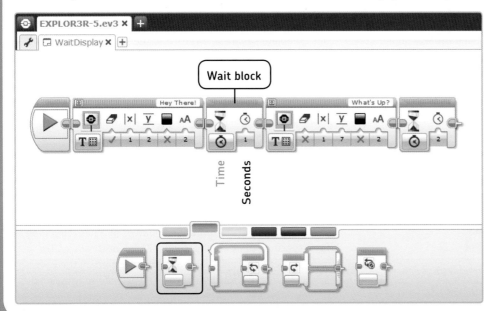

Figure 5-1: The WaitDisplay *program contains two Wait blocks in Time mode to pause the program after displaying text on the EV3 screen.*

understanding the WaitDisplay program

When you run the program, the text "Hey There!" should appear on the EV3 screen, and "What's Up?" should follow after one second. The second Wait block gives you time to read what is displayed on the EV3 screen. Without it, the program would end immediately after the text was displayed, making it impossible to read the message. Wait blocks are also useful for a robot with sensors, as you'll see in the next chapter.

DISCOVERY #14:
LEAVE A MESSAGE!

Difficulty: ▱ **Time:** 🕐🕑

Expand the *WaitDisplay* program to display a message that explains where you've gone next time you leave the house. Add Wait blocks between the text lines to make your message easy to read. Don't forget to leave some instructions for your family so they know how to access the message!

DISCOVERY #15:
BOARD GAME TIMER!

Difficulty: ▱▱ **Time:** 🕐🕑

Create a program that turns your EV3 into a timer that you can use while playing board games. The program should display a timer on the screen showing how much time is left for each turn (see Figure 5-2). When the time is up, your robot should say "Game Over!", indicating that it's the next player's turn. Or, you can use Sound blocks to make the robot say how much time is left.

HINT Use a series of Display blocks in Image mode, with Wait blocks in between. Look for *Timer 0*, *Timer 1*, and so forth in the list of images.

Figure 5-2: The board game timer in Discovery #15

the loop block

Imagine you're walking along a square-shaped path, like the one shown in Figure 5-3. As you walk, you follow a certain pattern over and over again: Go straight, then turn right, go straight, turn right, and so on.

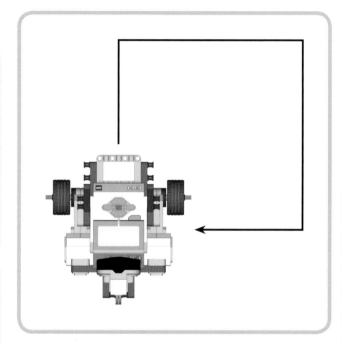

Figure 5-3: The EXPLOR3R moving in a square

To create this sort of behavior with your robot, you could use one Move Steering block to make it go straight and another block to make it turn right. To make your robot trace one complete square and return to the starting position, you would have to use each of these two blocks four times for a total of eight blocks.

Rather than use eight Move Steering blocks to create this program, it's much easier to use the Loop block, which lets you repeat sequences of blocks that are placed within it. Loop blocks are especially useful when you want to repeat certain actions many times.

using the loop block

The Loop block (see Figure 5-4) repeatedly runs the blocks you place within it. Depending on which mode you choose, it runs these blocks either for a specified number of times (*Count*) or for a specified number of seconds (*Time*), or it repeats the blocks indefinitely until you abort the program on the EV3 brick (*Unlimited*). (You'll learn to use many of the remaining modes in subsequent chapters.)

At the top of each loop, you can enter a *Loop Name* to describe the functionality of the blocks you place within it. You can resize the block manually if necessary, as shown in Figure 5-4. (There is also a *Loop Index* feature, which you'll learn about in Chapter 14, but you can ignore it for now.)

You place blocks in a Loop block by simply dragging one or more blocks into it, as shown in Figure 5-5.

seeing the loop block in action

To see the Loop block in action, complete the *OneSquare* program shown in Figure 5-6. When you run the program, the robot should play a tone, drive in a square-shaped pattern, play another sound, and then stop. If your robot doesn't make 90-degree turns when steering, try adjusting the number of degrees in the second Move Steering block, similar to what you did in Discovery #2 on page 39.

using loop blocks within loop blocks

The *OneSquare* program (see Figure 5-6) makes the EXPLOR3R drive in a square once. You can use another Loop block to repeat the square pattern so that the robot completes multiple squares. In *Unlimited* mode, the robot would keep driving in squares endlessly.

Try this out with the *InfiniteSquare* program shown in Figure 5-7. To create it, expand the program you've just made by picking a second Loop block from the palette and setting its mode to Unlimited. Drag the "Square" loop you already made and the second Sound block into the new loop. Now the robot will keep driving in a square, saying "Goodbye" after each round, until you quit the program with the back button on the EV3.

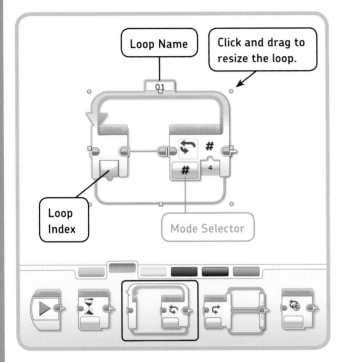

Figure 5-4: The Loop block in Count mode. In this configuration, the program will run any blocks placed within the loop four times. In the other modes, you can have the blocks in the loop repeat for a certain number of seconds or for an unlimited amount of time.

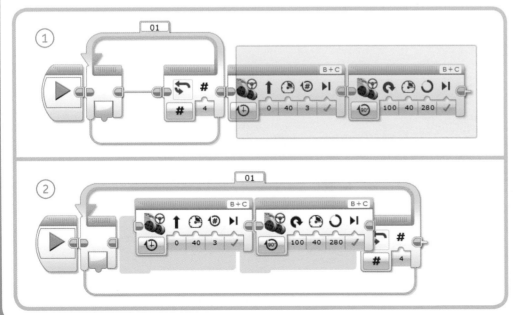

Figure 5-5: To place blocks inside a Loop block, first place all required blocks on the Programming Canvas (1). Next, select the blocks you want to move and drag them into the loop (2). The Loop block should resize automatically to create space for the blocks as you drag them in. When you drag a Loop block around, its contents remain inside it.

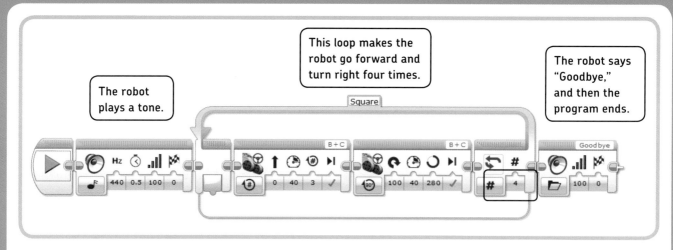

Figure 5-6: The OneSquare program contains a loop that runs four times. Once the two blocks inside the loop have run four times (resulting in a square), the program continues with the next block, a Sound block in this case. You can enter **Square** in the Loop Name field, as shown, to describe the function of the loop.

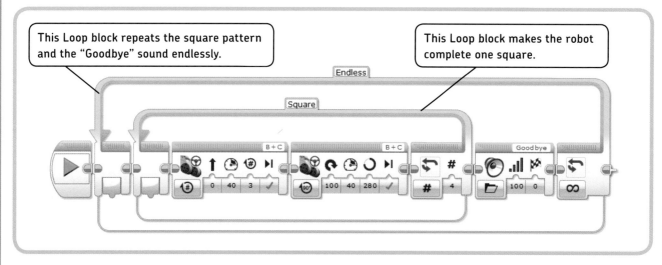

Figure 5-7: The InfiniteSquare program shows you how you can have one Loop block inside another. The inner Loop block makes the EXPLOR3R drive in a square-shaped pattern, and the outer loop makes the robot repeat the square and say "Goodbye" endlessly.

DISCOVERY #16:
GUARD THE ROOM

Difficulty: ⬜ **Time:** ⏱

Create a program to allow the EXPLOR3R to constantly move back and forth in front of your bedroom door, as if guarding it (see Figure 5-8). Use one Loop block in Unlimited mode, a Move Steering block to move forward, and another Move Steering block to turn around.

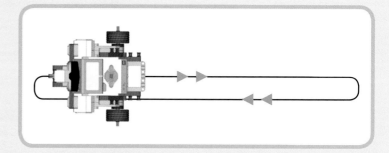

Figure 5-8: The path for the EXPLOR3R in Discovery #16

DISCOVERY #17: TRIANGLE!

Difficulty: ▢▢ **Time:** ⏱

You've created a program to make your robot drive in square-shaped patterns. How could you modify the *OneSquare* program to drive in a triangle-shaped pattern? How about a hexagon? Use additional Loop blocks to repeat each shape five times.

HINT In Discovery #2 on page 39, you found the number of degrees that the wheels should turn to have the robot make a 180-degree turn. Can you use that value to calculate the amount of degrees required for a 120-degree turn so the robot can drive in a triangular pattern?

making your own blocks: the my block

In addition to using ready-made blocks, you can make your own blocks, called *My Blocks*. Each My Block lets you combine multiple programming blocks into one. My Blocks are especially useful when you want to use a specific set of blocks in your program more than once. For example, you could create a My Block to make the robot say "Hello! Good morning!" and to change the status light color to red whenever you use that block.

Normally, it takes five programming blocks to make the robot do this. If you wanted this to happen multiple times, instead of inserting all five blocks each time, it would be much easier to make a My Block that combines all those blocks into one block that you can reuse. Also, using My Blocks can help you keep your programs looking organized because you'll see fewer blocks on the screen.

creating my blocks

To demonstrate the My Block functionality, let's write a program that makes the EXPLOR3R say "Hello! Good morning!", move forward, and then say the same thing again. Because the robot will greet you twice, you'll create a My Block called *Talk* to make it easier to repeat this action, as shown in Figures 5-9

to 5-11. Once you've created your My Block, you can place it in a program whenever you want the EXPLOR3R to tell you "Good morning!"

1. Create a new program called *MyBlockDemo*, and place and configure the five blocks for your robotic greeting on the canvas, as shown in Figure 5-9. Then, select these five blocks (selected blocks are outlined in blue) and click **Tools ▸ My Block Builder**.

2. The My Block builder now appears, as shown in Figure 5-10. Enter a name for your My Block, such as **Talk**, in the *Name* box. Use the *Description* area to describe your block so that you'll remember what it does if you want to reuse it later. Finally, choose an icon, such as the speaker, to help you remember that this My Block is used to make sounds. Click **Finish**.

3. Once you've finished creating your My Block, it should appear on the Programming Canvas, replacing the blocks that were originally there, as shown in Figure 5-11. Rearrange the blocks if necessary.

using my blocks in programs

Now that the My Block is ready, you can find it on the light blue tab of the Programming Palette, as shown in Figure 5-12. Add a Move Steering block, as well as another copy of the My Block, to complete the *MyBlockDemo* program. When you run the program, the robot should say "Hello! Good morning!", drive forward, and greet you again. The status light should be red while the sound plays and green while the robot is driving.

Notice how these My Blocks make this program much easier to understand or to explain to a friend. For this reason, it is sometimes useful to break your program into several My Blocks, even if you use some of them only once.

editing my blocks

You can edit the blocks inside My Blocks after you create them. To do so, double-click the My Block on the Programming Canvas to reveal its contents and then edit it as if it were a normal program. When you're finished, click **Save** and return to the program that uses the My Block.

To rename a My Block, double-click its name tab and enter a new name (just as you do for normal programs, as shown in Figure 3-11 on page 29).

managing my blocks in projects

You can use the My Blocks you create in any program within the same project. For example, you can use the Talk My Block in any program in the *EXPLOR3R-5* project file. But sometimes you'll want to use a My Block in other projects as well.

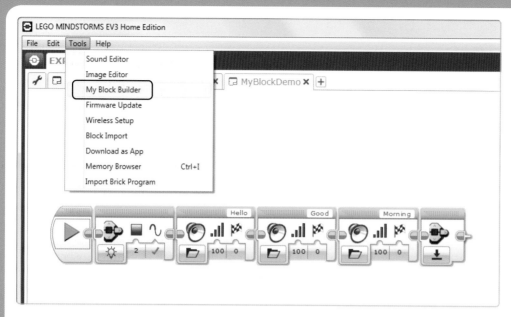

Figure 5-9: Place and configure the blocks as shown. Then draw a selection around all of them (except the orange Start block) and click **My Block Builder** in the menu.

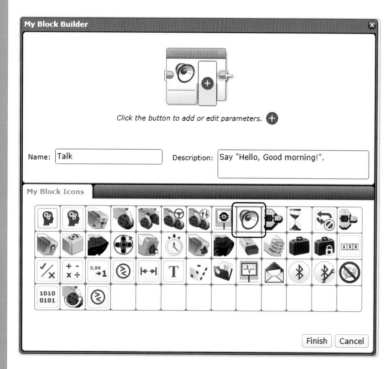

Figure 5-10: The My Block Builder. Choose a name, a description, and an icon for your My Block. Then click **Finish** to complete your block.

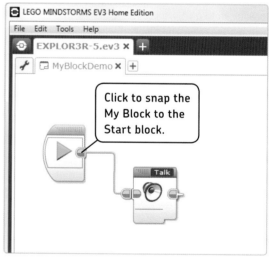

Figure 5-11: The completed My Block on the canvas. If your block is not correctly lined up with the Start block after its creation, click the left end of the Sequence Wire to snap the blocks back together.

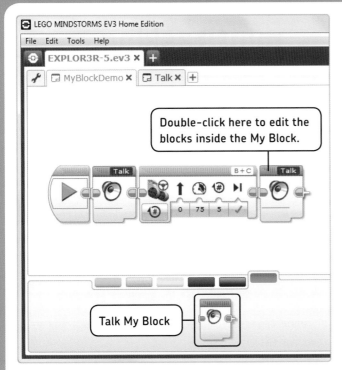

Double-click here to edit the blocks inside the My Block.

Talk My Block

Figure 5-12: The finished MyBlockDemo program

To copy the My Block you just made to another project (as shown in Figure 5-13), go to the Project Properties page (1), select **Talk.ev3p** on the My Block tab (2), and click **Copy** (3). Then, without closing the project, open the project you made before: *EXPLOR3R-4* (4). Go to its My Blocks on the Properties page (5) and click **Paste** (6). You should now be able to use the Talk My Block in the *EXPLOR3R-4* project.

Instead of using Copy, you can also click **Export**, which allows you to save a My Block to a file. You can email that saved file to a friend, who can then add it to a project using the **Import** button. To delete a My Block from a project, select the file and click **Delete**.

NOTE You can use the same method to share other files between projects, such as individual programs, custom sounds, and images.

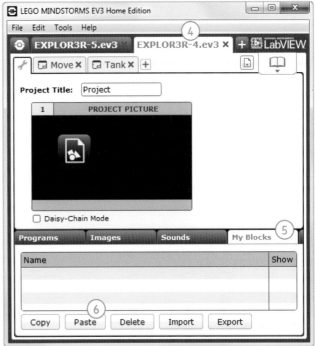

Figure 5-13: Copying a My Block from the EXPLOR3R-5 project to the EXPLOR3R-4 project.

DISCOVERY #18: MY SQUARE!

Difficulty: ▱   **Time:** ⏱

Open the *OneSquare* program from Figure 5-6 and turn the Loop block and its contents into a My Block called *MySquare*. Now you can use this block whenever you want your robot to drive in a square.

DISCOVERY #19: MY TUNE!

Difficulty: ▱   **Time:** ⏱

Remember the tune you made using Sound blocks in Discovery #6 on page 42? Convert that sequence of Sound blocks into a My Block so that you can easily use your favorite tune anytime in your programs.

multitasking

All the blocks that you've used so far are executed one at a time in the order in which they are lined up on the Programming Canvas. However, the EV3 can *multitask* by executing multiple blocks at the same time, using either multiple Start blocks or a split Sequence Wire. The methods are very similar, as you'll see.

using multiple start blocks

An easy way to have two sequences of blocks run *in parallel* (at the same time) is to add a second Start block, as shown in Figure 5-14. When you press Download and Run, both sequences start running simultaneously. The program ends once both sequences of blocks have finished running. To test one sequence individually, click the green arrow on its Start block.

When you run this program, the robot will move and play a sound at the same time.

splitting the sequence wire

Another way to have a robot multitask is to split the Sequence Wire, as shown in Figure 5-15. This is useful when you want two parallel sequences but you don't want them to start right at the beginning of the program. In the *MultiSequence* program shown, the robot plays a tone and then two actions occur in parallel: The robot drives forward while saying "Hello! Good morning!", using the Talk My Block that you made previously.

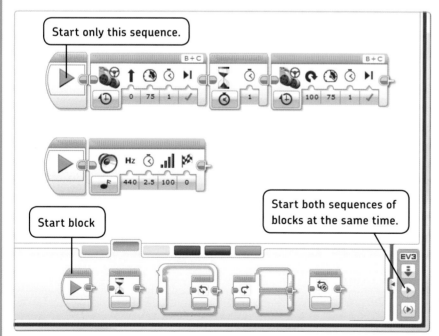

Figure 5-14: Multitasking with two Start blocks in the MultiStart program. You'll find the Start block with the Flow blocks (under the orange tab).

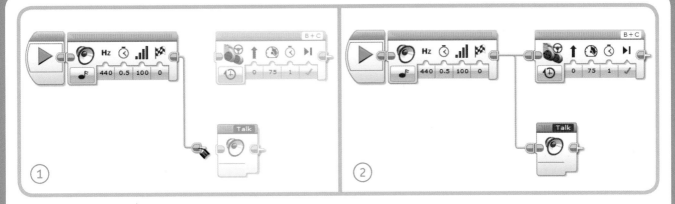

Figure 5-15: Multitasking with a split Sequence Wire in the MultiSequence *program. To create it, first place all required blocks on the canvas and then connect a Sequence Wire to each series of blocks. (This works even if the Sound block and the Move block are snapped together; the blocks will be separated automatically as you try to connect the Talk My Block.)*

avoiding resource conflicts

Just as you can't walk forward and backward at the same time, you can't program your robot to drive forward in one sequence while programming it to drive backward in another sequence. There is a *resource conflict* when two sequences of blocks try to control a single motor or sensor at the same time.

Unfortunately, the EV3 software won't tell you whether there's a resource conflict in your program; your program will probably still run, but the result will be unpredictable. For example, the robot might not drive in the expected direction. To avoid this potential problem, don't use the same motor or sensor in more than one sequence.

It's a good idea to avoid multitasking if possible because resource conflicts may arise unexpectedly. Sometimes you can accomplish the same functionality using just one sequence of blocks. For example, rather than moving and playing sounds using two parallel sequences, you can control both actions from a single sequence, as you'll see in Discovery #21.

further exploration

Having completed the first part of this book, you now have a solid foundation of several essential programming techniques. In this chapter, you learned how to use Wait and Loop blocks, how to create and edit My Blocks, and how to make your robots multitask.

In the next part of this book, you'll create robots that can interact with their environment with sensors. But before you do, practice a bit with what you've learned in this chapter by solving the following Discoveries.

DISCOVERY #20:
LET'S MULTITASK!

Difficulty: ▢▢ **Time:** ◷

Make the robot drive in a square-shaped pattern indefinitely and say repeatedly "LEGO, MINDSTORMS, EV3" at the same time.

DISCOVERY #21:
SINGLETASKING!

Difficulty: ▢ **Time:** ◷

The *MultiStart* program is a simple example that illustrates the principle of multitasking, but it's not always necessary to use multiple sequences to move and play sounds at once. Can you create a new program that does the same thing as the *MultiStart* program, with only one sequence of blocks?

For more of a challenge, do the same for the *MultiSequence* program.

HINT You learned how to do this in Chapter 4. What did the Play Type setting in the Sound block do?

DISCOVERY #22:
COMPLEX FIGURES!

Difficulty: ▭▭ **Time:** ⏱⏱⏱

Create a program that makes the EXPLOR3R drive in the pattern shown in Figure 5-16 while making different sounds.

HINT If you look carefully, you'll see that you can divide the track into four identical parts, so you have to configure a set of Move blocks for only one of these parts. Then you can place these blocks in a Loop block configured to repeat four times.

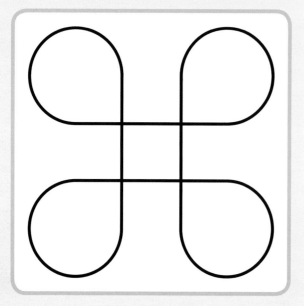

Figure 5-16: The drive track for Discovery #22

DESIGN DISCOVERY #3:
MR. EXPLOR3R!

Building: ✺ **Programming:** ▭

Add the Medium Motor to create a waving hand for your EXPLOR3R. Use other LEGO parts to further decorate your robot and turn it into your own Mr. EXPLOR3R. Make it continuously wave its hand and say a greeting like "Good morning!" at the same time.

PART II

programming robots with sensors

6

understanding sensors

The LEGO MINDSTORMS EV3 set includes three types of sensors: Touch, Color, and Infrared. You can use these sensors to make your robot respond to its environment. For example, you can program your robot to make a sound when it sees you, avoid obstacles while driving, or follow colored lines. This part of the book will teach you how to create working robots that use these sensors.

To learn how to work with sensors, we'll expand the EXPLOR3R robot by adding a bumper that detects obstacles using the Touch Sensor, as shown in Figure 6-1. Once you have a handle on how to make programs that use the Touch Sensor, you'll learn how to use the other sensors in subsequent chapters.

Figure 6-1: The EXPLOR3R uses a bumper with a Touch Sensor to detect objects it runs into.

what are sensors?

LEGO MINDSTORMS robots can't actually see or feel the way humans do, but you can add *sensors* to them so they can collect and report information about their environment. By designing programs that can interpret this sensor information, you can make your robots seem intelligent by having them respond to their environment. For instance, you could create a program that makes the robot say "blue" when one of its sensors sees a piece of blue paper.

understanding the sensors in the EV3 set

Your EV3 set contains three sensors that you can attach to your robot (see Figure 6-2) as well as some built-in sensors. The *Touch Sensor* detects whether the red button on the sensor is pressed or released. The *Color Sensor* detects the color of a surface and the intensity of a light source, as you'll see in Chapter 7. The *Infrared Sensor* (covered in Chapter 8) measures the approximate distance to a nearby object, and it receives signals from the infrared remote.

In addition, each motor in the EV3 set has a built-in *Rotation Sensor* to measure the motor's position and speed, and the EV3 brick can detect which of its *Brick Buttons* are pressed (see Chapter 9).

Figure 6-2: The EV3 set comes with a Touch Sensor (left), a Color Sensor (middle), and an Infrared Sensor (right).

understanding the touch sensor

The Touch Sensor allows your robot to "feel" by detecting whether the red button on the sensor is currently pressed or released, as shown in Figure 6-3. The EV3 retrieves this information from the sensor and can use it in programs. For example, you could make your robot say "hello" whenever you press the Touch Sensor.

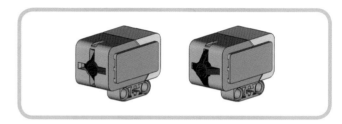

Figure 6-3: The Touch Sensor detects whether the red button is pressed (left) or released (right).

Despite its simplicity, the Touch Sensor is useful for many applications. For example, robots can use the Touch Sensor to detect obstacles in front of them. You can also use the Touch Sensor to detect that a mechanism in your robot has reached a certain position. In Chapter 18, for example, you'll use the sensor to detect whether a robotic arm is lifted all the way.

creating the bumper with the touch sensor

If you build a bumper and attach it to the Touch Sensor, then anytime the EXPLOR3R runs into an object with its bumper, the sensor will become pressed. The program on your EV3 can use this information to decide to move the robot in a different direction. Build the bumper and attach it to the robot as shown on the following pages. Be sure to connect the Touch Sensor to input port 1 using a short cable.

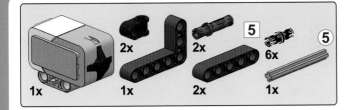

1

2

5

3

4

5

5

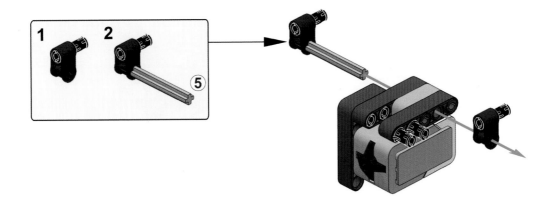

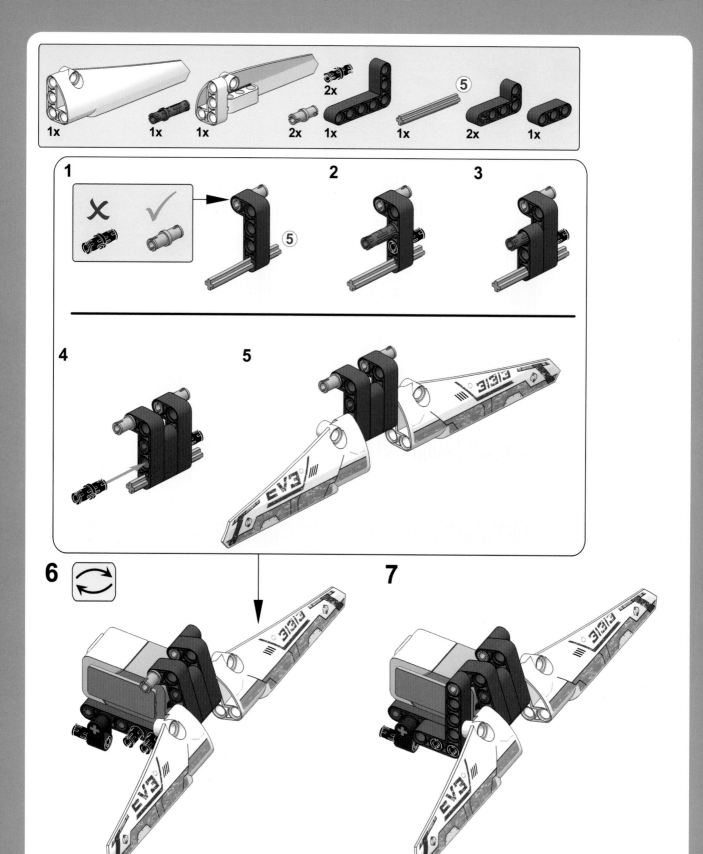

8

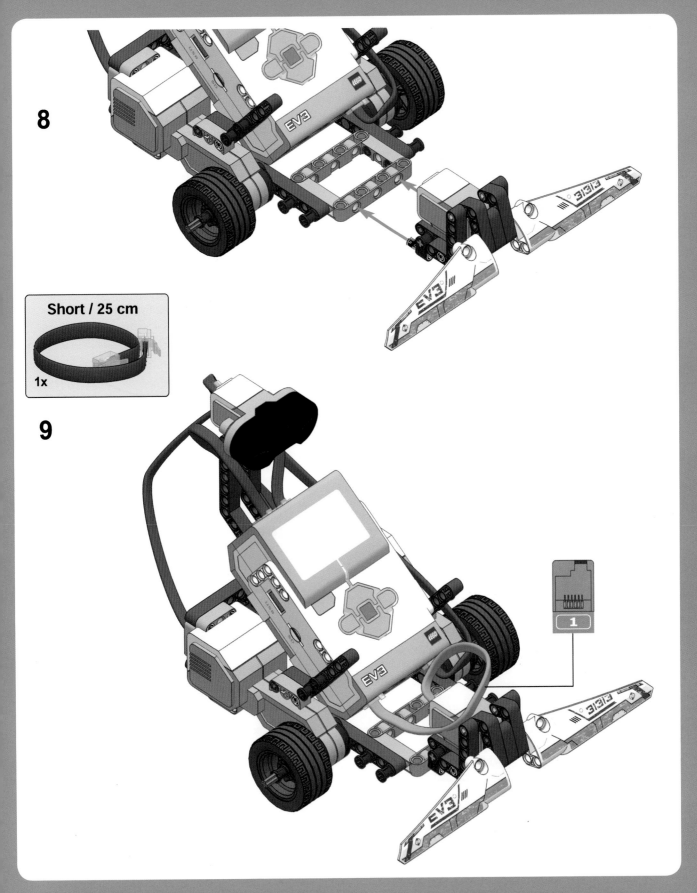

Short / 25 cm

1x

9

1

viewing sensor values

You can see the measurements reported by each sensor by opening the **Port View** application on the Brick Apps tab of your EV3 brick, as shown in Figure 6-4. For the Touch Sensor, a measurement of **1** means that it's pressed, while **0** indicates that it's released.

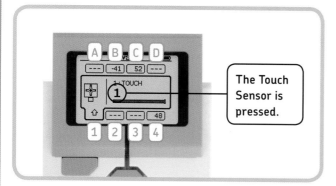

Figure 6-4: The Port View application on the Brick Apps tab. The EV3 brick automatically determines which sensors you've connected to the EV3 and displays their measurements on the screen. Use the buttons (Left, Right, Up, and Down) to see more details regarding each sensor.

You can use the buttons on the EV3 to navigate to measurements of the other sensors. At the bottom right of your screen (port 4), you should see the distance measurement of the Infrared Sensor (48% in this example). The two values at the top (–41 and 52) indicate the positions of the motors on ports B and C on your robot.

Some sensors can take more than one type of measurement. To see other measurements of the Infrared Sensor, for example, navigate to port 4, press the Center button, and choose a sensor mode. You'll learn more about the meaning of each value as you read on.

If your robot is connected to the computer, you can also view sensor measurements on the Hardware Page in the EV3 software, as shown in Figure 6-5. Just use the method you find most convenient.

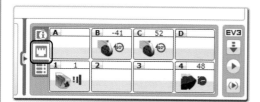

Figure 6-5: You can also view sensor measurements from within the EV3 software. If the values aren't being updated continuously, download a program to the robot to refresh the connection. Click one of the sensors to choose which type of measurement you want to see.

programming with sensors

Now let's look at how you can use these measurements in your programs. Let's try using the Touch Sensor in a program that has the robot play a sound when the Touch Sensor is pressed.

Several programming blocks let you use sensors in your program, including the Wait, Loop, and Switch blocks. In this chapter, you'll learn how each of these blocks work with the Touch Sensor, and the same principles apply to the other sensors in the EV3 set as well.

sensors and the wait block

Earlier, you used a Wait block to pause the program for a set amount of time (say, five seconds). But you can also use a Wait block to pause a program until a sensor is triggered. For example, you can configure a Wait block to pause until the Touch Sensor is pressed by selecting the *Touch Sensor* mode, as shown in Figure 6-6.

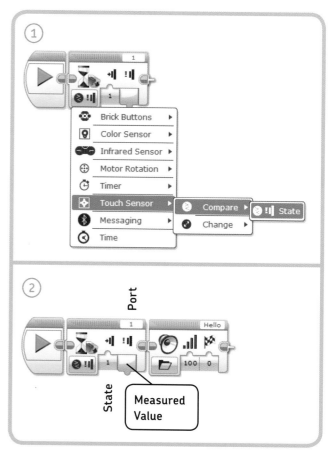

Figure 6-6: The WaitForTouch program makes the robot play a sound when the Touch Sensor is pressed.

After selecting this mode, you must also choose between Compare and Change mode. In *Compare* mode, you specify with the *State* setting whether the program should wait until the sensor is released (0), pressed (1), or bumped (2). If you choose *bumped*, the program waits for a press followed by a release.

In *Change* mode, the program waits until the state of the sensor changes: If the sensor is pressed when the block begins running, the program waits until it's released. If it's released at first, the program pauses until the sensor is pressed.

The *Port* setting lets you specify which input port your sensor is connected to (in this case, port 1). Finally, the *Measured Value* plug allows you to use the last sensor measurement later on in your program (we'll get back to this in Part V of the book).

sensors and the wait block in action

Create a new project called *EXPLOR3R-Touch* with a program called *WaitForTouch*, as shown in Figure 6-6. The mode of the Wait block is set to **Touch Sensor – Compare – State**. When you run the program, nothing will happen at first, but when you press the Touch Sensor (by pushing the bumper), the robot should say "Hello."

Now keep the Touch Sensor pressed as you start the program again. The sound plays immediately because the Wait block has nothing to wait for—the sensor is already pressed.

DISCOVERY #23:
HELLO AND GOODBYE!

Difficulty: 🖵🖵 **Time:** ⏱
Can you create a program that has the robot say "Hello" when you press the bumper on the robot and then "Goodbye" when you release the bumper?

HINT Add another pair of Wait and Sound blocks to the *WaitForTouch* program (see Figure 6-6). The first Wait block should wait for a press, and the second should wait for a release. Where do you place these new blocks?

---

avoiding obstacles with the touch sensor

Now that you're familiar with the Touch Sensor and the Wait block, you're ready to make some more exciting programs. The next program, *TouchAvoid*, will make the EXPLOR3R robot drive around a room and turn around when it feels something, such as a wall or chair, with its bumper. You can see an overview of the program in Figure 6-7.

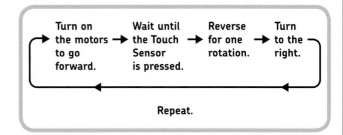

Turn on the motors to go forward. → Wait until the Touch Sensor is pressed. → Reverse for one rotation. → Turn to the right.

Repeat.

Figure 6-7: The program flow in the TouchAvoid *program. After turning right, the program returns to the beginning and repeats.*

You can accomplish each action in this diagram with one programming block. You'll use a Move Steering block in **On** mode to turn on the motors and then use a Wait block to wait for the sensor to be pressed. (Note that while the program waits, the robot keeps moving forward.)

Once the sensor is triggered, you use a Move Steering block to reverse and then another to turn around, both in **On For Rotations** mode. After the robot turns around, the program returns to the beginning, which is why you must place the four blocks you use inside a Loop block configured in **Unlimited** mode. Create the program now, as shown in Figure 6-8.

DISCOVERY #24:
AVOID OBSTACLES
AND A BAD MOOD!

Difficulty: 🖵 **Time:** ⏱
Expand the *TouchAvoid* program by making it display a happy face on the EV3's display as the robot moves forward and a sad face when it's reversing and turning.

HINT Place two Display blocks somewhere in the Loop block.

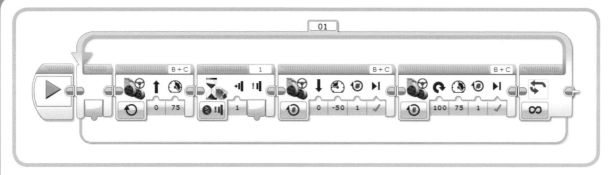

Figure 6-8: The TouchAvoid program. The Wait block is configured in Touch Sensor – Compare – State mode.

For most programs, it's best to use the Compare mode because doing so makes predicting your robot's behavior easier. Regardless of the initial state of the sensor, the robot will always wait to change its behavior until the Touch Sensor reaches the state of your choice.

sensors and the loop block

As you learned in Chapter 5, you can configure a Loop block to loop a certain number of times, loop for a specified number of seconds, or loop endlessly. You can also program a Loop block to stop repeating based on sensor input. For example, you can make your robot drive back and forth until the Touch Sensor is pressed. To configure the Loop block like this, select the **Touch Sensor – State** mode, as shown in Figure 6-10. As before, choose **1** in the Port setting.

Create the *LoopUntilTouch* program and run it to see how it works. You should notice that the program checks the sensor measurement only once each time the blocks inside the loop have completed. For the loop to end, the sensor will need to

using change mode

So far we've used the Wait block in Compare mode to make the program pause until the Touch Sensor reaches a state of your choice (pressed or released). Now we'll create a program with a Wait block in *Change* mode that will pause the program until the sensor state changes (either from released to pressed or from pressed to released). Create and run the *WaitForChange* program, as shown in Figure 6-9.

If the bumper is released when the program starts, the robot should drive until it hits an object and then stop. If the bumper is already pressed when the program starts, the robot should continue to try to go forward until the sensor is released and then stop.

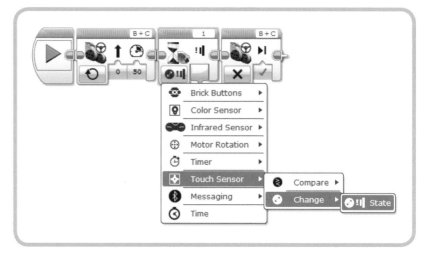

Figure 6-9: The WaitForChange program

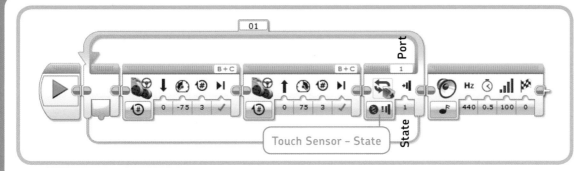

Figure 6-10: The LoopUntilTouch *program. To configure the Loop block, click the Mode Selector and choose* **Touch Sensor – State**.

be pressed just after the robot moves forward. If the sensor is not pressed at this point in the loop, the robot moves back and forth once more before checking the state of the Touch Sensor again.

This is the expected behavior for the Loop block, but sometimes you'll want the loop to end even if you don't press the sensor at exactly the right time. To accomplish this, set the state setting to *Bumped* (2) and run the program again. In this configuration, the loop doesn't check whether the Touch Sensor is pressed at the *end* of the loop; rather, it checks whether you bump (press and release) the sensor at any time *during* the loop. If you do, the blocks should stop repeating after they complete the current run. (The EV3 continuously monitors the state of the Touch Sensor while the loop runs so that you don't have to worry about it.)

DISCOVERY #26: HAPPY TUNES!

Difficulty: 🖭 **Time:** ⏲
Use a Loop block to make the robot play a tune until the robot's bumper is pressed, at which point the robot should scream and quickly turn around.

HINT You can use the My Block that you made in Discovery #19 on page 56 for your tune. If you have yet to create your own tune, simply select a sound file from the list in a Sound block.

sensors and the switch block

You can use a *Switch block* to have a robot make a decision based on a sensor measurement. For example, you can make your robot drive backward if the Touch Sensor is pressed or say "No Object" when it's not pressed, as shown in Figure 6-11.

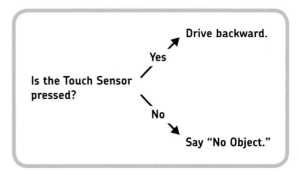

Figure 6-11: A robot can make a decision based on a sensor measurement.

The Switch block checks whether a given *condition* (such as "The Touch Sensor is pressed") is true or false, as shown in Figure 6-12.

The Switch block in this example contains two sequences of blocks; the switch decides which sequence to run based on whether the condition is true or not. If the condition is true, the block in the upper part of the switch is run, and the robot moves backward; if the condition is false, the lower blocks are run, and the robot should say "No Object."

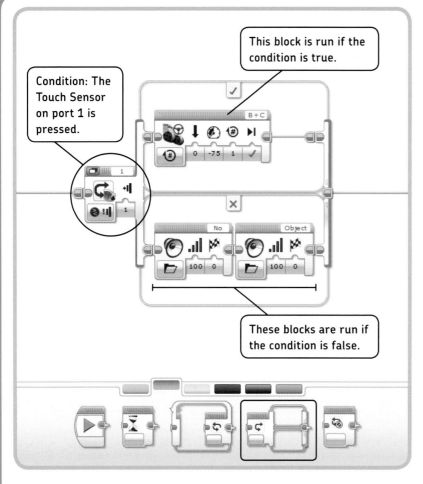

Condition: The Touch Sensor on port 1 is pressed.

This block is run if the condition is true.

These blocks are run if the condition is false.

Figure 6-12: The Switch block checks whether the condition is true or false and runs the appropriate blocks. You specify the condition using the mode and settings on the Switch block.

configuring a switch block

You define the condition by configuring the mode and settings of the Switch block. Once the program arrives at the Switch block, the robot checks whether the condition is true. Then, it decides which set of programming blocks in the switch to run.

There's a mode for each sensor; in this case, you'll choose the one for the Touch Sensor, namely **Touch Sensor – Compare – State** (the only available option). Once you've chosen this mode, you can specify in the state setting whether the Touch Sensor must be pressed (1) or released (0) for the condition to be true. As before, set Port to **1** to specify how the Touch Sensor is connected to your EV3.

sensors and the switch block in action

The *TouchSwitch* program you'll now create makes the robot drive forward for three seconds. Then, if the Touch Sensor is pressed, the robot reverses for a short while. If the sensor is not pressed, the robot instead says "No Object." Finally, regardless of the Switch block's decision, the robot plays a tone. Now create the program, as shown in Figure 6-13.

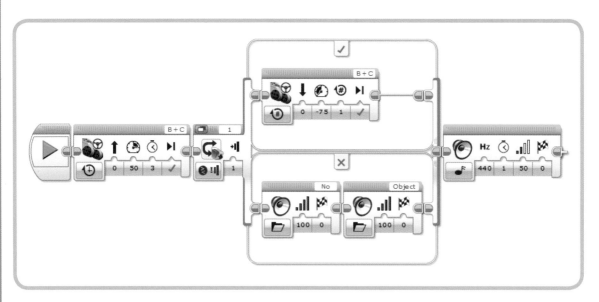

Figure 6-13: The TouchSwitch program has the robot decide what to do based on a sensor reading.

Try running this program a few times, and determine when you need to press the Touch Sensor to make the robot go backward. Your experiments should show that the robot takes a measurement when the Switch block runs and that it uses this single measurement to determine whether the condition is true. In this program, the sensor measurement is taken just after the robot finishes going forward. When either the reverse action or the "no object" action is complete, the tone plays.

adding blocks to a switch block

There's no limit to the number of blocks you can place inside a Switch block. If one part of a switch has multiple blocks, they're simply run one by one. You can also leave one of the two parts of a Switch block empty, as shown in Figure 6-14.

Run this modified program to see what happens. If the condition is true (the bumper is pressed), the robot should say "Object" and move backward, and the program should continue by playing the tone. If the condition is false (the sensor is not pressed), the program will find no blocks in the lower part of the switch and instantly move on to the Sound block after the switch.

DISCOVERY #27:
STAY OR MOVE?

Difficulty: 🔲 **Time:** ⏱
Make the robot stand still for three seconds. Then, if the Touch Sensor is released, the robot should turn around and drive forward for five wheel rotations. But if the sensor is pressed, the robot should do nothing, and the program should end immediately.

DISCOVERY #28:
DIFFICULT DECISIONS!

Difficulty: 🔲 **Time:** ⏱⏱
Let's practice with the Switch block! Create a program to implement the decision tree shown in Figure 6-15. How do you configure the Switch block, and why do you have to put a Wait block at the end of the program?

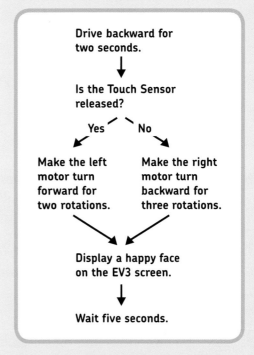

Figure 6-15: The program flow for Discovery #28

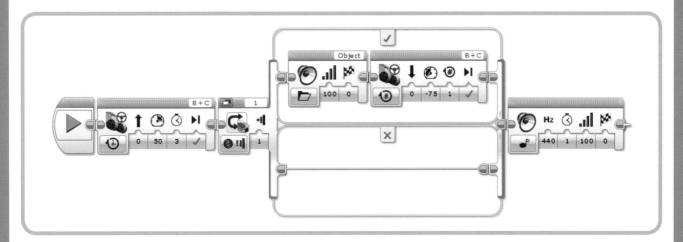

Figure 6-14: A modified version of the TouchSwitch *program. The switch does not have any blocks to run if the condition is false, so the program immediately plays a tone after moving forward if the sensor is not pressed.*

using flat view and tabbed view

Normally, you see the complete Switch block on your screen in *Flat View*. When you make large programs containing Switch blocks, it's easy to lose track of how your program works. In such cases, you can display the block in *Tabbed View* to decrease the size of the Switch block, as shown in Figure 6-16. Both parts of the switch are still in the program, but they're on separate tabs, which you can open by clicking them.

repeating switches

Every time your program arrives at the Switch block, it checks the state of the Touch Sensor to decide whether the blocks in the true or false part of the switch should be run.

To have a robot check a condition more than once, you can drag a Switch block into a Loop block. For example, you could program a robot to say "Yes" if the Touch Sensor is pressed and to say "No" otherwise. If you place a Switch block with this configuration in a Loop block, the robot will keep checking the sensor reading and continue to say "Yes" or "No" accordingly.

Now create the *RepeatSwitch* program shown in Figure 6-17.

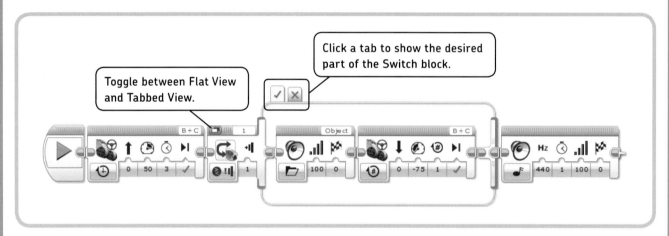

Figure 6-16: Decrease the size of the Switch block by changing to Tabbed View. This option just changes your view of the block; it won't affect the way the program works.

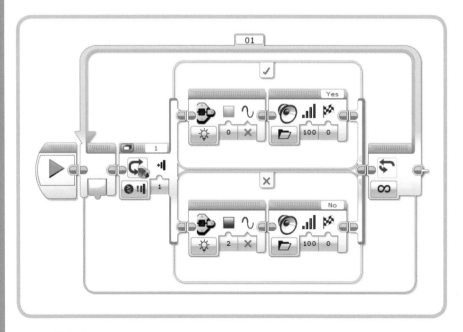

Figure 6-17: The RepeatSwitch program

understanding compare, change, and measure mode

If you use sensor input in any programs with Wait, Loop, and Switch blocks, you often have to choose among Compare, Change, and Measure when configuring a block's mode. You'll see more examples of these modes as you read on, but let's take some time to look at them side by side to see how they work.

compare mode

Compare mode (⊘) makes your robot take a sensor value and check it against the *condition* as specified in the block's settings. A condition is a statement like "The Touch Sensor is pressed," "The measured light intensity is less than 37%," or "The Color Sensor sees red or blue."

* A *Wait block in Compare mode* keeps taking new sensor measurements until the condition becomes true. When it's true, the program moves on to the next block in line (see Figure 6-6).
* A *Loop block in Compare mode* checks the condition against a new sensor value each time it's done running the blocks inside the loop. If the condition is true, the program moves on to the block after the loop; if it's false, the loop runs again (see Figure 6-10). Loop blocks are always in Compare mode.
* A *Switch block in Compare mode* runs the sequence of blocks at the top if the condition is true; it runs the blocks at the bottom if it's false (see Figure 6-13).

change mode

Change mode (⊘) is available only in Wait blocks. A *Wait block in Change mode* takes an initial measurement and keeps taking new measurements until it finds one that's different from the first one. For example, if the Touch Sensor is pressed when the block starts, it waits until the Touch Sensor is released. Then, the program continues with the next block (see Figure 6-9).

measure mode

Measure mode (⊞) is available only in Switch blocks. A *Switch block in Measure mode* contains a set of blocks to run for each possible sensor value. You'll see how this works in Chapter 7 when you create a program that does something different for each color the Color Sensor can see.

configuring the modes

The text usually makes clear which mode you should use for the example programs in this book, but all of the information you need is also visible in the programming diagrams. If you're not sure which mode to choose, just look at the icons on each block, as shown in Figure 6-18.

Once you've selected the sensor (1) and made a choice of Compare, Change, or Measure (2), you choose the *sensor operation mode* (3). The Touch Sensor measures just one thing (the state of the red button), but the Color Sensor has three operation modes, as shown in Figure 6-18. As you practice with each mode in the next chapters, you'll be able to create your own programs with sensors in no time.

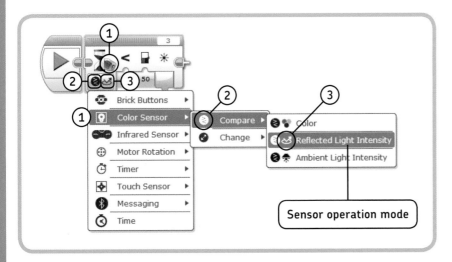

Figure 6-18: When you see Wait, Loop, or Switch blocks in the example programs in this book, choose the menu items to match the icons shown on the blocks. Begin by choosing the correct sensor (1) and then choose Compare, Change, or Measure (2). Finally, choose the sensor operation mode (3).

further exploration

In this chapter, you've learned that robots use sensors to gather input from their environment. You've also learned how to create programs for robots with sensors using Wait, Loop, and Switch blocks.

You've been using only the Touch Sensor so far, but you can use all the programming techniques you've learned in this chapter when working with the other sensors. You'll continue with the Color Sensor in Chapter 7, you'll learn about the Infrared Sensor in Chapter 8, and you'll meet the built-in Rotation Sensors and the Brick Buttons in Chapter 9. Before you continue, practice your sensor-programming skills by solving these Discoveries.

DISCOVERY #29: CHOOSING DIRECTIONS!

Difficulty: Time:
Expand the *TouchAvoid* program from Figure 6-8 to make the robot go right after running into the first object and left when the sensor is pressed again. The next obstacle should result in a right turn again, and so on.

HINT Duplicate the four blocks inside the loop so there are eight blocks in the loop, and change the steering setting in the second set of blocks.

DISCOVERY #30: WAIT, LOOP, OR SWITCH?

Difficulty: Time:
Program the robot to wait for a Touch Sensor press. Then, if the sensor is still pressed after five more seconds, make the robot say "Yes." Otherwise, make it say "No."

HINT You'll need a combination of Switch and Wait blocks.

DISCOVERY #31: BRICK BUTTONS!

Difficulty: Time:
As you'll see in Chapter 9, you can use most of the Brick Buttons the same way you've used the Touch Sensor. Without skipping ahead, can you make the robot say "Left" if you press the Left button on the EV3 Brick and "Right" if you press the Right button? Can you make it say "Up" and "Down" when you push the Up and Down buttons as well?

DESIGN DISCOVERY #4: INTRUDER ALARM!

Building: **Programming:**
Can you turn your robot into an alarm that alerts you to intruders entering the room? Use the Touch Sensor as a switch that gets activated when someone breaks into the room. Play a loud beeping sound when this happens.

HINT Place something heavy, like a book, in front of your robot, pressing the bumper. Then, design a contraption that automatically pulls the book away when someone enters the room. Your robot should play the sound as soon as the Touch Sensor is no longer pressed.

DESIGN DISCOVERY #5: LIGHT SWITCH!

Building: **Programming:**
Design a robot that toggles the light switch in your room whenever you press the Touch Sensor. Each time you press the sensor, the robot should switch the light on or off. You can add the Medium Motor to the EXPLOR3R to do this, or you can design a completely new robot dedicated to this task.

7

using the color sensor

In this chapter, you'll learn to use the Color Sensor by adding it to the EXPLOR3R (see Figure 7-1) so the robot can detect colored paper, follow lines, and respond to light signals.

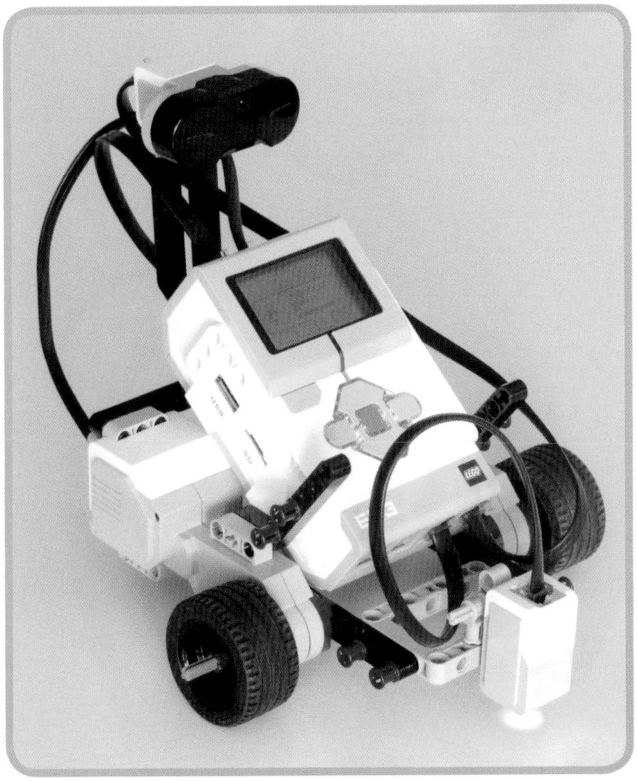

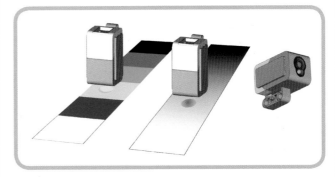
Figure 7-1: Using the Color Sensor, the EXPLOR3R can detect colors and follow lines.

The *Color Sensor* can detect the color of a surface (in Color mode), the amount of light reflected by a surface (in Reflected Light Intensity mode), or the brightness of ambient light (in Ambient Light Intensity mode), as shown in Figure 7-2.

You'll create programs for the EXPLOR3R to try out each of these modes using the Wait, Loop, and Switch blocks, much as you've already done with the Touch Sensor. You'll see more applications for this sensor as you build the robots later in the book—for example, LAVA R3X in Chapter 19, which measures reflected light intensity to detect your handshake.

Figure 7-2: The three operation modes of the Color Sensor: Color mode (left), Reflected Light Intensity mode (middle), and Ambient Light Intensity mode (right). The sensor on the right points upward to measure the intensity of light in a room.

attaching the color sensor

Before you begin programming, remove the Touch Sensor attachment from the robot (don't take it apart though; you'll need it later). Then, connect the Color Sensor to the robot using the instructions on the next page.

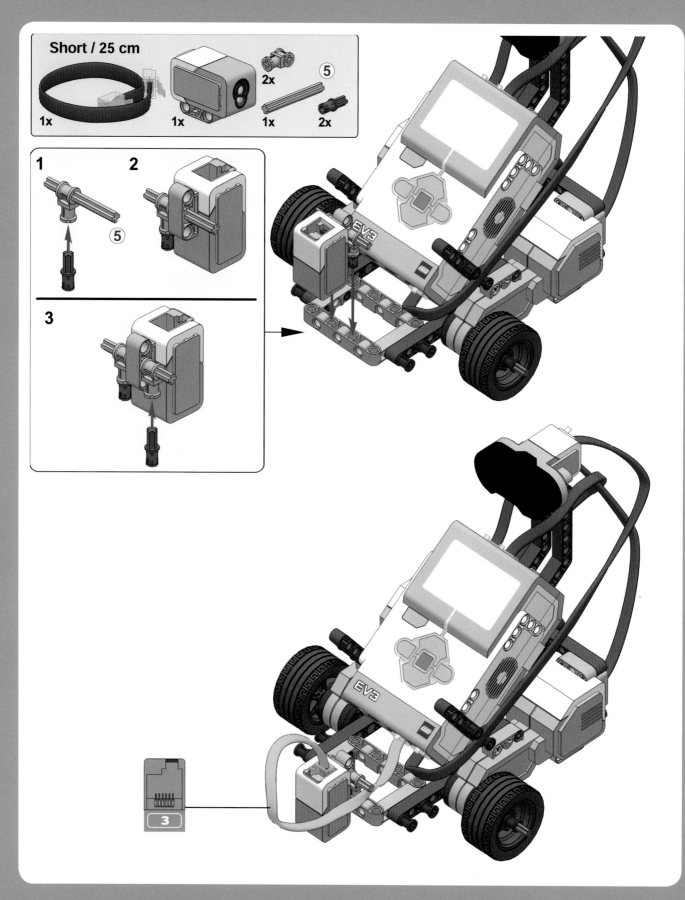

Short / 25 cm

1x   1x   2x   ⑤   1x   2x

1

2

⑤

3

3

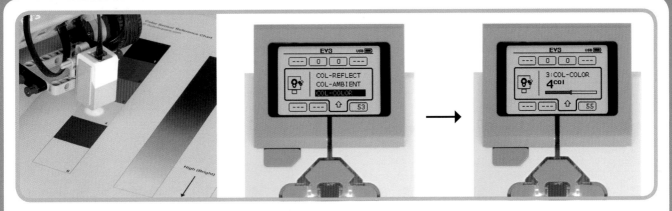

Figure 7-3: Go to the Port View app to see the sensor measurement. Navigate to the sensor on input port 3, press the **Center** button, select **COL-COLOR**, and press **Center** again. You should now see a number between 0 and 7, representing a color. The sensor value is **4** in this case (representing yellow).

color mode

The first operation mode you'll use is *Color mode*, in which the sensor can detect the color of surfaces about 1 cm (0.4 inches) away. The sensor is mounted pointing straight down so it can detect the color of the surface beneath the robot.

To test the color measurement, download and print the color reference chart from *http://ev3.robotsquare.com/color.pdf* and position the robot on top of it (if you don't have a color printer or if the sensor does not detect the colors properly, try using the Mission Pad that comes with the EV3 set). Then, go to the Port View app on the EV3 brick to see the detected color, displayed as a number (see Figure 7-3).

The sensor can distinguish among black (1), blue (2), green (3), yellow (4), red (5), white (6), and brown (7). A **0** measurement indicates that the sensor is not able to detect any color, which may mean that the surface is too far away from the sensor or too close to it.

staying inside a colored line

Programs can use Wait, Loop, and Switch blocks to make decisions based on the sensor value. For example, you can make the robot drive around without going outside the black outline shown in Figure 7-4. To do this, the robot should drive forward until it sees the black line. Then, it should back up, turn around, and move forward in another direction.

creating the test track

First, you need to create the circular test track that you'll use to test the sensor. The track is composed of a set of tiles printed on standard A4 or US Letter paper. Download and print the tiles from *http://ev3.robotsquare.com/testtrack.pdf*, cut them to size along the dashed line, and use some tape to keep them together so your test track looks just like the one shown in Figure 7-4.

If you're unable to print the test track, create your own track using some black tape and a light-colored surface, such as white kitchen tiles or a large sheet of plywood.

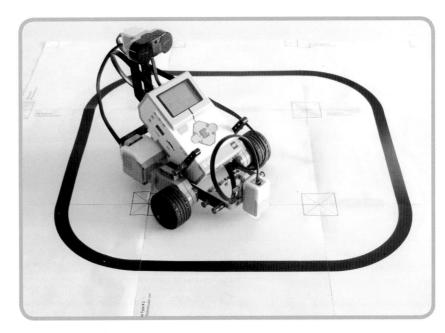

Figure 7-4: The EXPLOR3R drives around without leaving the black shape on the test track.

creating the program

Figure 7-5 shows the program flow we'll need to make the robot drive around without leaving the black shape. It's similar to the wall-avoidance program you made for the Touch Sensor (see Figure 6-7 on page 67), only this time the program waits until it sees the black line rather than for a button press.

For this program, you'll use a Wait block in **Color Sensor – Compare – Color** mode. In this configuration, you can choose a combination of colors that the sensor should look for using the *Set of Colors* setting. When it sees any of them, the block stops waiting, and the program moves on to the next block. Figure 7-6 shows the finished *StayInCircle* program with a Wait block configured to wait for a black line. Place the robot inside the black outline on the test track, and run the program to see it in action.

(see Figure 6-7 on page 67)

> ## DESIGN DISCOVERY #6: BULLDOZER!
>
> **Building:** ☀ **Programming:** ▭
>
> The *StayInCircle* program will keep the robot moving around the circle in different directions. If you put some LEGO pieces in the circle, you can make your robot push them out. But first you need to give EXPLOR3R a bulldozer blade. Can you build such a blade with your LEGO pieces?

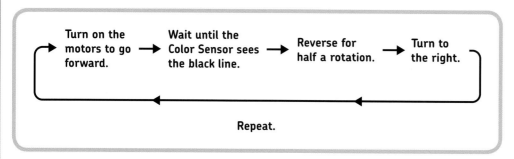

Figure 7-5: The program flow of the StayInCircle program

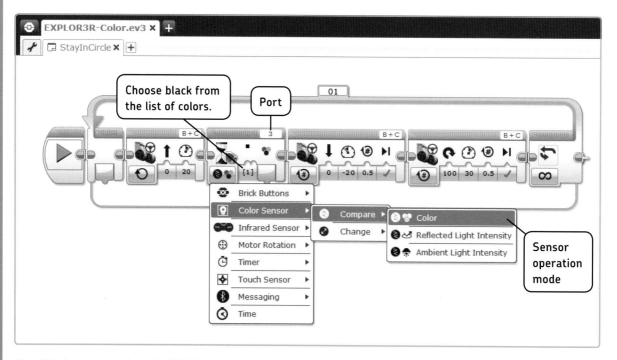

Figure 7-6: Create a new project called EXPLOR3R-Color with one program called StayInCircle. To configure the Wait block, first choose the mode; then click the Set of Colors setting and choose black (1) from the list of colors that appears.

following a line

In your next project, you'll use the Color Sensor to create a line-following robot—a robot that follows the line on your test track. Let's look at the strategy behind this program.

When the robot is following a black line on a white mat, the sensor will always detect one of two colors: white or black. Therefore, to create a line-following program for a black-and-white environment, you can use a Switch block that looks for the color black. When the sensor sees black, the Switch block triggers a Move Steering block to perform one movement; if it sees another color (white), it performs a different movement, as shown in Figure 7-7.

If the robot sees white, it won't know which side of the line it's on. You need to make sure it always stays on only one side of the line; otherwise, it will stray off the line into the white area. You can do this by always driving EXPLOR3R right when it sees black and left when it sees white. If you make the robot drive forward a little as it steers and if you repeat this behavior indefinitely, the robot follows the line. The *ColorLine* program implements this strategy, as shown in Figure 7-8.

NOTE If your robot strays off the line in corners or on sharp curves, make it drive slower (choose 20% instead of 25% speed).

Now let's have a look at the robot's behavior as it follows the line. Notice that the robot actually follows the *edge* of the line. The program keeps adjusting the robot's steering to the sensor measurement, causing it to zigzag across the edge: As soon as it sees the black line, it tries to move away from it by going right; as soon as it sees the white area, it tries to move back to the line by going left.

As a result, the robot keeps the line to the left of the sensor. This means that if you place the robot on the line such that it follows the circle clockwise, it traces the inner edge of the circle; if you make it drive around counter-clockwise, it follows the outer edge. To see this, turn the robot around manually while the program runs to make it follow the line the other way; it should start following the other edge of the line.

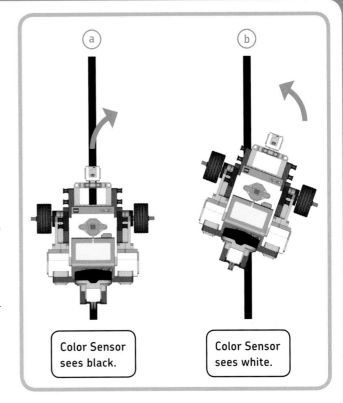

Figure 7-7: The EXPLOR3R steers right if it sees the black line (a) and steers left if it sees the white area (b). As it steers, it moves forward, so if you have the program repeat this behavior, you end up with a line-following robot.

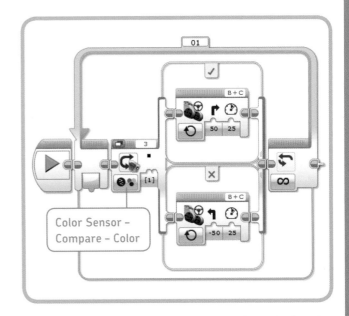

Figure 7-8: The ColorLine program. Note that the Move Steering blocks are in On mode. Once the robot starts to turn, it instantly goes back to the beginning of the program to see whether a different color has been detected or whether it should keep turning in the same direction. The On mode just switches on the motors and has the program continue.

the switch block in measure mode

The Switch block in the *ColorLine* program makes the robot go right if the sensor sees the black line. If it sees any other color, such as green or red, it goes left. By changing the Switch block's mode to **Color Sensor – Measure – Color**, you can configure the block to do something different for each color.

The *ShowColor* program in Figure 7-9 changes the brick status light color based on the Color Sensor measurement. The Switch block in this program has four *cases*, each containing one or more blocks. Each case corresponds to a different color measurement: The status light turns green if the sensor

measures green, red if it measures red, and orange if it measures yellow. If no color is detected (☐), the status light turns off. The program runs the blocks belonging to whichever case it detects when it gets to the Switch block.

But what happens if the sensor sees black, blue, white, or brown? If none of the cases match the sensor value, the switch runs the *default case*, which is marked by a dot, as shown in Figure 7-9. Here, it runs the same blocks that run when no color is detected.

Create the program now and move the sensor over the color reference chart to test it.

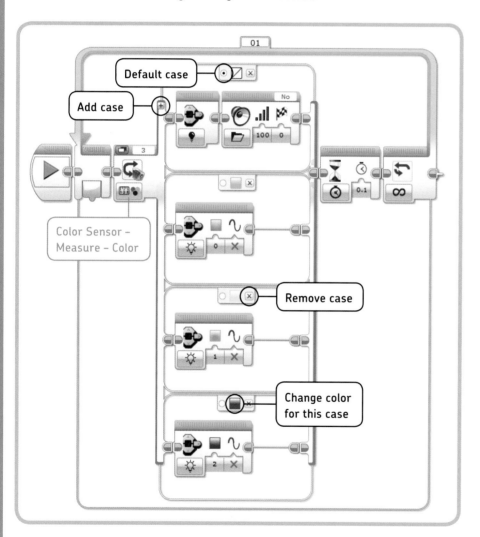

Figure 7-9: The ShowColor *program. To configure the Switch block with its cases, first select the* **Color Sensor – Measure – Color** *mode and add two cases by clicking the **+** sign twice. Now that your switch has four cases, select the proper color for each. Don't forget to mark the no-color case as the default by ticking the indicated field.*

DISCOVERY #32: CREATE YOUR OWN TRACK!

Difficulty: 🔲 **Time:** ⏱⏱⏱

The test track you just made is good to begin with, but the EXPLOR3R can handle much more challenging tracks. Go to *http://ev3.robotsquare.com/lines.pdf* to create your own custom track. You can choose from 30 types of tiles, including ones with straight lines, corners, and junctions. Print out the tiles you like, cut them to size along the dashed lines, and use some tape to put them together.

To begin, print four corners (four copies of page 3), a zigzagged line (page 15), and a straight line with a blue line across (page 18) to create the track shown in Figure 7-10. Run the *ColorLine* program you made to test the EXPLOR3R on your new track.

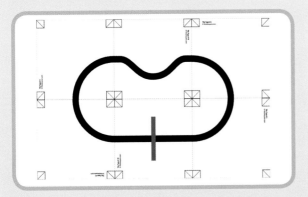

Figure 7-10: The line-following track for Discoveries #32 and #33

TIP Some printers allow you to print all required tiles at once by entering the page numbers in the Page field of your printer settings as follows: 3,3,3,3,15,18

DISCOVERY #33: STOP AT THE BLUE SIGN!

Difficulty: 🔲🔲 **Time:** ⏱

Modify the *ColorLine* program so it follows the black line on the test track you made in Discovery #32 until it comes across the blue line. When it sees blue, the robot should stop and make a sound.

HINT Change the mode of the Loop block to look for blue.

DISCOVERY #34: SAY COLOR!

Difficulty: 🔲🔲 **Time:** ⏱⏱

Create a program that has the robot tell you which color it sees. Make it say "Blue" if the sensor sees blue, and so on. Make it say "No Object Detected" if it doesn't see any color. When you're ready, turn the Switch block that determines the color into a My Block called *SayColor*. You can use this block anytime you want to know which color the robot sees.

HINT Your program will be similar to the *ShowColor* program.

reflected light intensity mode

The Color Sensor can also measure the brightness of a color in *Reflected Light Intensity mode*. For example, the sensor can see the difference among white, grey, and black paper by shining a light on the paper and measuring how much light is reflected. The Reflected Light Intensity is measured as a percentage from 0% (very low reflectivity: dark) to 100% (very high reflectivity: light).

Black paper doesn't reflect much light, resulting in a measurement below 10%. White paper can result in a measurement higher than 60%. To verify these values, go to Port View on the EV3 brick, choose port 3, and select **COL-REFLECT** (see Figure 7-3 for instructions). Place the robot on the color reference chart that you printed, and observe how the sensor value changes as you move the robot over the bar with various tones of grey.

DISCOVERY #35: SUPER REFLECTOR!

Difficulty: 🔲 **Time:** ⏱

You should be able to find at least one material that results in a Reflected Light Intensity value as high as 100%. Which material is this, and why is the sensor value so high?

setting the threshold value

In Color mode, the sensor could see that the test track was either black or white. Now observe the reflected light measurement as you drag the robot with your hands (very slowly!) from the black line onto the white area of the test track. You'll notice that the value gradually increases from around 6% (black) to around 62% (white). When the sensor partly sees the black line and partly sees the white paper, the value will be between these two extremes—as if the sensor were seeing a grey surface. You can use this more detailed measurement to improve your line-following robot.

To tell your robot what measurement you consider to be the white area or the black line, you'll define white and black in your program using a *threshold value*. You'll consider a measured value greater than this threshold as white and a measurement lower than this threshold as black. In other words, you take dark grey tones to be black and light grey tones to be white, as shown in Figure 7-11.

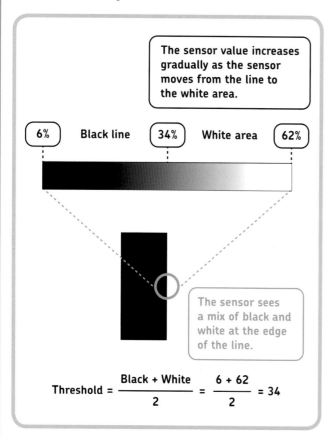

Figure 7-11: *At the edge of the line, the sensor sees a mix of black and white, resulting in a measurement between these extremes, as if it were seeing a grey surface. The threshold is the average of the sensor value found with the sensor on the black line (a low number) and the one found on the white area (a bigger number). To calculate the average, add both values and divide the total by 2.*

comparing sensor values with a threshold

You'll now create a new line-following program that uses Reflected Light Intensity mode and a threshold value to determine whether the sensor is on the black line or on the white area. As before, the robot should turn right if it sees the black line and left if it sees white.

To accomplish this, use a Switch block in **Color Sensor – Compare – Reflected Light Intensity** mode, as shown in Figure 7-12. Enter the previously calculated threshold in the Threshold Value setting. The *Compare Type* setting specifies which sensor values make the condition *true*—that is, which sensor values will make the blocks in the top of the Switch run. You can choose to run the upper blocks when the sensor value meets one of these conditions:

0. Equal To (=) the threshold

1. Not Equal To (≠) the threshold

2. Greater Than (>) the threshold

3. Greater Than Or Equal To (≥) the threshold

4. Less Than (<) the threshold

5. Less Than or Equal To (≤) the threshold

You want the robot to turn to the right if the sensor sees black, which is when the sensor value is *less than* the threshold, so you can choose that as the condition that triggers the upper set of blocks. Create and run the *ReflectedLine1* program now (see Figure 7-12) and test that it performs the same way as the previous program.

NOTE Be sure to calculate your own threshold values rather than using the values given in the diagrams. You may find different measurements for black and white, depending on factors such as the light level in the room, the robot's battery level, and the type of paper you use.

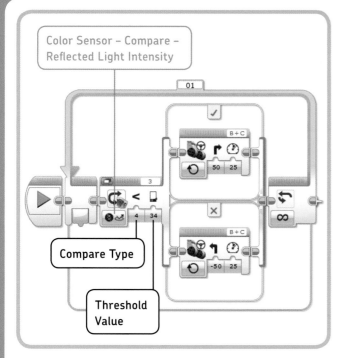

Figure 7-12: The ReflectedLine1 program makes the EXPLOR3R follow the line on the test track using Reflected Light Intensity mode. Less than (<) is option 4 in the list of Compare Types.

following a line more smoothly

The advantage of the Reflected Light Intensity mode is that the robot can measure not only black and white but also a mix of black and white as it moves across the edge of the line. If the robot sees white in the *ReflectedLine1* program, it makes a sharp turn to the left to get back to the line.

But sharp turns aren't necessary if the sensor is close to the line. If the robot measures light grey, a soft left turn might be sufficient to return to the line. This makes the robot follow the line more smoothly, rather than bouncing left and right repeatedly. Soft turns aren't always enough to return to the line, though, so the robot still needs to take a sharp turn when it's far off the line (when it sees white).

To determine whether the robot measures black, dark grey, light grey, or white, you need two additional threshold values, midway between the three known values, as shown in Figure 7-13.

The diagram in Figure 7-14 shows how the robot should determine which direction to turn and whether to make a sharp or soft turn. Run the *ReflectedLine2* program (see Figure 7-15) to verify that the robot follows the line more smoothly than before. The Brick Status Light blocks help you see whether the robot is making a sharp turn (red) or a soft turn (green) when the program runs.

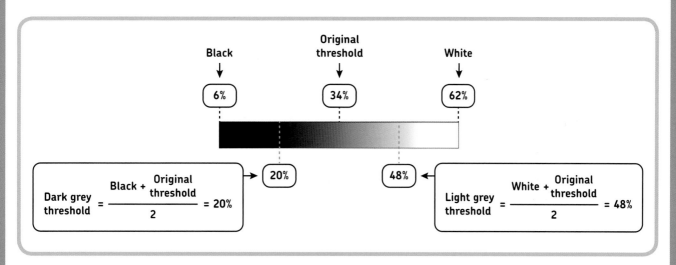

Figure 7-13: You need to calculate two additional thresholds to distinguish black, dark grey, light grey, and white. As before, each threshold is the average of two known values. For example, the dark grey threshold (20%) is the average of black (6%) and the original threshold (34%).

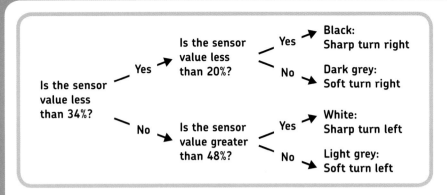

Figure 7-14: The flow diagram of the ReflectedLine2 program

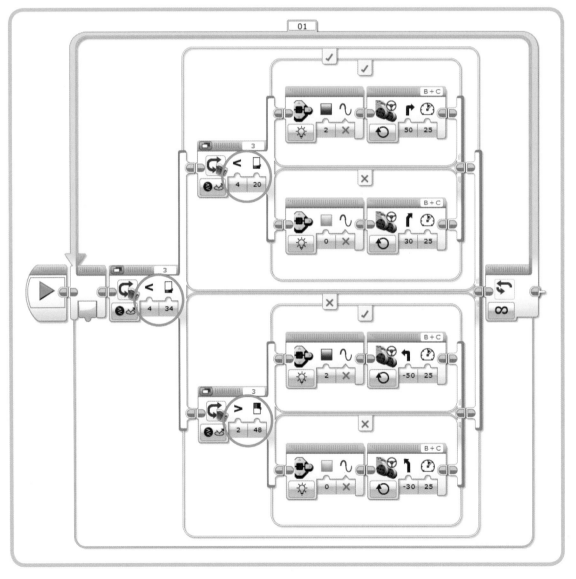

Figure 7-15: The ReflectedLine2 program. Note that the Switch block at the bottom is configured to run the blocks at its top branch if the sensor value is greater than (>) the light grey threshold value. For greater than (>), choose option **2** in the Compare Type setting.

ambient light intensity mode

The Color Sensor can be used to detect the light level in a room or the brightness of a light source using *Ambient Light Intensity mode*. You'll use this measurement to see whether the light in your room is on or off. To give the sensor a better view of its surroundings, mount the Color Sensor attachment as shown in Figure 7-16.

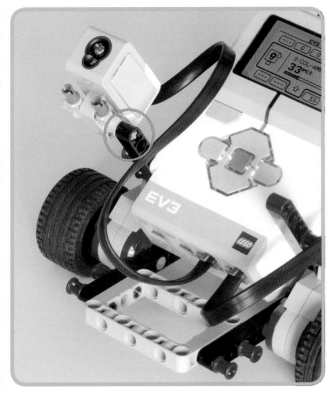

Figure 7-16: Remove the Color Sensor from the front of your robot and reattach it to the side. You won't need any additional LEGO elements. The cable remains connected to input port 3.

measuring ambient light intensity

Use Port View on the EV3, and choose **COL-AMBIENT** to see the measurement for ambient light intensity. The sensor value ranges from 0% (very dark) to 100% (very bright). If you cover the sensor with your hand, for example, it should report a value below 5%, while you might get a value near 70% if you hold it near a lamp.

You can measure ambient light intensity in a program with the same methods you used to measure reflected light, but now you'll use **Color Sensor – Compare – Ambient Light Intensity** mode. To detect whether the lights in a room are on, for example, the robot can use a Switch block to see whether the sensor value is greater than the threshold value. In this situation, the threshold should be the average of the sensor value measured when the lights are on and when they are off.

NOTE The sensor emits a blue light in Ambient Light Intensity mode, but the light actually turns off for a very short period each time it does a measurement. This way, reflected light is not measured; the sensor sees only the light coming from the robot's environment.

a morse code program

Now you'll create a program that lets you control your robot with light signals in a dark room. You'll program the robot to drive to the right if it sees a light for more than two seconds and to the left if it sees a shorter light signal. This allows you to control the robot's direction by switching the lights in the room on and off. This is a simplified form of *Morse code*, a communication method used before the invention of the telephone.

The robot first uses a Wait block to wait for the light to come on. Then, after waiting two more seconds with another Wait block, the program uses a Switch block to determine whether the light is still on. If so, it turns right; otherwise, it turns left.

Create the *MorseCode* program, as shown in Figure 7-17, and run the program in a dark room. Alternatively, you can try running it in a room with the lights on while sending light signals with a bright flashlight.

NOTE In my room, the sensor value is 2% when the lights are off and 16% when the lights are on. Therefore, I chose a threshold value of 9% for the *MorseCode* program. However, you should determine your own threshold value.

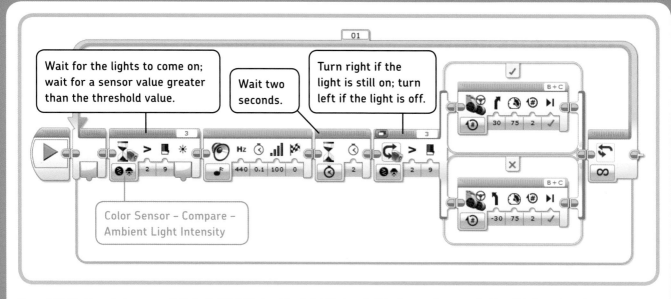

Figure 7-17: The MorseCode program. Both the first Wait block and the Switch block are in Color Sensor – Compare – Ambient Light Intensity mode.

DISCOVERY #36: MORNING ALARM!

Difficulty: ▢ **Time:** 🕐🕐

Can you turn your robot into an alarm that goes off when the sun rises? Place your robot near a window and have a Wait block pause the program until the ambient light intensity goes above a threshold value that you've calculated. Then, the robot should repeatedly play loud tones until you press the Touch Sensor, which acts as a snooze button.

TIP Your robot normally turns itself off if you don't use it for 30 minutes, so it won't wake you up in the morning unless you go to the Settings tab on the EV3 brick, select *Sleep*, and choose *Never*. Remember to change the Sleep setting back to 30 minutes the next day so it doesn't drain the batteries if you forget to switch it off.

further exploration

The Color Sensor allows your robot to sense its environment by detecting color, reflected light intensity, and ambient light intensity. Robots use measurements from this sensor to perform various kinds of tasks. For example, you've made the EXPLOR3R follow lines and respond to light signals.

You've also learned to calculate threshold values and compare them to sensor values to have your robot detect changes in its environment. The robot used a threshold to determine whether it saw a black line and whether the light in a room was on. Threshold values are also useful for other sensors, and we'll use them again later.

The Color Sensor is a versatile device, and there are many more cool ways to use it. Try solving some of the Discoveries to see what you can invent!

DISCOVERY #37:
COLOR TAG!

Difficulty: ▫▫ **Time:** 🕐🕐

Place the Color Sensor on your robot, as shown in Figure 7-16, and have the robot drive in different directions as you hold differently colored objects near the sensor. Each movement should last three seconds.

DISCOVERY #38:
FINGERPRINT SCANNER!

Difficulty: ▫▫ **Time:** 🕐🕐

Can you make the robot turn left if you press the Touch Sensor and turn right if you "press" the Color Sensor? Remove both sensors from their attachments so you can hold them in your hands. Then connect them to the EV3 brick with the longest cables you have. How do you make the Color Sensor detect the touch of your fingers?

HINT What is the Reflected Light Intensity when you place your finger on the sensor?

DISCOVERY #39:
COLOR PATTERN!

Difficulty: ▫▫▫ **Time:** 🕐🕐

Expand the program you made in Discovery #37 to make the robot respond to different color *patterns*. For example, make the robot steer left if it sees red for two seconds, but make it steer right if it sees red for one second and then blue for one second.

HINT Create a program similar to the *MorseCode* program.

DISCOVERY #40:
OBSTACLES ON THE LINE!

Difficulty: ▫▫▫ **Time:** 🕐🕐🕐

Can you make the robot follow the line on the test track and turn around if there is an obstacle in its path? Reconnect the Touch Sensor attachment to the robot and place the Color Sensor attachment just to the left of it, as shown in Figure 7-18. (Only one of the two blue pins will be connected, but that's fine.)

HINT Modify the Loop block in your line-following programs to repeat until the Touch Sensor is pressed. Then, make the robot turn around, find the line, and follow it in the other direction.

Figure 7-18: The EXPLOR3R with both the Color Sensor (port 3) and the Touch Sensor (port 1) attached. Note that the line-following programs you created earlier should still work with this configuration.

DISCOVERY #41:
CRAZY TRACK!

Difficulty: ⬜⬜⬜⬜ **Time:** ⏲⏲⏲

Go to *http://ev3.robotsquare.com/lines.pdf* and print two corners (two copies of page 3), a three-way junction (page 33), a turning point (page 17), a yellow face (page 26), and a green star (page 28) to create the track shown in Figure 7-19. Make the robot follow the line until it finds the green star, regardless of its starting position.

HINT Make the robot turn around and follow the line in the other direction if it sees the yellow face.

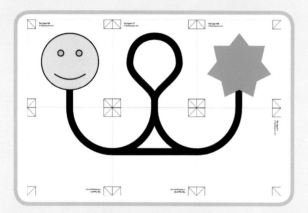

Figure 7-19: The line-following track for Discovery #41

DESIGN DISCOVERY #7:
DOORBELL!

Building: ❋ **Programming:** ⬜

Can you make the EV3 brick play a sound when someone steps through the doorway? Mount the Color Sensor on one side of the door frame, and position a flashlight on the other side, pointing directly at the Color Sensor. How can you program the robot to detect when someone steps through the door?

HINT The ambient light intensity should drop when someone blocks the light from the flashlight by stepping through the doorway.

DESIGN DISCOVERY #8:
SAFE DEPOSIT!

Building: ❋❋❋ **Programming:** ⬜⬜⬜

Can you design a safe deposit box that you can unlock with a colored security card, as shown in Figure 7-20? Use one motor to move each colored square on the security card past the Color Sensor, and use another motor to open and close the deposit box upon scanning the card successfully. You can scan this particular card using these steps:

1. Rotate the wheel until the Color Sensor sees either red, yellow, green, or blue.

2. Eject the card if the color is not red. Proceed otherwise.

3. Turn the motor to reach the next color.

4. Eject the card if the color is not yellow. Proceed otherwise.

5. Turn the motor to reach the next color.

6. Etc.

HINT Use one Switch block for each of the colored squares. Use the first switch to determine whether the first color is red. If so (true), rotate the motor and use another switch to determine whether the next color is yellow. If so (true), rotate the motor, and so on. The false tab of each switch should contain blocks to eject the card.

Figure 7-20: You can print a copy of this card from http://ev3.robotsquare.com/securitycard.pdf.

8

using the infrared sensor

The *Infrared Sensor* lets your robot "see" its surroundings by measuring the approximate distance to an object using infrared light. In addition, the sensor gathers information from the *infrared remote* (also called the *beacon*). The sensor can detect which buttons on the remote you press, approximately how far away the remote is, and the relative direction, or heading, from the robot to the remote.

You can implement each of these features in your programs using the Infrared Sensor in four different modes: Proximity, Remote, Beacon Proximity, and Beacon Heading (see Figure 8-1). You can actually see the infrared light from the beacon by viewing it with a digital camera, such as the one in a smartphone. You can also see a faint light coming from the sensor if it's in Proximity mode. In this chapter, you'll learn how each mode works as you create programs for the EXPLOR3R to make it avoid obstacles, respond to remote control commands, and find the beacon.

proximity mode

The robot measures the distance from the sensor to an object with *Proximity mode*. Rather than measuring this distance in inches or centimeters, the sensor gives you the distance as a percentage from 0% (very close) to 100% (very far). Go to Port View on the EV3 brick, choose input port 4, and select **IR-PROX** (short for *infrared proximity*) to see the sensor value.

The sensor value is determined by measuring how much of the infrared light it emits gets reflected back. The closer an object is to the sensor, the more light is reflected back to the sensor. Some surfaces reflect infrared light better than others, which makes them seem closer. For example, a white wall might appear to be closer than a black wall, even though they are actually the same distance from the sensor.

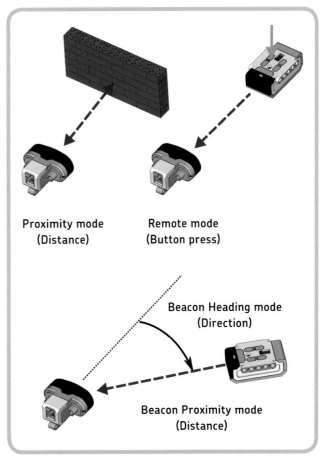

Proximity mode (Distance)

Remote mode (Button press)

Beacon Heading mode (Direction)

Beacon Proximity mode (Distance)

Figure 8-1: The operation modes of the Infrared Sensor. The red dashed lines represent the invisible rays of infrared light. If you block the path between the sensor and the remote, the sensor won't be able to get a correct measurement.

avoiding obstacles

Although the sensor doesn't measure the exact distance to an object, it's good at seeing whether or not an obstacle is in the way. The sensor value should be 100% when the sensor doesn't see anything, and the value should drop to around 30% if the robot gets close to a wall. If you program the robot to drive forward but then move away if it sees something closer than 65% (the threshold value), you'll have an obstacle-avoiding robot. The *ProximityAvoid* program in Figure 8-2 uses a Wait block to look for measurements less than 65%.

combining sensors

You're not limited to using just one sensor in your program. In fact, you can use all of the sensors connected to your EV3 in a single program. This can make your robot more *reliable*.

For example, the Infrared Sensor doesn't always see small objects in the robot's path, but the Touch Sensor is good at detecting these. On the other hand, the Touch Sensor won't detect some soft obstacles, like a curtain, but the Infrared Sensor will. If you combine the sensor measurements, the EXPLOR3R is less likely to get stuck somewhere when it drives around.

One way to combine sensor measurements is to use a Switch block within a Switch block, resulting in the program flow in Figure 8-3. Create and run the *CombinedSensors* program that implements this decision tree, as shown in Figure 8-4.

NOTE Don't forget to reconnect the Touch Sensor attachment you made in Chapter 6 to your robot. The Touch Sensor should be connected to input port 1.

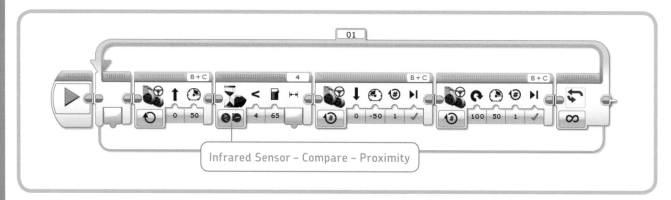

Infrared Sensor – Compare – Proximity

Figure 8-2: Create a new project called EXPLOR3R-IR with a program called ProximityAvoid, and configure the blocks as shown. Notice the similarity with the TouchAvoid program you made in Chapter 6.

DISCOVERY #42: CLOSE UP!

Difficulty: ▢ **Time:** ◷

Make the robot repeatedly say "Detected" if it sees an object closer than 50%, and make it say "Searching" otherwise. Then try out other threshold values, such as 5% or 95%, to see how close and how far the sensor can detect objects reliably. The sensor doesn't measure an exact distance, and you'll find that the results vary depending on what kind of object you're trying to detect.

HINT You'll need to place a Switch block in a Loop block.

DISCOVERY #43: THREE SENSORS!

Difficulty: ▢▢ **Time:** ◷

Expand the *CombinedSensors* program with a third sensor. Make the robot stand still if the Color Sensor sees something blue, and make it continue avoiding obstacles when the blue object is removed.

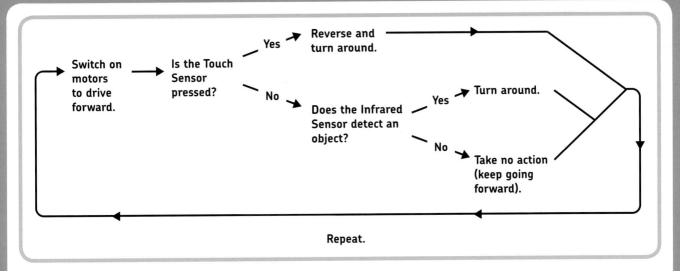

Figure 8-3: *The flow diagram for the* CombinedSensors *program*

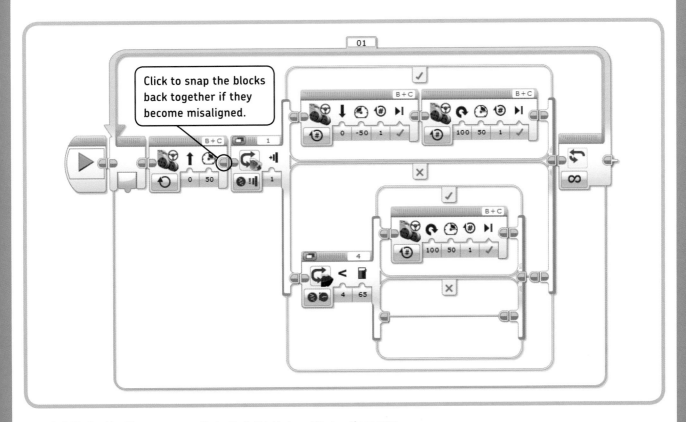

Figure 8-4: *The* CombinedSensors *program. Resize the Switch blocks and the loop if necessary.*

remote mode

In *Remote mode*, the Infrared Sensor detects which buttons on the infrared remote you press, allowing you to make programs that do something different for each button. This is how you were able to remotely control your robot in Chapter 2. The IR Control application you used on the EV3 brick is actually a program that makes the robot steer in different directions according to the buttons you press. The sensor can detect 12 button combinations, or *Button IDs*, as shown in Figure 8-5.

A Switch block in Measure mode lets you choose a set of blocks to run for each sensor value—each Button ID, in this example. The *CustomRemote* program uses such a switch to make the EXPLOR3R drive forward if the two top buttons are pressed (Button ID 5). It steers left when the top-left button is pressed (ID 1) and right when the top-right button is pressed (ID 3), and it stops if no buttons are pressed (ID 0). Stopping is the default, so the robot also stops if you press an unspecified button combination.

Because the program is configured to listen to the remote on channel 1, your remote should be on channel 1 as well (see Figure 2-10). If you have another EV3 robot, switch it to a different channel (2, 3, or 4) to avoid interference. Create and run the program now (see Figure 8-6), and drive your robot around with the remote.

This technique is especially useful (and fun!) because now you can make a custom remote control program for all of your robots. For example, in Chapter 12 you'll build a Formula EV3 Race Car that drives and steers differently than the EXPLOR3R. The normal IR Control application won't work for that robot, but you can solve this problem by making your own program to drive and steer the robot.

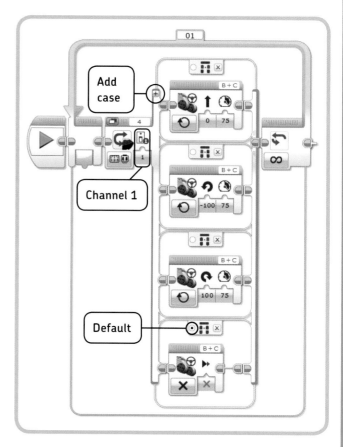

Figure 8-6: The CustomRemote *program*

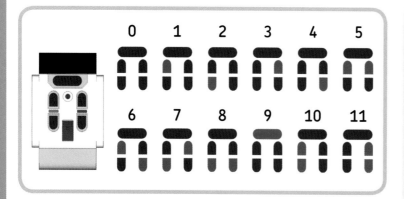

Figure 8-5: The Infrared Sensor can detect 12 button combinations (Button IDs) on the remote control. Pressed buttons are shown in red.

DISCOVERY #44: UNLOCK THE REMOTE!

Difficulty: ☐ **Time:** ⏱

Can you protect your program with a secret button combination? Add two Wait blocks just before the Loop block in the *CustomRemote* program. Make these blocks wait for you to press Button IDs 10 and 11, respectively, before the rest of the program will run. (As an extra challenge, try making the code more secure using the technique you learned in Design Discovery #8 on page 88.)

beacon proximity mode

In addition to detecting which button you've pressed on the infrared remote, the Infrared Sensor can detect the signal strength and the direction the signal is coming from when you press any of the buttons. The robot can use this information to locate the remote, or *beacon*, and drive toward it.

In *Beacon Proximity mode*, the sensor uses the beacon's signal strength to calculate a relative distance ranging from 1% (the beacon is very close to the sensor) to 100% (it's very far). For best results, hold the beacon at roughly the same height as or just above the sensor and point it at the eyes of the robot (see Figure 8-1).

The little green light on the beacon indicates it's busy sending a signal. It doesn't matter which button you press when it comes to beacon proximity or heading, but it's handy to use the button at the top (Button ID 9). The green light turns on when you press it once, and it turns off when you press it again, so you don't have to keep the button pressed all the time.

Now you'll make the *BeaconSearch1* program that has the robot repeatedly say "Searching" until the Infrared Sensor measures a beacon proximity of less than 10%. In other words, it will keep saying it's searching until you hold the beacon very close to the sensor. You can accomplish this with a Loop block, configured as shown in Figure 8-7. After the loop finishes, the robot says "Detected."

NOTE The sensor can also tell you whether it sees a signal at all. You could make the robot stop moving if no signal is detected, for example. This requires the use of a few blocks you haven't seen yet, but we'll get back to that in Chapter 14.

beacon heading mode

The Infrared Sensor can measure the direction of the beacon's signal using *Beacon Heading mode*. This heading gives the robot a rough sense of the angle between the beacon and the sensor. The measurement is a number between –25 and 25, as shown in Figure 8-8. The sensor doesn't provide a precise angle, but it's good enough to determine whether the beacon is to the robot's left (negative values) or right (positive values).

The sensor is able to see the beacon in all directions—even if it's behind the sensor—but the measurement is most reliable in the green region of Figure 8-8. The heading value is zero if the beacon is directly in front of the sensor or right behind it, but it's also zero when the sensor detects no signal at all.

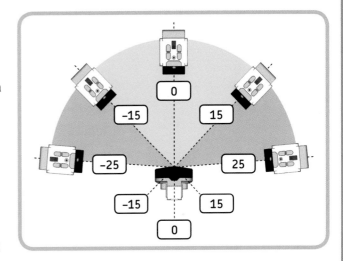

Figure 8-8: The Beacon Heading sensor value ranges between –25 and 25. Negative values indicate that the beacon is to the left of the sensor; positive values mean it's to the right. A value near 0 means that the robot sees the beacon straight ahead or right behind it.

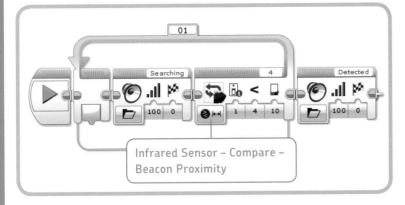

Figure 8-7: The BeaconSearch1 program

Information about whether the beacon is to the robot's left or right is all you need to make the robot drive toward the beacon. When the sensor detects the beacon to the left, the robot should drive to the left. If the beacon is on the right, the robot should steer right. You can use a Switch block to see whether the Beacon Heading sensor value is less than (<) 0, indicating that the beacon is to the left.

If you continuously adjust the steering to the sensor measurement and drive forward at the same time, your program will make the robot find the beacon. Remove the Sound blocks from your previous program and add a Switch block with two Move Steering blocks to complete the *BeaconSearch2* program, as shown in Figure 8-9. The loop makes the robot keep searching for the beacon until beacon proximity is below 10%; then the program ends.

NOTE Remember to make the beacon transmit a signal continuously, either by keeping one of the buttons pressed or by toggling the button at the top (Button ID 9) so the green indicator light is permanently on.

DISCOVERY #45: SMOOTH FOLLOWER!

Difficulty: 🔲🔲 **Time:** ⏱

Can you expand the *BeaconSearch2* program to make the robot drive toward the beacon more smoothly? Have the robot make soft turns (25% steering) if the beacon is in the green area of Figure 8-8 and sharp turns (50% steering) if it's in the red area.

HINT Use the techniques you learned in "Following a Line More Smoothly" on page 83. You don't need to calculate threshold values; they're given in Figure 8-8.

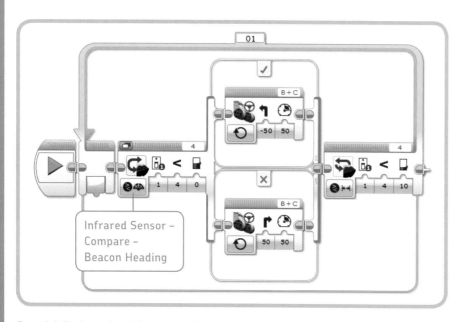

Infrared Sensor –
Compare –
Beacon Heading

Figure 8-9: The BeaconSearch2 *program makes the robot drive toward the beacon and stop when it's very close to the beacon. If you move around with the beacon, the robot will follow you.*

combining sensor operation modes

Combining multiple operation modes of the Infrared Sensor in one program can make your robot behave unexpectedly because the sensor needs time to switch from one mode to the other. (Beacon Proximity and Beacon Heading are the only operation modes of the Infrared Sensor you can use together without delays.)

For example, say you wanted to change the Loop block in the *CustomRemote* program (see Figure 8-6) to loop until the sensor detects a proximity below 10%. A Switch block would detect the pressed button in Remote mode, and then the Loop block would try to read the proximity value in Proximity mode. But because the sensor has to change from one mode to the other each time, the program would run very slowly, and your robot may not respond properly to the remote control. (Try this out if you like).

If timing isn't critical, though, it's fine to use different modes in one program. For example, the *MultiMode* program in Figure 8-10 works as expected. First, it waits until you press the top-right button on the remote (Button ID 3), after which you'll hear a beep. Then, it says "Yes" if the proximity measurement is lower than 30%; otherwise, it says "No." You'll hear a gap between the beep and the spoken word—that's the delay caused by switching from Remote mode to Proximity mode.

further exploration

The Infrared Sensor lets your robot detect objects in its environment from a distance. When combined with the infrared remote, the sensor can act as a remote control receiver and a beacon detector. You've also seen how to combine the Touch Sensor and the Infrared Sensor to make the robot avoid obstacles more reliably. You can, of course, add the Color Sensor for even more sophisticated programs. Before you go on, practice your building and programming skills with these Discoveries!

DISCOVERY #46: FOLLOW ME!

Difficulty: ▢▢ **Time:** 🕐🕐

Can you make the EXPLOR3R follow you in a straight line while keeping a fixed distance? Use the Infrared Sensor in Proximity mode to detect the distance to your hand (keep it in front of the robot). The robot should follow as you move your hand away; it should reverse if you move your hand closer. Make the robot stand still if it sees your hand at a distance between 35% and 45%.

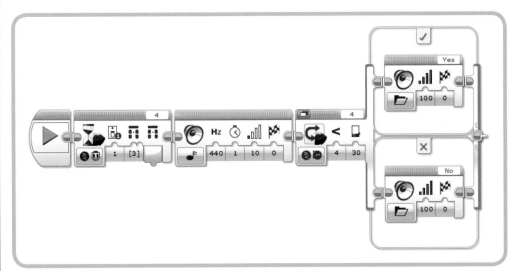

Figure 8-10: The MultiMode *program. Timing isn't critical here, so it's okay to use both Remote mode and Proximity mode in the same program.*

DISCOVERY #47:
SONAR SOUNDS!

Difficulty: ▭▭▭ **Time:** ◷◷

Can you make the EV3 play sounds to guide you toward the beacon with your eyes closed? Make it play tones at different frequencies and volumes based on the location of the beacon. Play low tones (400 Hz) if the beacon is to the left of the sensor and high tones (1000 Hz) if the beacon is to the right. The closer you are to the beacon, the louder the sound should be.

HINT First, use a Switch block to determine whether the beacon is to the left or to the right. Next, in both parts of this switch, place a Switch block to determine whether the beacon is close or far away. Then, you'll have four spots to place a Sound block, each configured to play one of these tones: low and loud, low and soft, high and loud, and high and soft.

DESIGN DISCOVERY #9:
RAILROAD CROSSING!

Building: ✹ **Programming:** ▭

Can you design an automated railroad crossing for LEGO model trains? Use a motor to move a barrier that stops cars from crossing the railway when a train passes. Use the Infrared Sensor or the Color Sensor to spot when a model train approaches and when the barriers should be lowered as well as when they should be raised again.

DESIGN DISCOVERY #10:
FOOLPROOF ALARM!

Building: ✹ **Programming:** ▭▭

Can you use all three sensors in the EV3 set to create an intruder alarm that never fails? Use the Touch Sensor to detect a door that's being opened (Design Discovery #4 on page 74), use the Color Sensor to detect people stepping through the doorway (Design Discovery #7 on page 88), and use the Infrared Sensor in Proximity mode to detect movements near an object of interest, such as a phone.

TIP Design your robot and your program in such a way that you (only you!) can still enter the room without having the alarm go off.

9

using the brick buttons and rotation sensors

In addition to the Touch, Color, and Infrared Sensors, the EV3 contains two types of built-in sensors: Brick Buttons and Rotation Sensors. You can use the *Brick Buttons* on the EV3 brick to control or influence a program while it's running. For example, the program can ask you to press one of the buttons to choose what the robot should do next.

Each of the EV3 motors has a built-in *Rotation Sensor* that determines the position of the motor, allowing you to precisely control wheels or other mechanisms. The sensor also measures the motor speed, making it possible to detect when a motor is moving slower or faster than intended.

using the brick buttons

You can use the EV3 brick's Up, Down, Left, Right, and Center buttons in your programs just as you use the Touch Sensor. You can make your robot respond by playing a sound when you press a particular button, for example. You can also make the robot wait for the button to be released or *bumped* (a press followed by a release).

One interesting way to use multiple buttons in a program is to create a menu on the EV3 screen, letting you choose the next action in the program. The *ButtonMenu* program in Figure 9-1 plays one of three sounds based on which button the user presses.

Two Display blocks show a simple menu on the screen, asking the user to choose whether the robot should say "Hello," "Okay," or "Yes." Then, a Wait block (in **Brick Buttons – Compare** mode) pauses the program until the user presses either the Left, Center, or Right button.

DISCOVERY #48: LONG MESSAGE!

Difficulty: 🖵 **Time:** ⏱
When you display a long message on the EV3 screen, you might find that the screen is too small to display it entirely. Create a program that lets you use the Down button to scroll through your message.

HINT Make the robot display some new text on the screen each time you press the button.

DISCOVERY #49: CUSTOM MENU!

Difficulty: 🖵 **Time:** ⏱⏱
Can you expand the *ButtonMenu* program to make your robot do useful things besides playing sounds? Take three programs you made previously, turn them into My Blocks, and place them in the switch of the *ButtonMenu* program. Reconfigure the Display blocks to describe what happens as you press each button.

TIP This technique is often used in robotics competitions because it provides a way to start different programs very quickly. To change the actions of each *sub program*, simply modify the blocks in each My Block.

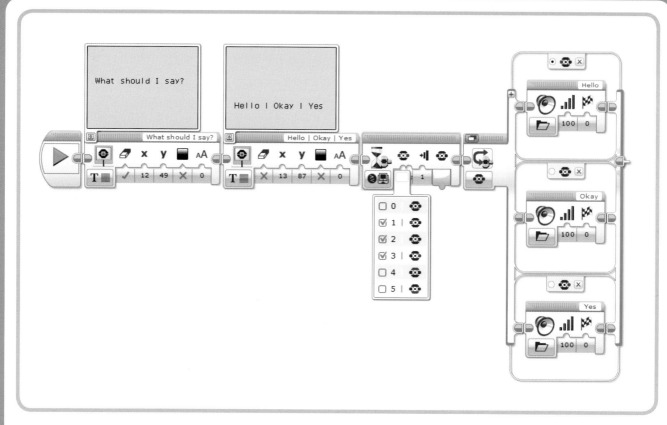

Figure 9-1: The ButtonMenu *program. The Wait block is configured in* **Brick Buttons – Compare – Brick Buttons** *mode, and the Switch block is in* **Brick Buttons – Measure – Brick Buttons** *mode.*

Next, a Switch block (in **Brick Buttons – Measure** mode) determines which button is being pressed, and the robot plays the requested sound. After the Wait block completes, the Switch block runs so quickly that the button is still pressed by the time the Switch checks the button state, even if you release it right away.

using the rotation sensor

When you tell the robot to move forward for three rotations with the Move Steering block, the vehicle knows when to stop moving because the Rotation Sensor in each EV3 motor tells the EV3 how much it has turned. The program can also tell you how fast a motor is currently turning by measuring how fast the motor position changes.

You can use Wait, Loop, and Switch blocks in Motor Rotation mode to measure motor position (Degrees mode or Rotations mode) and motor speed (Current Power mode).

motor position

The motor position tells you how much a motor has turned since you started the program. Use the Port View app on the EV3 brick, navigate to motor B or C, and rotate the motors with your hands to see the sensor values change.

When you first start Port View (or your own program), the sensor value is 0. The value becomes positive when you rotate a motor forward; it becomes negative if you turn it backward past 0, as shown in Figure 9-2. For example, if you rotate the motor forward by 90 degrees and then backward for one rotation (360 degrees), the motor should report a position of –270 degrees.

You can use the position measurement to create a program that plays a sound if you turn one wheel 180 degrees forward by hand, as shown in Figure 9-3. A Wait block in **Motor Rotation – Compare – Degrees** mode waits until the Rotation Sensor value is greater than or equal to (≥) 180 degrees. Because these sensors are built into the EV3 motors, they are always connected to output ports (you use the motor on output port B in this program).

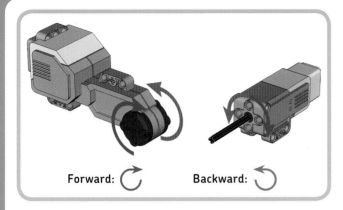

Forward: ⟳   Backward: ⟲

Figure 9-2: If you program a Large or Medium Motor to go forward, it turns in the direction of the blue arrow and the Rotation Sensor value becomes positive.

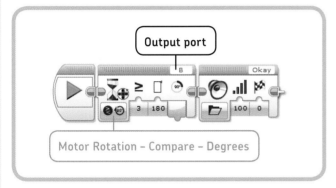

Figure 9-3: The HandRotate program makes the robot say "Okay" once you rotate motor B forward by 180 degrees. Note that the Wait block would do the same thing if you used Motor Rotation – Compare – Rotations mode with a threshold value of 0.5.

resetting the motor position

Now suppose you want to repeat the actions in the *HandRotate* program with a Loop block so that the sound plays again when you rotate the wheel another 180 degrees. The first sound plays when you rotate the motor forward by 180 degrees. But during the second run of the loop, the sound would play immediately because the sensor value is already greater than 180 degrees, which is not what you want.

The solution is to reset the Rotation Sensor value to 0 at the beginning of the Loop, using a *Motor Rotation block* in *Reset* mode, as shown in Figure 9-4. (You'll explore the other features of this block and the other Sensor blocks later in Chapter 14.)

Run the program and verify that you hear the sound once each time you rotate the wheel 180 degrees forward.

rotational speed

The Rotation Sensor calculates how fast a motor turns as a value between –100% and 100% based on the rate at which the motor position changes. The value is positive when the motor turns forward (blue arrow in Figure 9-2), negative when the motor turns backward (green arrow), and 0 when the motor is not turning.

For the Large Motor, a *Current Power* sensor value of 50% corresponds to a *rotational speed* of 85 rotations per minute (rpm). You can reach this speed by rotating a motor with your hands or by using any of the Move blocks with its Power setting at 50%.

NOTE Current Power mode measures rotational speed; it does not measure current or power consumption!

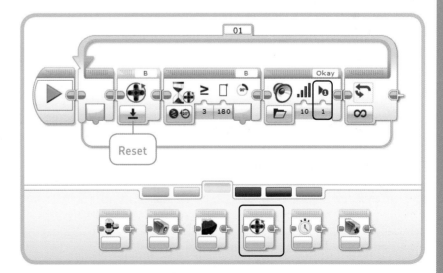

Figure 9-4: The HandRotateReset program sets the sensor value to 0 at the beginning of each loop with the Rotation Sensor block in Reset mode. Note that the Play Type setting in the Sound block is Play Once (1) so that the program doesn't wait for the sound to finish.

DISCOVERY #50:
BACK TO THE START!

Difficulty: 🔲🔲 **Time:** ⏱

Can you make a program that returns a motor to the position it was in when the program started? The robot should give you five seconds to turn the motor to a random position manually, and then the motor should return to its starting point. Use the decision tree shown in Figure 9-5 as a guide for your program.

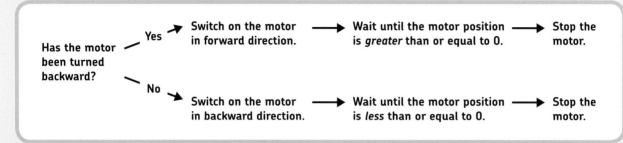

Figure 9-5: The flow diagram for Discovery #50. How does the robot determine that a motor has been turned backward?

calculating the rotational speed

You can calculate the speed measured in rotations per minute (rpm) using the Current Power value as follows:

large motor
rotational speed (rpm) = sensor value × 1.70

medium motor
rotational speed (rpm) = sensor value × 2.67

For example, if the Current Power value of a Large Motor is 30%, the motor rotates at 30 × 1.70 = 51 rotations per minute. Because one rotation per minute is equivalent to six degrees per second, you can calculate the rotational speed in degrees per second (deg/s) as follows:

rotational speed (deg/s) = rotational speed (rpm) × 6

In this example, you get 51 × 6 = 306 degrees per second.

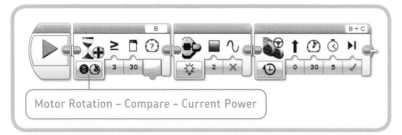

Motor Rotation – Compare – Current Power

Figure 9-6: The PushToStart program

measuring rotational speed in a program

To measure rotational speed in a program, you use the Current Power mode of the Rotation Sensor, as shown in Figure 9-6. The *PushToStart* program uses a Wait block configured to pause the program until motor B reaches a sensor value of 30% (51 rpm). Then, a Move Steering block takes over at the same speed. Run the program and push EXPLOR3R forward with your hands until it begins to move by itself.

DISCOVERY #51:
COLORED SPEED!

Difficulty: 🔲🔲 **Time:** ⏱⏱

Create a program that continuously changes the brick status light color to green if motor B is rotating forward, orange if it's rotating backward, and red if the motor stands still. Rotate motor B with your hands to test your program.

HINT You'll need a Loop block, two Switch blocks, and three Brick Status Light blocks.

understanding speed regulation

So far, you've been using several kinds of green Move blocks to make your robot move. These blocks make the motors turn at a constant, *regulated speed*. When the motors slow down because of an obstacle or an incline, the EV3 supplies some extra power to the motor to keep it going at the desired speed. The Power setting on these blocks actually specifies the *speed* that the motors try to maintain. That is, a Large Motor turning at 20% speed (34 rpm) while performing a heavy task might consume more power than a motor doing a light task at 40% speed (68 rpm).

When you don't want the EV3 to supply that extra power to maintain constant speed, you can use *unregulated speed*.

seeing speed regulation in action

To see the difference between regulated speed and unregulated speed, you'll create a program that makes the robot drive up a slope, such as a table with one end lifted up in the air. First, the robot drives at *unregulated* speed for three seconds, and then it drives at *regulated* speed for three seconds.

You need two *Unregulated Motor blocks* (one for each motor) from the advanced tab of the Programming Palette, with their Power setting at 20%, as shown in Figure 9-7. After

waiting for three seconds, you stop the motors by setting their power to 0%. To drive at a *regulated speed* of 20% (34 rpm), the program uses the Move Steering block that you've seen before.

Place the EXPLOR3R on a tilted surface and run the *SteepSlope* program. You should find that the robot drives quite slowly up the slope during the first three seconds and that it goes faster during the next three seconds.

During the first part of the drive, the EV3 switches on both motors and leaves them alone for the next three seconds. The robot runs slowly because it takes more power to drive up a hill than to drive on even terrain. During the second part, the Rotation Sensors tell the robot that it's going slowly, prompting the EV3 brick to supply some extra power to get the robot back up to speed.

stopping a stalled motor

If you try to slow down one of the wheels when a Move Steering block runs, you'll feel that the robot tries harder to get back up to speed. This is fine for wheeled vehicles, but it's often undesired for mechanisms that don't make complete turns, such as a claw mechanism.

To avoid this problem, you can use the Unregulated Motor block and the Rotation Sensor to detect when the motor is *stalled* (blocked), as demonstrated by the *WaitForStall* program (see Figure 9-8). The program switches on motor B at 30% power and waits for the speed to drop below 5%, indicating that the motor is stalled. If you run this program on the EXPLOR3R, your robot should drive in circles until you slow down the robot by blocking its path.

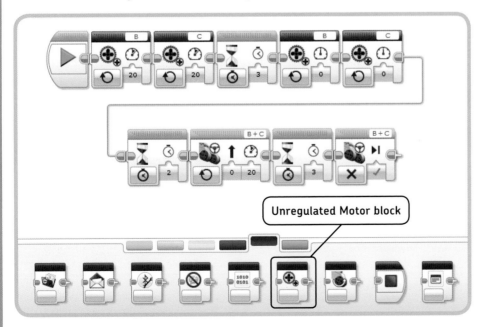

Figure 9-7: The SteepSlope *program. Note that I used a Sequence Wire to split the program in two for better visibility, but you won't need to do this in your program.*

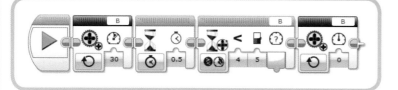

Figure 9-8: The WaitForStall program turns motor B until it is stalled. Note that the first Wait block is required to give the motor some time to get up to speed—otherwise, the rotational speed would be 0 when the speed is first measured, causing the program to end immediately.

further exploration

Now that you've learned how to work with all of the sensors in the EV3 set, you can create robots that interact with their environment. The EXPLOR3R is, of course, only one example. As you continue reading this book, you'll build several robots with sensors, each of which will use sensors differently.

So far, you've learned to use the components that are essential to create a working robot: the EV3, the motors, the sensors, and the programming software. The following chapters will explore each of these subjects in more detail so that you'll be able to create increasingly sophisticated (and fun!) robots. In the next chapter, you'll begin looking at how you can use the Technic building elements in the EV3 set to construct your own robots.

The following Discoveries will help you explore more possibilities with the sensors you've seen in this chapter.

DISCOVERY #52: BRICK BUTTONS REMOTE!

Difficulty: ▨ **Time:** ⏱⏱

Remove the EV3 brick from your robot (leaving the cables connected), and create a program that lets you move the robot around by pressing the EV3 buttons. Make the robot drive forward if you press the Up button, go left if you press the Left button, and so on.

HINT Use a Switch block in *Brick Buttons – Measure – Brick Buttons* mode.

DISCOVERY #53: LOW SPEED OBSTACLE DETECTION!

Difficulty: ▨▨▨ **Time:** ⏱⏱

Can you make your robot drive around the room and move away from obstacles without the use of the Touch, Color, or Infrared Sensors? Make the robot drive forward using Unregulated Motor blocks until it detects an obstacle. The robot should then reverse, turn around, and continue driving in a new direction.

HINT The rotational speed of a motor drops when the robot runs into an obstacle.

DESIGN DISCOVERY #11: AUTOMATIC HOUSE!

Building: ✹✹✹ **Programming:** ▨▨

Have you ever built houses from regular LEGO bricks? Now that you know how to work with motors, how to use sensors, and how to make working programs, how about trying to build a robotized house with the EV3?

TIP Use a motor to automatically open the door when someone presses the doorbell (the Touch Sensor), and set an intruder alarm that sounds when the Infrared Sensor sees someone. Use another motor to close and open the shutters based on the light level measured with the Color Sensor.

PART III

robot-building techniques

10

building with beams, axles, connector blocks, and motors

You've already learned a lot about programming robots, but learning to build robots is just as important. Building experience comes with practice, but this part of the book aims to give you a solid introduction to building robots with your EV3 set. In this chapter, you'll begin looking at how you can create sturdy structures for your robots using beams, frames, pins, connector blocks, and axles, as shown in Figure 10-1. You'll also learn how the LEGO unit grid can help you design your own constructions of beams and connector blocks. In Chapter 11, you'll learn how gears work.

Each of the examples presented in this chapter can be built using the pieces in your EV3 set, though not all at the same time. Try out the examples to get a sense of how sturdy each construction is and which ones will be useful for your robots.

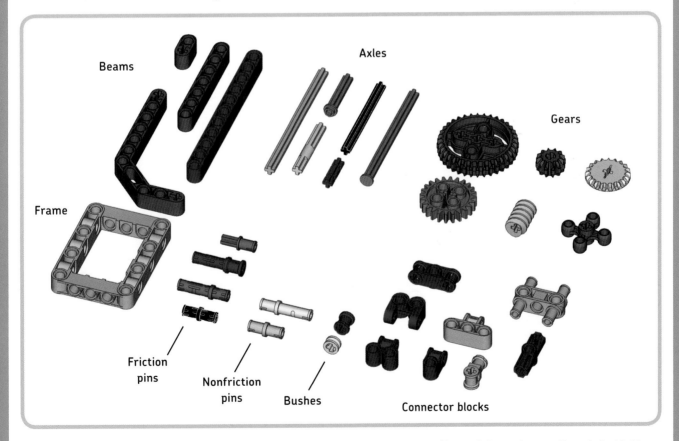

Figure 10-1: The EV3 set contains many different beams, frames, axles, gears, connector blocks, and pins. (You can find a complete parts list on the back inside cover.)

using beams and frames

So far you've learned a lot about using motors, sensors, and the EV3 brick in your robots. You use *beams* to create structures that hold all of these elements together. The length of beams and other elements is measured in *LEGO units*, sometimes called *modules*, as shown in Figure 10-2. The shortest straight beam in the EV3 set is two LEGO units in size, or 2M; the longest beam is 15M.

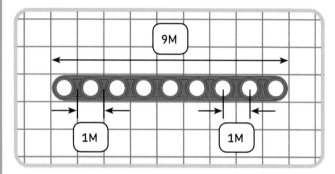

Figure 10-2: The length of beams and other elements is measured in LEGO units. The distance between the center points of two holes is exactly one LEGO unit, or 1M. Consequently, the distance between the center points of the leftmost hole and the rightmost hole of this 9M beam is 8M.

extending beams

You can extend beams in length or width by joining multiple beams using *friction pins*. You'll need at least two pins for a rigid connection, and it's best to have at least three holes overlap, as shown in Figure 10-3.

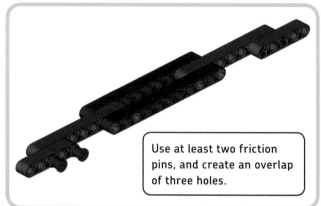

Use at least two friction pins, and create an overlap of three holes.

Figure 10-3: You can extend beams using friction pins (the red, blue, and black pins in the set).

using frames

The EV3 set contains two types of *frames*, as shown in Figure 10-4. Frames make it easy to create large structures with many attachment points for other elements, such as beams and motors. Frames are also useful for connecting beams at a right angle.

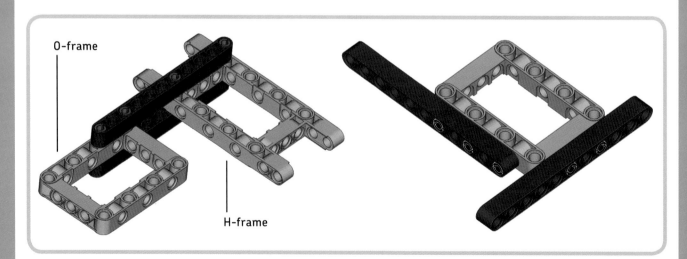

Figure 10-4: Frames can be used to create large and sturdy constructions with many connection points for other elements (left) and to connect beams at a right angle (right).

using beams to reinforce structures

Beams are useful not only to create new structures but also to *reinforce*, or strengthen, existing structures. For instance, consider the top structure consisting of two frames in Figure 10-5. It's easy to pull the two frames apart because doing so breaks only two pin connections.

The bottom structure is reinforced with two 3M beams, making it very hard to pull the two frames apart. (You could make the structure even stronger by replacing the 3M beams with 11M beams that span the full width of the frame.)

using angled beams

The EV3 set contains many *angled beams* of different types and angles. The set contains four types of angled beams with a 90-degree angle, or *right angle*, as shown in Figure 10-6. You'll use these beams to join straight beams at a right angle.

In addition to right-angled beams, your set contains two types of beams with a 53.13-degree angle, as shown in Figure 10-7. This might seem like a strange angle, but it's actually quite useful because it can form the corner of a certain common triangle. Specifically, you can use this angle to create a Pythagorean (right-angled) triangle whose sides are 3M, 4M, and 5M, as shown in Figure 10-8.

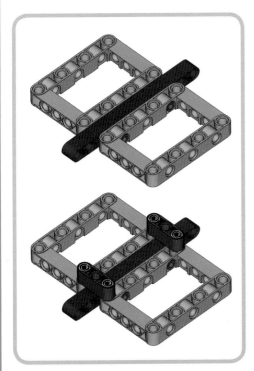

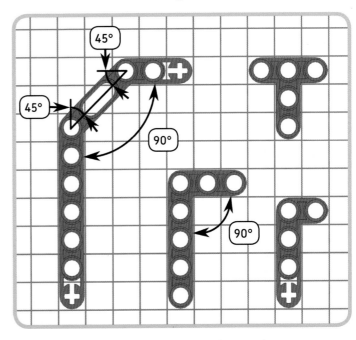

Figure 10-6: Four types of beams with a right angle (90 degrees)

Figure 10-5: You can reinforce a structure with beams. Compare the strength of both structures by trying to pull the two frames apart. You should find that the structure on the bottom is much sturdier.

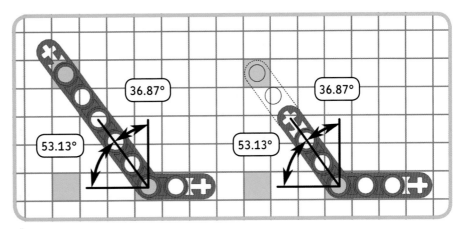

Figure 10-7: Two beams with a 53.13-degree angle. Because the angles of both beams are the same, you can extend the shorter one (right) with a straight beam to achieve similar building possibilities to those provided by the larger one (left).

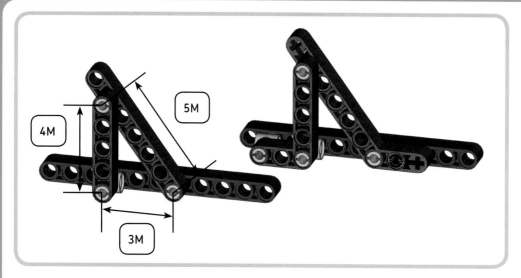

Figure 10-8: Creating two of the same right-angled triangles with straight beams (left) and with angled beams (right). The pins aren't actually green, but they are colored to match the connection points marked in green in Figure 10-7. Note that the triangle sides are measured *between the center points* of the holes at each corner instead of by counting the number of holes on the beam.

DISCOVERY #54: BIGGER TRIANGLES!

Difficulty: ☀ **Time:** ⏱

There is another useful right-angled triangle that you can build using the elements in the EV3 set. In fact, it is twice the size of the triangle shown in Figure 10-8. Can you build this triangle? What about a triangle that is three times the size?

using the LEGO unit grid

The left of Figure 10-9 is a grid of 1M-sized squares. This *LEGO unit grid* can help you design sturdy robots in a structured way. If you attach new elements to the beams so that their holes align with this grid, it will be easier to add more elements later (a).

If you stray off the grid by placing pieces at an angle, connecting more elements becomes difficult because LEGO parts come only in a limited number of fixed sizes (b). This type of construction is not recommended for the main structure of your robot, though it may be fine for decorative parts, such as the tail of an animal robot.

Building off the grid can cause beams to stretch or bend and damage your LEGO elements (c). You should always avoid this type of construction. In general, avoid connections that you can't make without bending or stretching elements slightly. If you're not sure whether an angled construction causes pieces to bend or not, it's best to stick to the unit grid by using right angles.

You can stay on the grid with 53.13-degree angled beams by using the connection holes marked green in Figure 10-7. In this way, you can use angled beams to build right-angled constructions, as shown in Figure 10-10.

NOTE You can print a copy of the LEGO unit grid from *http://ev3.robotsquare.com/grid.pdf* to use as a reference for your own designs. Be sure to select *actual size*, or *100%*, in the print size settings. The 15M beam on the printed chart should be the same size as an actual 15M beam. If it's not the same size, try adjusting the scale setting and print the grid again.

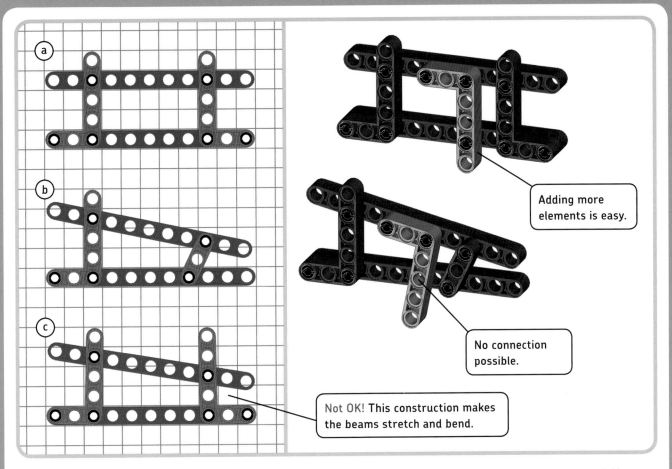

Figure 10-9: Building on the grid is recommended (a). You can build off the grid if necessary (b) as long as you do not stress the beams to make them fit (c).

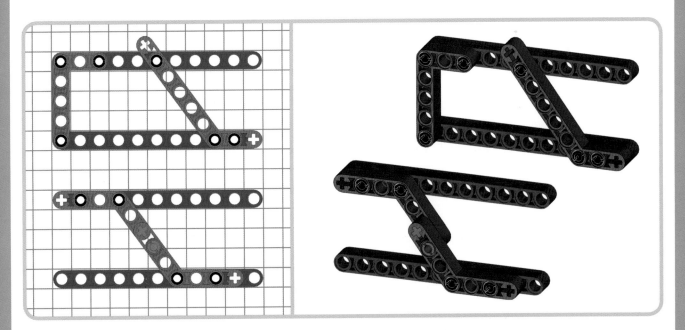

Figure 10-10: Two ways to use 53.13-degree angled beams while staying on the unit grid

DISCOVERY #55: ANGLED COMBINATIONS!

Difficulty: ☀   **Time:** ⏱

Can you find a combination of two 53.13-degree angled beams to connect the two parallel 11M beams shown in Figure 10-11?

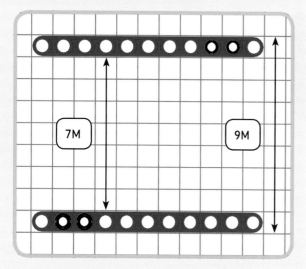

Figure 10-11: The two straight beams of Discovery #55

using axles and cross holes

An axle is a shaft to which you can mount rotating elements, like wheels or gears. Axles spin freely in *round holes*, but you can use them to create rigid connections by mounting them in *cross holes*, as shown in Figure 10-12. The shortest axle in the EV3 set is 2M; the longest axle is 9M.

You can prevent an axle from falling out of a round hole by adding a *bush* to secure it or by using an *axle with a stop*, as shown in Figure 10-13.

The *axle pin with friction*, shown in Figure 10-14, is a useful connector. One end is a pin with friction, which can be mounted in a round hole, where it will spin with some resistance. The other end is an axle, which can be mounted in a cross hole, where it will remain stationary. Some LEGO sets contain a similar tan or grey *axle pin without friction*; its pin rotates smoothly in a round hole.

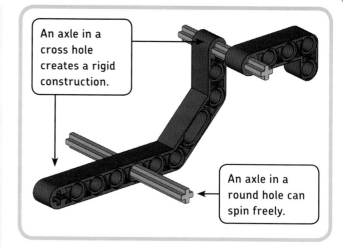

An axle in a cross hole creates a rigid construction.

An axle in a round hole can spin freely.

Figure 10-12: Axles spin freely in round holes, but they form a rigid connection in cross holes.

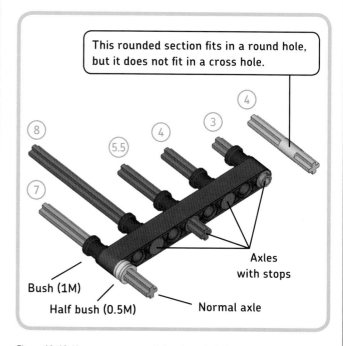

This rounded section fits in a round hole, but it does not fit in a cross hole.

Axles with stops

Bush (1M)

Half bush (0.5M)   Normal axle

Figure 10-13: You can secure an axle in a beam hole by using bushes. If an axle has a stop at one end, you'll need just one bush.

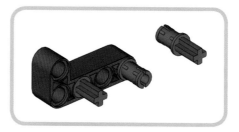

Figure 10-14: The blue axle pin with friction connects a round hole to a cross hole.

using connector blocks

You use *connector blocks* to join beams, axles, motors, and sensors at various angles. Each type of connector block in the EV3 set can be used in many ways, but this section provides some useful examples to get you started.

extending axles

Some connector blocks can be used to extend two axles, as shown in Figure 10-15. Doing so allows you to join two axles at an angle or combine axles to make them longer.

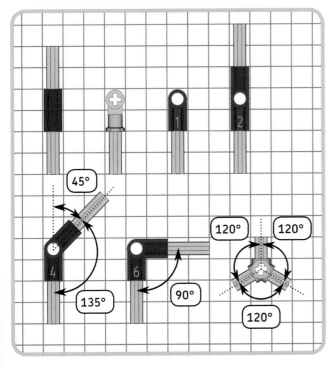

Figure 10-15: Extending axles with connector blocks

connecting parallel beams

You can use frames or beams to connect two parallel beams, but you can also use a combination of connector blocks, as shown in Figures 10-16 and 10-17. This is useful when space is limited or when you've run out of beams or frames. Use the unit grid as a reference for your own designs. For example, if you need to bridge a 3M gap between two parallel beams, you can use option *f* in Figure 10-16.

In Figures 10-16 through 10-18, you'll see how to combine certain connectors on the left, and you'll see examples of these combinations in use on the right.

NOTE The examples show how to connect two beams using connector blocks, but you can apply the same principles to other elements with beam holes. For example, you can use these combinations of connector blocks to attach a sensor to the beam holes of a motor or the EV3 brick.

connecting beams at right angles

Many connector blocks have round holes or cross holes positioned perpendicular to one another. This makes it possible to connect beams at a right angle or to connect parallel beams whose holes are placed perpendicular to one another, as shown in Figure 10-18.

securing parallel beams

You've seen earlier that you can strengthen structures using beams (see Figure 10-5). But depending on the orientation of the beam holes, you may need to add connector blocks before you can add beams for reinforcement, as shown in Figure 10-19.

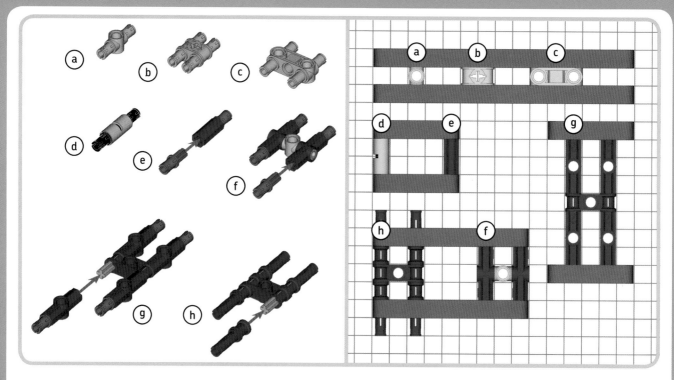

Figure 10-16: Connecting parallel beams with their beam holes facing each other

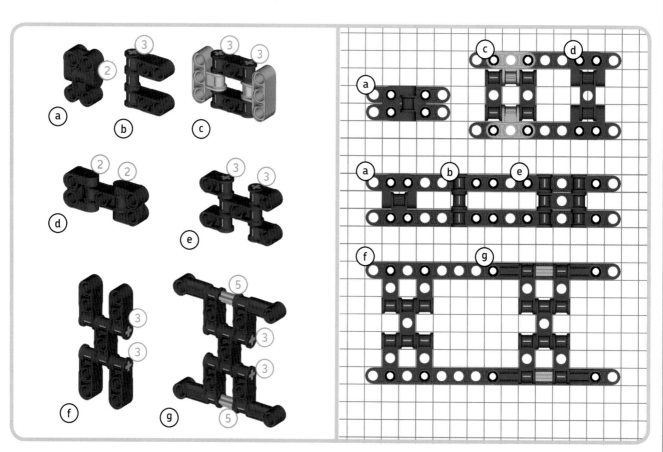

Figure 10-17: Connecting parallel beams with their flat sides facing each other. The circled numbers denote the length of the axles used in the constructions.

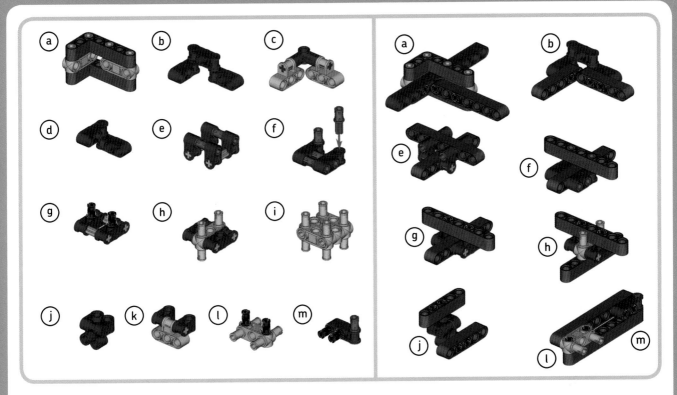

Figure 10-18: Using connector blocks to connect beams at a right angle. Each of the grey axles are 3M in size.

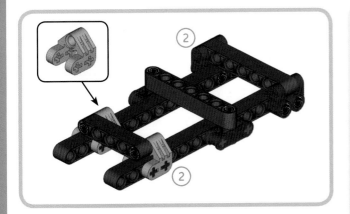

Figure 10-19: You can use connector blocks to create attachment points for beams that reinforce a construction. Note that these structures aren't rigid when used on their own; you should use these techniques to strengthen structures like the ones in Figure 10-16.

DISCOVERY #56:
CONSTRUCTIVE CONNECTORS!

Difficulty: ✻ **Time:** ⏱⏱

Can you combine connector blocks to make sturdy connections between beams, as shown in Figure 10-20? Expand the examples from Figures 10-16 through 10-19 or create your own combinations.

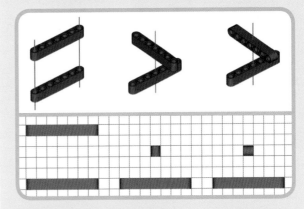

Figure 10-20: The green lines indicate how the two beams are positioned relative to one another.

using half LEGO units

Certain combinations of connector blocks result in a 0.5M offset from the unit grid, as shown in Figure 10-21. When used properly, this technique gives you more building options without compromising sturdiness. For example, you can create constructions that are 7.5M in size rather than 7M or 8M.

Note that you cannot easily strengthen such structures with beams because the distance between two beam holes should always be a whole number. That is, in this particular construction, you can't use a beam to connect the top beam to the one in the middle.

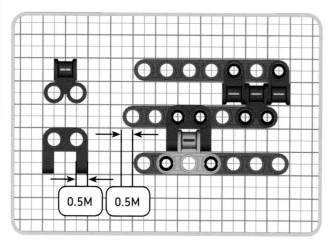

Figure 10-21: Some combinations of connector blocks result in a 0.5M offset from the unit grid. The beam in the middle has a 0.5M offset relative to the other two beams.

DISCOVERY #57:
HALF-UNIT BEAMS!

Difficulty: ☀ Time: ⏱
Can you join two beams to create an 18.5M beam using connector blocks?

HINT Use the connector blocks that are shown in Figure 10-21.

---

using thin elements

Most of the elements in the EV3 set are exactly 1M in width, but some *thin* elements are just 0.5M wide, as shown in Figure 10-22. Thin elements can be used when there is no space for larger elements. (The EV3 set does not contain many thin elements; you'll need to use thin elements from other LEGO Technic sets to take full advantage of this technique.)

One element to highlight here is the *cam*, which is especially useful for creating a rotating mechanism that presses the Touch Sensor once per rotation, as you'll see in Chapter 13.

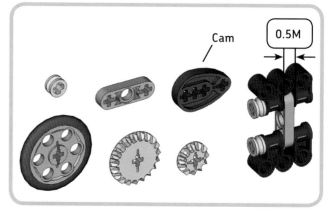

Figure 10-22: The thin elements in the EV3 set

creating flexible structures

You use nonfriction pins (the grey and tan pins in the EV3 set) to create hinges and flexible mechanisms instead of rigid constructions. For example, the nonfriction pins in the mechanism of Figure 10-23 make it easy to turn the gear.

The EV3 set contains two types of *steering links* (6M and 9M), normally used for steering mechanisms in LEGO Technic cars. These links can be used to replace beams in certain constructions. While a link creates a less sturdy connection than a beam, it can be used to connect elements that are not in the same plane. That is, if you widen the moving beam of the previous mechanism, you cannot use a beam to connect it to the gear, but you can use a steering link, as shown in Figure 10-24.

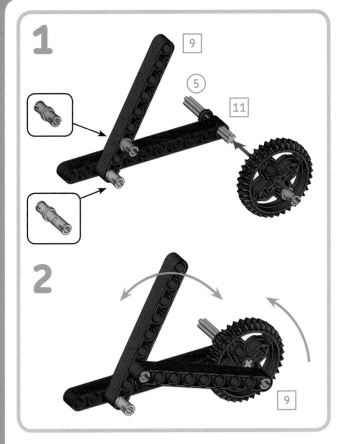

Figure 10-23: This dynamic structure uses nonfriction pins so that it's easy to turn the gear that makes the beam move back and forth. For comparison, replace the pins with friction pins, and you will find that it's much harder to turn the gear.

Figure 10-24: This modified version of the dynamic structure uses a steering link instead of a beam. You connect a steering link to a round hole or a cross hole using tow ball pins, as shown.

building with motors and sensors

We'll now look at how you can employ the size and shape of motors to use them as central components of your robots. Because motors are large and have many attachment points for pins and axles, it's often practical to use a motor as a starting point for a mechanism such as a robotic claw or a tank drive.

This method makes it possible to test each mechanism, or *module*, on its own. Once you've verified that all of the modules work well on their own, you can combine them into a single, sturdy robot.

building with the large motor

The geometry of the Large Motor (see Figure 10-25) makes it easy to connect two motors using a frame and friction pins. Doing so gives you a head start in creating a vehicle robot, as shown in Figure 10-26. All you need to do is add wheels or treads and the EV3 brick.

connecting wheels and treads

The Large Motor is strong and fast enough to drive wheels directly, as shown in Figure 10-27. You can also directly drive tank treads by bracing them with two 13M beams. Figure 10-28 shows the essential geometry, and you'll see another example when building the SNATCH3R in Chapter 18. (You'll learn how to connect gears to the Large Motor in the next chapter.)

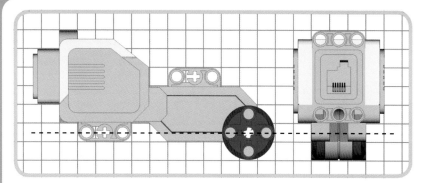

Figure 10-25: The geometry
of the Large Motor

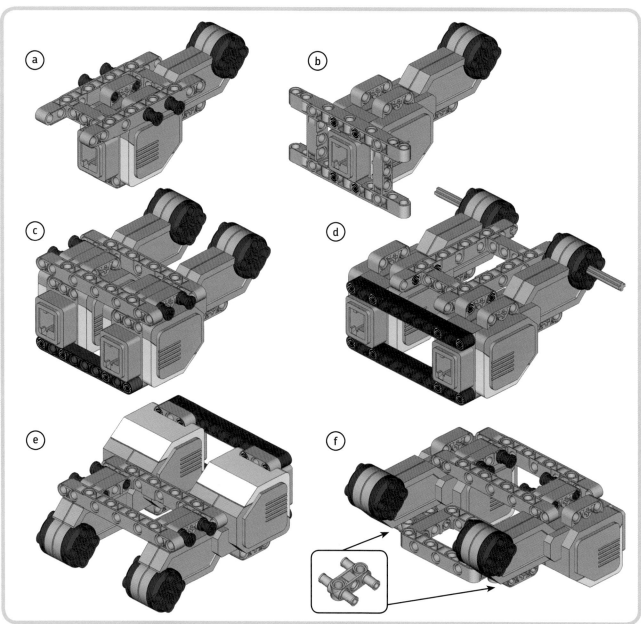

Figure 10-26: Adding frames to a motor makes it easier to mount the motor in a robot. For example, you can create a base for a vehicle robot by joining two Large Motors using frames, beams, and friction pins in various ways.

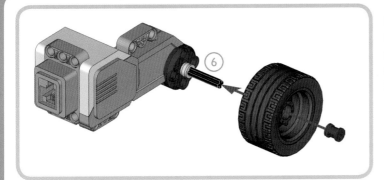

Figure 10-27: You can connect wheels directly to the Large Motor using a 6M axle, a half bush to create some space between the motor and the wheel, and a regular bush to prevent the wheel from sliding off the axle.

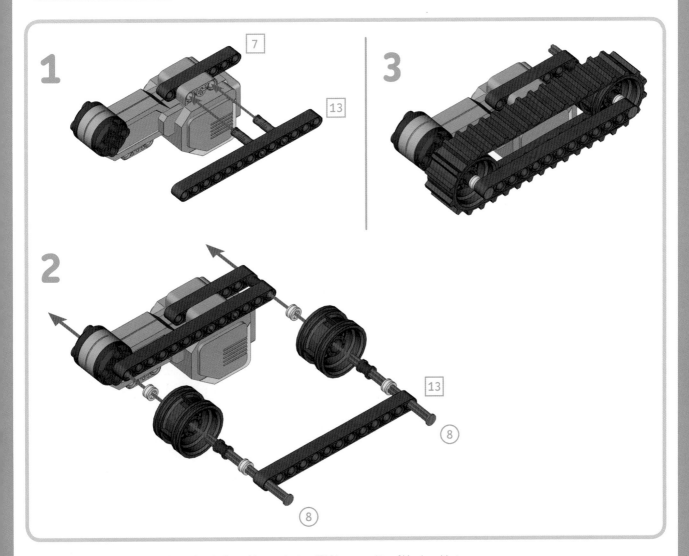

Figure 10-28: You can connect a tank tread to the Large Motor using two 13M beams and two 8M axles with stops.

connecting beams to the motor shaft

You can connect wheels and gears to the red motor shaft on the Large Motor using the cross hole, but you can also make beams and other elements rotate by connecting them to the round holes on the shaft. For example, you can make the motor continuously rotate a 3M beam to create a reciprocating mechanism, as shown in Figure 10-29.

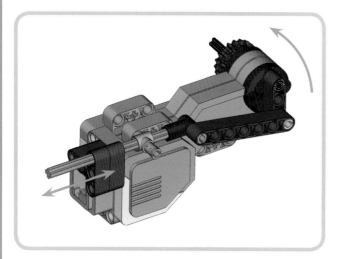

Figure 10-29: You can connect gears and wheels to the motor shaft with an axle or connect beams using the round pin holes. This mechanism makes the grey 9M axle move back and forth each time the motor shaft makes one rotation.

building with the medium motor

The Medium Motor (see Figure 10-30) is more compact than the Large Motor, allowing you to use it in small mechanisms, such as the steering mechanism of a race car. The motor has round mounting holes near the front, and you can create more connection points in the back by adding a frame, as shown in Figure 10-31. You can also add a frame as shown in Figure 10-32 so you can mount it between the two Large Motors of a vehicle robot, for example, to motorize a forklift mechanism.

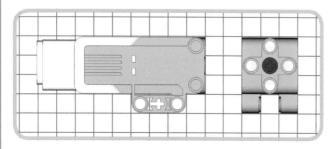

Figure 10-30: The geometry of the Medium Motor

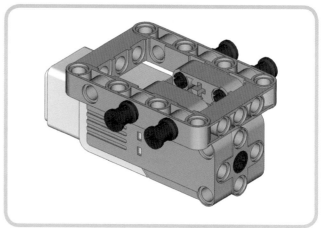

Figure 10-31: Adding connection points to the Medium Motor using a frame

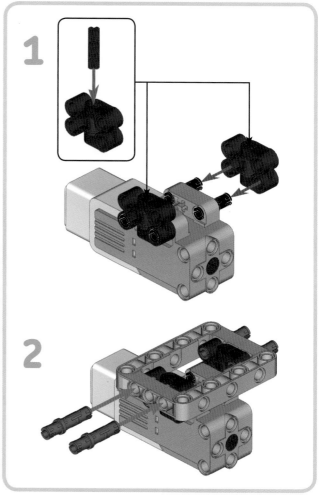

Figure 10-32: You can attach the Medium Motor to a frame so that it's easy to mount between the two Large Motors of a vehicle robot. For example, you can remove the frame in example e in Figure 10-26 and put this construction in its place.

building with sensors

Each of the sensors in the EV3 set has attachment points for one axle and two pins, as shown in Figure 10-33. In addition, the Infrared Sensor has two round holes in the back. To create a rigid connection, you'll need to use either two pins and a beam or an axle and a beam with a cross hole.

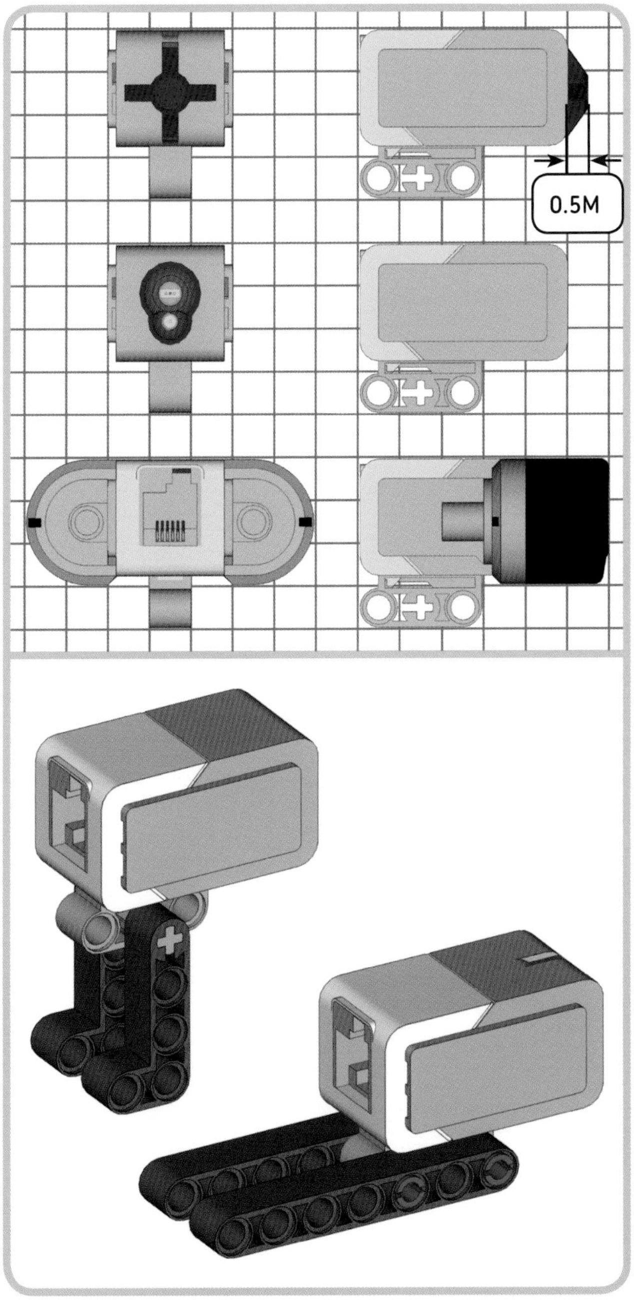

Figure 10-33: The geometry of the sensors in the EV3 set (top) and attaching them to your robot (bottom)

miscellaneous elements

In addition to the pieces discussed in this part of the book, the set contains various other elements, such as swords and monster teeth, which you can use to decorate your robots. Finally, your set includes a ball shooter and ball magazine component. You can find instructions on how to build a shooter with these components by following the building steps for EV3RSTORM in the EV3 software (see Figure 3-2 on page 26).

further exploration

In this chapter, you've learned the essentials of using beams, frames, pins, axles, connector blocks, and motors to create components for your robots. You've also learned how the LEGO unit grid can help you design your own sturdy constructions. There is no recipe for creating a perfect robot. Instead, the best way to gain experience in designing your own robots is just to try things out. You can begin by building the designs presented in this book and then modify them to create your own robots using the Design Discoveries throughout.

In the next chapter, you'll look at how gears work and how you can use them with the EV3 motors.

DESIGN DISCOVERY #12: TANK DRIVE!

Building: ☼ **Programming:** ▭
The EXPLOR3R moves around on two wheels and a support wheel in the back. Can you create a version of the EXPLOR3R that drives around on tank treads? Test your creation by driving it around with the infrared remote.

HINT **Use the tank tread example in Figure 10-28. Why should you remove EXPLOR3R's support wheel?**

DESIGN DISCOVERY #13: TABLETOP CLEANER!

Building: ✸✸ **Programming:** ▭▭

Can you create a robot that drives around a tabletop without driving off the table? Make the robot sweep away any LEGO elements in front of it with the Medium Motor so the robot cleans the tabletop as it drives around.

Add the Infrared Sensor to the front of the robot, about 25 cm (10 inches) ahead of the robot's front wheels, and make it point downward so it sees the tabletop. How do you make the robot detect the approaching table edge?

TIP If you've already built the robot from Design Discovery #12 on page 119, you can use it as a starting point for this Design Discovery.

DESIGN DISCOVERY #14: CURTAIN OPENER!

Building: ✸✸ **Programming:** ▭▭▭

Can you design a robot that automatically opens the curtains when the sun rises and closes the curtains when the sun sets? Use the Color Sensor to measure the ambient light intensity, and add an option to override the automatic behavior with the infrared remote control.

TIP If you use just one threshold value, the robot may end up opening and closing the curtains repeatedly when the light level fluctuates around the threshold. To avoid this problem, you can use two different threshold values. What should the robot do when the light level is in between these values?

building with gears

You can use *gears* to transfer motion from one rotating axle to the next. For example, you can transfer the motion of a rotating motor to the wheels of a robot to make it drive. Gears can also be used to change the output speed and torque of a rotating axle.

A series of gears used to transfer motion is called a *gear train*. In this chapter, you'll begin learning how gears work as you experiment with a basic gear train. You'll then see how the gear ratio controls the performance of the gear train. Finally, you'll explore each of the gears in the EV3 set and discover how you can use them effectively in your own robots.

gearing essentials

To begin, create a mechanism with two gears, as shown in the following steps. You'll use this mechanism to experiment with the essentials of gears. Be sure to try out the other examples as you read on—it's the best way to really understand how gears work.

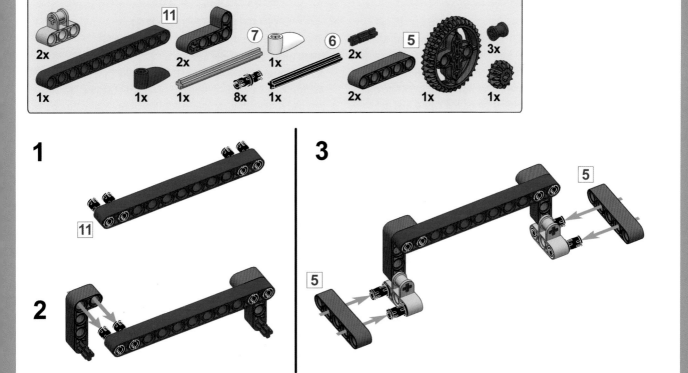

4

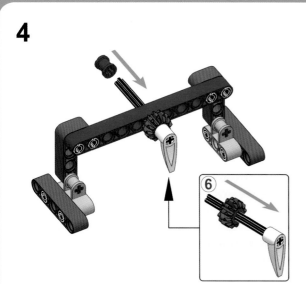

Do not attach the bushes too tightly; the gears should turn smoothly.

5

Make sure that both dials point downward as shown before you add the bushes.

Before we look at gears in detail, let's rotate the gears manually and observe what happens:

* Turning one gear makes the other gear turn. Regardless of which gear you turn, the other gear always rotates in the *opposite* direction.
* For every turn of the red dial, the white dial completes precisely *three* turns. (To see this, have both dials point down at first and then count how many times the white dial goes round as you rotate the red dial once.)
* The small gear always rotates faster than the big gear. In fact, the small gear turns three times as fast as the big gear.
* If you try to block the grey axle (attached to the big gear) with your hand, you'll find that you can still turn the black axle (attached to the small gear) with some effort. On the other hand, if you block the black axle, it's very difficult to turn the grey axle.

You'll find explanations for each of these observations as you read on.

taking a closer look at gears

If you look at the gears in our example more closely, you'll see that the small gear has 12 *teeth* (we'll call it a *12T* gear), while the big gear has 36 teeth (*36T*). At the *contact point*, the teeth of both gears *mesh*, as shown in Figure 11-1. If you turn the

small gear manually, its teeth force the teeth of the big gear to follow, causing the big gear to turn, in the opposite direction. We'll refer to the gear that we turn manually as the *input gear*. The gear that follows as a result is the *output gear*.

For each tooth of the small gear passing through the contact point, there is one tooth of the big gear that follows. As the small gear (12T) makes three complete rotations, each of its teeth passes through the contact point three times so that a total of 36 teeth pass through the contact point (3 × 12 = 36). During this time, all 36 teeth of the big gear (36T) are pushed through the contact point so that the big gear completes one turn.

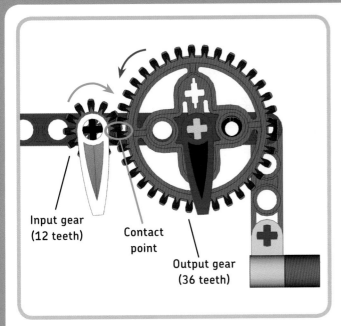

Input gear
(12 teeth)

Contact
point

Output gear
(36 teeth)

Figure 11-1: If you look at the gearing mechanism closely, you'll see that the teeth of both gears mesh. The input gear makes the output gear turn by pushing the teeth of the output gear at the contact point in the direction of the arrow.

calculating the gear ratio for two gears

As you've just seen, three rotations of the 12T gear (the white dial) result in one rotation of the 36T gear (the red dial). You can describe this configuration with the *gear ratio*. The gear ratio is the factor by which *output speed decreases* relative to the input speed. The gear ratio is also the factor by which *output torque increases* relative to the input torque. (More torque makes it easier for a vehicle to drive up a hill. We'll talk more about what torque means in a moment.)

You calculate the ratio as follows:

$$\text{Gear ratio} = \frac{\text{Number of teeth on the } output \text{ gear}}{\text{Number of teeth on the } input \text{ gear}} \qquad 1$$

The output has 36 teeth and the input has 12 teeth, so in our example, the formula gives us 36 ÷ 12 = 3. The gear ratio is the factor by which the output speed decreases so that the output gear spins 3 times as slow as the input gear. In other words, 3 rotations of the input result in just 1 rotation of the output.

Gear ratios are sometimes written as the number of teeth on the output and the input separated by a colon—in this case, for example, 36:12. If you simplify this ratio to its lowest terms, you get 3:1, which has the same meaning. (Reading from left to right, you see again that 3 rotations of the input result in 1 rotation of the output.)

calculating output speed

Once you've calculated the gear ratio of an existing design, you can use the ratio to calculate the output speed if you know the input speed:

$$\text{Rotational } output \text{ speed} = \frac{\text{Rotational } input \text{ speed}}{\text{Gear ratio}} \qquad 2$$

If the gear ratio is 3 and if you rotate the input gear at 30 rotations per minute (rpm), the output gear will turn at 30 ÷ 3 = 10 rpm, which confirms that the speed decreases by a factor of 3.

calculating the required gear ratio

You can rewrite the previous formula to calculate the required gear ratio for your design if you know the input speed and the output speed you want to achieve:

$$\text{Gear ratio} = \frac{\text{Rotational } input \text{ speed}}{\text{Rotational } output \text{ speed}} \qquad 3$$

For example, if you want an output gear with a wheel to rotate at 120 rpm while you have a motor rotate the input gear at a constant speed of 72 rpm, you need the following gear ratio: 72 ÷ 120 = 0.6. You can accomplish this ratio with a 20T gear as the input and a 12T gear as the output (12 ÷ 20 = 0.6).

Not every gear ratio can be realized with the gears in the EV3 set, so you may want to use one of the gear combinations given in this chapter and use formula [2] to calculate whether the resulting output speed is satisfactory for your design.

NOTE Be sure to use the same units for the rotational input and output speeds. (If you measure the input speed as rotations per minute, you should measure the output speed in rotations per minute, too.)

decreasing and increasing rotational speed

Now let's look at how the example gear train could be used in a robot. You can use gears to change the rotational speed of an output, such as a wheel, relative to the speed of an input, such as a motor. To *gear down*, or decrease the speed, the output gear should have more teeth than the input gear so that the gear ratio is greater than 1, as shown in Figure 11-2.

This configuration decreases the wheel speed by a factor of 3. As a result, the output torque is increased by a factor of 3.

Now let's look at what happens if you interchange the two gears, as shown in Figure 11-3. The 36T gear is the input driven by a motor, and the 12T gear is the output connected to a wheel.

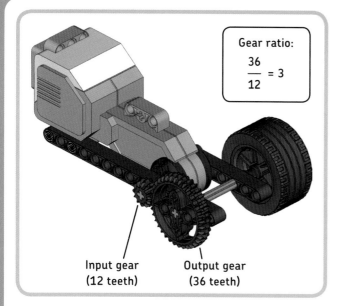

Gear ratio:
$$\frac{36}{12} = 3$$

Input gear
(12 teeth)

Output gear
(36 teeth)

Figure 11-2: Decreasing the rotational output speed by a factor of 3 while increasing the torque by a factor of 3. The gear ratio is 3 (or 3:1).

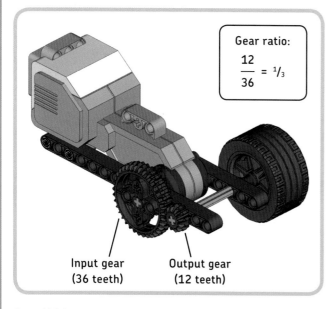

Gear ratio:
$$\frac{12}{36} = \frac{1}{3}$$

Input gear
(36 teeth)

Output gear
(12 teeth)

Figure 11-3: Increasing the rotational output speed by a factor of 3 while decreasing the torque by a factor of 3. The gear ratio is $\frac{1}{3}$ (or 1:3).

The gear ratio is $12 \div 36 = \frac{1}{3}$, or approximately 0.333. Therefore, the speed *decreases* by a factor of $\frac{1}{3}$, but that's the same as saying that the speed *increases* by a factor of 3. (If you rotate the input at 30 rpm, formula [2] gives you $30 \div 0.333 = 90$ rpm as the output speed, which is indeed three times as fast.)

Increasing the output speed is called *gearing up*. Increasing the output speed means that the output torque decreases, so it will be harder to drive up a hill.

DISCOVERY #59: GEARING MATH!

Difficulty: ☀ **Time:** ○

What is the gear ratio of each set of gears shown in Figure 11-4? If you turn the input gears at 10 rpm, what will be the rotational speed of the output gears?

TIP You can verify your answer by building the gear trains. Add dials to the gear axles to make it easier to see how much each gear turns.

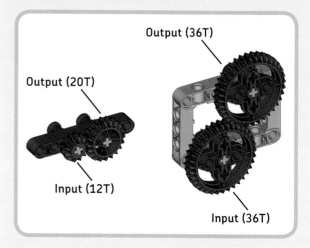

Output (36T)

Output (20T)

Input (12T)

Input (36T)

Figure 11-4: What are the gear ratios of these gear trains?

If you use two gears with the same number of teeth, the gear ratio is 1, and both speed and torque remain unchanged.

what is torque?

You've just seen how you can increase torque, but what exactly is torque? Why can increasing it be useful? To experience the concept of torque, replace the red dial with a weight consisting of two wheels, as shown in Figure 11-5. Now try to lift the weight by turning the *grey* axle manually. When you do this, your hand has to apply *torque* to the axle in order to counterbalance the torque created by the weight of the wheels placed at a distance from the axle.

Torque is the product of a force and the distance between the force and the axle. In this case, the force is gravity acting on the wheels. If you increase the weight by adding more wheels, or if you place the wheels farther from the axle using a longer beam, the torque created by the wheels will increase, and you'll need to apply a greater torque to the grey axle to lift them.

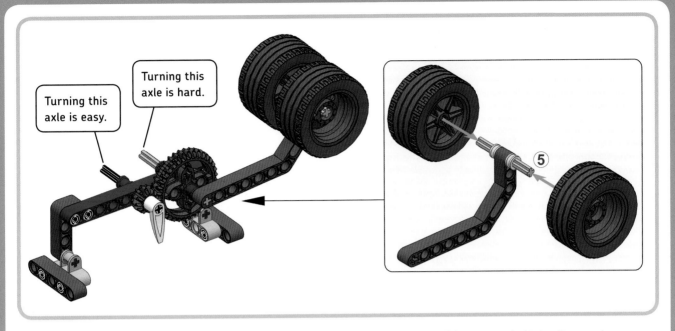

Figure 11-5: Lifting the weight by turning the black axle is easier because the required torque is just a third of the torque required to turn the grey axle.

Now try to lift the weight by turning the *black* axle. When you do this, the gears increase the torque you apply with your hands by a factor of 3 so that lifting the weight is considerably easier. However, you have to turn the black axle farther, compared to the grey axle, to fully lift the weight. (You have to turn it three times as far to accomplish the same effect.)

when do you increase torque?

Increasing torque with gears is useful if a motor cannot provide sufficient torque for a certain application, such as lifting a heavy weight. If your motor has difficulty making certain movements required in your program, you can use gears to increase the output torque and reduce the load on the motor (see Figure 11-2).

The maximum torque provided by the Large Motor is about three times the maximum torque provided by the Medium Motor, so it's easier to use Large Motors for heavy-duty tasks. If both of the Large Motors are already in use—say, to drive your robot—you can use gears to increase the output torque of the Medium Motor. Visit *http://ev3.robotsquare.com/* for details about the relationship between torque and rotational speed in EV3 motors.

decreasing torque

Sometimes you'll want to decrease a motor's output torque to protect a fragile mechanism. While you can decrease the output torque with gears using the technique shown in Figure 11-3, it's easier to limit the output torque in your program by using an Unregulated Motor block at a low power level,

such as 30%, as you learned in Chapter 9. When you use this block, the motor does not significantly increase torque when an external force slows it down.

creating longer gear trains

The gear train you've seen so far has only two gears, but you can extend it with more gears to transfer motion across a greater distance. For example, add a 20T gear, as shown in Figure 11-6, and use it as the input. Let's look at this new mechanism in more detail.

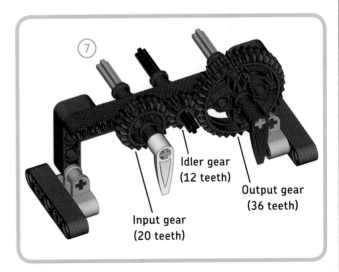

Figure 11-6: Remove the white dial from the black axle, and add a 7M axle with a 20T gear as shown. Make sure that both dials point downward.

The gear train now consists of an input gear (20T), an output gear (36T), and an *idler gear* (12T) in the middle. The idler gear transfers the motion of the input gear to the output gear. In addition, it reverses the direction in which the output gear rotates so that the input and the output rotate in the same direction (see Figure 11-7).

NOTE Each gear in a gear train rotates in the direction opposite to the one next to it. Consequently, the input and output of a gear train with an *odd* number of gears rotate in the *same* direction; the input and output of a gear train with an *even* number of gears rotate in the *opposite* direction.

calculating the compound gear ratio

You can determine the relationship between the input speed and the output speed of a gear train by calculating the overall gear ratio, or *compound gear ratio*. To calculate it, first determine the gear ratio of each pair of adjacent gears and then multiply these ratios, as shown in Figure 11-7.

The example mechanism has two pairs of adjacent gears. First, the input gear transfers motion to the idler gear with a gear ratio of 0.6. Next, the idler gear transfers motion to the output gear with a gear ratio of 3. Notice that the idler gear serves as the *output* of the first pair, while it serves as the *input* of the second pair.

If we calculate the compound ratio here, we get 0.6 × 3 = 1.8. This means that the output speed is decreased by a factor of 1.8 and the torque is increased by a factor of 1.8.

Therefore, 1.8 rotations of the input gear equals 1 rotation of the output. To see this, make both dials point downward, as shown in Figure 11-7, and rotate the white dial 9 times. The red dial should rotate 5 times (9 ÷ 1.8 = 5), after which both dials should point downward again.

Interestingly, because the central gear is an idler gear, it does not affect the compound gear ratio. You get the same ratio by dividing only the output and the input gear, which gives you 36 ÷ 20 = 1.8. This is because the number of teeth on the idler gear (12T) cancels out when calculating the compound gear ratio:

$$\frac{\cancel{12}}{20} \times \frac{36}{\cancel{12}} = \frac{36}{20} = 1.8$$

further increasing torque and decreasing speed

Sometimes the torque increase provided by two gears is not enough. You can further increase the gear ratio, and therefore torque, by coupling multiple pairs of gears that have a gear ratio greater than 1. To see how this works, modify your gear train so it looks like the one shown in Figure 11-8.

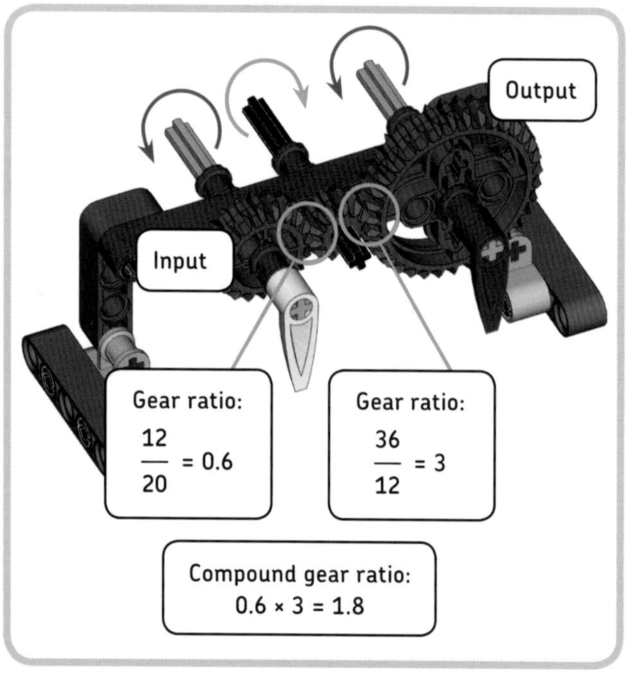

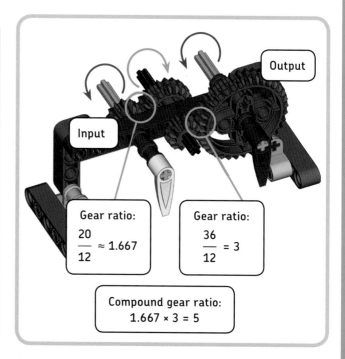

Figure 11-7: Calculating the compound gear ratio

Figure 11-8: Coupling two pairs of gears to attain a larger gear ratio. If you don't round the intermediate gear ratio values, the compound gear ratio is exactly 5 in this example.

In the example, the first pair of gears has a gear ratio of 20 ÷ 12 ≈ 1.667. The second pair has a ratio of 36 ÷ 12 = 3. You get the compound gear ratio of the gear train by multiplying these numbers, which gives you 1.667 × 3 = 5. Consequently, the output is 5 times as slow as the input. Put another way, rotating the white dial 5 times makes the red dial complete 1 rotation.

As a result, the torque increases by a factor of 5. If you replace the red dial with the weight you built in Figure 11-5, it should be easy to lift the weight by turning the white dial because of the increased torque.

balancing speed and torque

If you use the gear with the red dial in Figure 11-8 as the mechanism's input, the white output dial will rotate five times as fast. In principle, you can increase the output speed further by adding even more gears, but doing so reduces the output torque, too. Eventually, you'll reach the point where the output torque is no longer sufficient to overcome the friction in the gear train and the gears won't move at all. Similarly, you can't use gears to increase the speed of a race car indefinitely because there would not be enough torque to make the car accelerate from standstill.

In general, you'll have to experiment with various gear combinations to find the proper balance between torque and speed in your design:

* First, consider whether it's necessary to use gears at all. You may be able to accomplish the required speed and torque by changing the Power settings of the blocks in your program.
* If the maximum speed of your motor isn't high enough, try increasing the speed using a gear ratio less than 1, while making sure that enough torque remains for your robot to work properly.
* If a motor struggles to perform a heavy task, you can increase the torque using a gear ratio greater than 1, at the cost of a reduction in speed.

DISCOVERY #60:
PREDICTABLE MOVEMENT

Difficulty: ☀ **Time:** ◷

Can you analyze the gear train shown in Figure 11-9 before you actually build it? How fast does the red dial on the right turn compared to the white dial? And how fast does it turn compared to the red dial on the left? In which direction does each dial turn? Once you think you know the answers, build the gear train and verify your predictions.

Figure 11-9: Gear train with a 36T gear (left), a 12T gear (middle), and another 36T gear (right)

DISCOVERY #61:
COMPOUND DIRECTION!

Difficulty: ☀ **Time:** ◷

What is the compound gear ratio of the gear train in Figure 11-10? How is this gear train different from the gear train of Figure 11-1? Why might adding these 24T gears be useful?

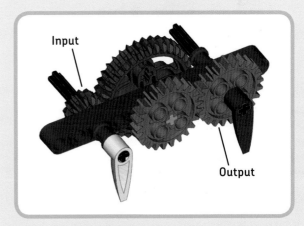

Figure 11-10: What is the compound gear ratio of this gear train?

friction and backlash

There are two important aspects of gears that can reduce the performance of your gear train. First, each gear introduces some *friction* to the mechanism. Friction causes rotating objects to slow down as they slide against other objects, and it reduces the output torque. You can experience friction by pushing the gears and bushes in Figure 11-8 tightly against the beam. You should find that it now takes more torque to turn the axles than when the bushes and gears were connected loosely. You can reduce friction in your design by bracing gears between two beams, as you'll see in "Constructing Sturdy Gear Trains" on page 134.

Second, each gear introduces some play, or *backlash*, as shown in Figure 11-11. Even if you block the gear on the left, the gear on the right can still turn by a tiny amount because of the space between the teeth. This means that you lose some control over the exact position of the output. Regardless of how accurately you move the input gear, the output gear can always move back and forth a little. The longer your gear train is, the more backlash you will see.

Figure 11-11: Backlash is caused by the existence of some play between meshing gear teeth.

The output shafts of the EV3 motors also experience some backlash, caused by the gear train inside each motor.

using the gears in the EV3 set

The EV3 set contains spur gears, bevel gears, double-bevel gears, knob wheels, and worm gears, as shown in Table 11-1. *Spur gears* can be used to transfer motion between parallel axles, while *bevel gears* can be used to transfer motion between perpendicular axles. *Double-bevel gears* can be used for both parallel and perpendicular configurations. (*Perpendicular* means that the axles are placed at a right angle to one another.)

table 11-1: the technic gears

Category	Qty	Gear	Teeth	Radius
Spur gear	0*		8	0.5M
	0*		16	1M
	2		24	1.5M
	0*		40	2.5M
Bevel gear	1		12	N/A
	1		20	N/A
Double-bevel gear	2		12	0.75M
	4		20	1.25M
	5		36	2.25M
Knob wheel	4		N/A	N/A
Worm gear	2		1	N/A

* These gears are not included in the EV3 set, but knowing their properties can be handy when combining your EV3 set with other Technic sets.

working with the unit grid

When you combine gears to create a gear train, it is essential to place them at the right distance to make their teeth mesh properly. If you place the gears too close together, then they can't turn; if you place them too far apart, then the teeth will *slip*. When a gear slips, it turns without catching the teeth from the other gear, and you'll hear a rattling sound.

Assuming you space them properly, you can use any combination of two spur gears to create a gear train. Similarly, you can combine all of the double-bevel gears. In fact, you can combine spur gears with double-bevel gears.

The required *distance between the center points* of two gears is the sum of their radii, as shown in Figure 11-12. The radii of spur gears and double-bevel gears, measured in LEGO units (M), are given in Table 11-1. For example, the *radius* of a 12T gear is 0.75M, and the radius of a 36T gear is 2.25M, so the distance between the center points of the gears must

be 0.75M + 2.25M = 3M. Because 3 is a whole number, it's easy to mount these gears on a beam using axles, exactly 3M apart.

gears and half units

Adding the radii of two gears will sometimes result in a number between two whole numbers, such as 1.5M or 2.5M. For example, the distance between two 20T gears is 1.25M + 1.25M = 2.5M. You can use connector blocks to achieve such a distance, as shown in Figure 11-13.

NOTE Instead of calculating the required distance between gears yourself, you can use the gear ratio calculator at *http://gears.sariel.pl/*. You can choose where you'll place the gear axles on the unit grid, and the calculator will tell you which gears you can use to bridge this distance.

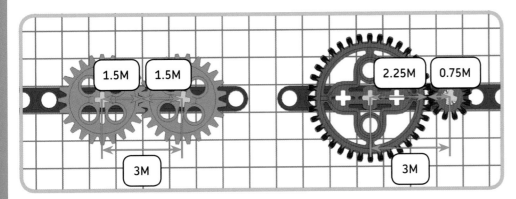

Figure 11-12: Calculating the required distance between the center points of two gears. If the sum of the two radii is a whole number, you can add the gears to a beam.

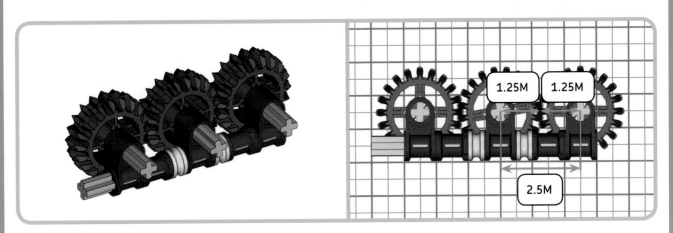

Figure 11-13: If the sum of the radii is a number between two whole numbers, you'll have to use connector blocks to create a 0.5M offset, as discussed in Chapter 10.

gear trains around a corner

You can extend a gear train along the corner of an angled beam by placing one gear at the corner hole, as shown in Figure 11-14. When you do this, it's best to brace the gear train between two beams, as you'll see later in this chapter.

using improper combinations

You can combine spur gears with double-bevel gears, but placing them at the proper distance can be tricky because their radii do not add to a whole number or a number halfway between two whole numbers. For example, the 12T double-bevel gear and the 24T spur gear must be placed 0.75M + 1.5M = 2.25M apart.

This distance cannot be accomplished on the unit grid or with a 0.5M offset, but you can get close to this distance using the corner of a right-angled beam, as shown in Figure 11-15. You can calculate the distance between holes on an angled beam using the Pythagorean theorem or measure the distance with a ruler (1M equals 8 mm, or approximately 5/16 inches).

You should always test your gear train carefully if the gears are not placed at precisely the proper distance. Gears should turn smoothly, and they should never slip, even if you block one gear with your hands. If you're not sure whether a particular improper combination will work, it's better to simply choose a combination of gears whose radii add up to a whole number.

using bevel and double-bevel gears

You can use bevel gears and double-bevel gears to transfer motion between two perpendicular axles, as shown in Figure 11-16.

A double-bevel gear is actually a combination of a spur gear and two bevel gears. So far, we've used only the spur gear teeth to transfer motion between two parallel axles, but you can use the bevel teeth on either side to create perpendicular connections. In fact, the 20T bevel gear is the same as the beveled section of the 20T double-bevel gear, and the 12T bevel gear is the same as the beveled section of the 12T double-bevel gear, so each combination in Figure 11-16 has the same gear ratio.

perpendicular connections on the unit grid

Every combination of bevel and double-bevel gears can be made to fit on the unit grid, but Figure 11-17 shows some particularly useful combinations. To transfer motion between perpendicular axles in your design, you can choose one of these examples and use the unit grid as a reference to design a construction that holds the axles in place.

NOTE The method for calculating the distance between two gears using their radii works only for parallel gear configurations, as shown in Figure 11-12. For perpendicular configurations, use the unit grid, as shown in Figure 11-17.

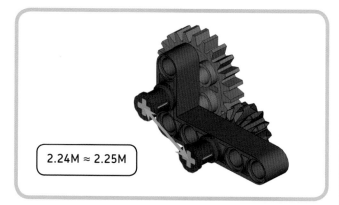

2.24M ≈ 2.25M

Figure 11-14: Creating a gear train along the corner of a beam. Both 20T gears turn in the same direction and at the same speed. The 12T idler gear has no effect on the gear ratio.

Figure 11-15: While you cannot easily accomplish a 2.25M distance between two gears, this configuration achieves a distance of 2.24M, which is close enough. This combination of gears is useful because the gear ratio is exactly 2, allowing you to double the torque and reduce the speed by a factor of 2 (or vice versa).

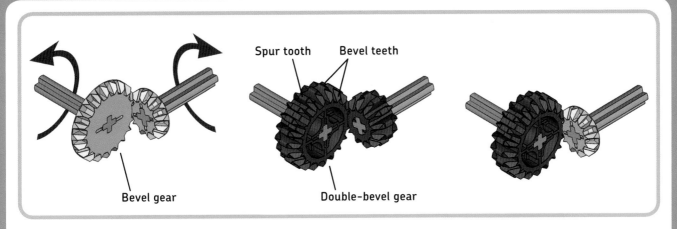

Figure 11-16: You use bevel and double-bevel gears to transfer motion between two perpendicular axles. The gear ratio of each combination shown above is the same. If you use the 20T gear as the input and the 12T gear as the output, the gear ratio is 12 ÷ 20 = 0.6.

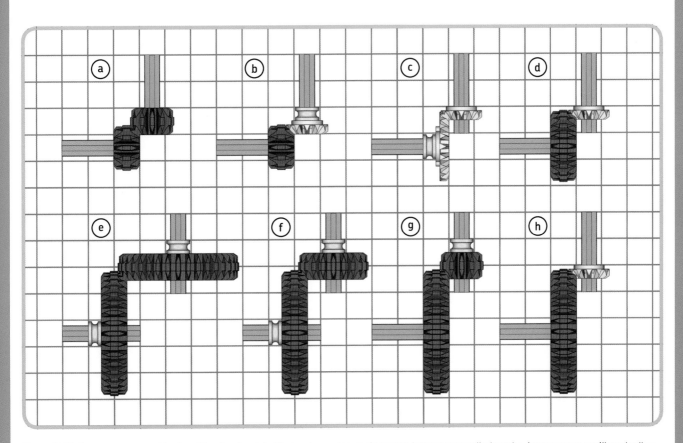

Figure 11-17: You can use any combination of two bevel and double-bevel gears to transfer motion between perpendicular axles. In some cases, you'll need yellow bushes to create a 0.5M offset.

connecting perpendicular axles

When transferring motion between two perpendicular axles, it is important to create a sturdy construction so that the gears do not slip. Figure 11-18 demonstrates how you can accomplish this using angled beams and connector blocks.

The EV3 set also contains a specialized element for connecting two small gears on perpendicular axles, as shown in Figure 11-19. This element can easily be connected to the Medium Motor (see Figure 11-28b, later in the chapter). You can also securely connect perpendicular axles using a frame, as shown in Figure 11-20.

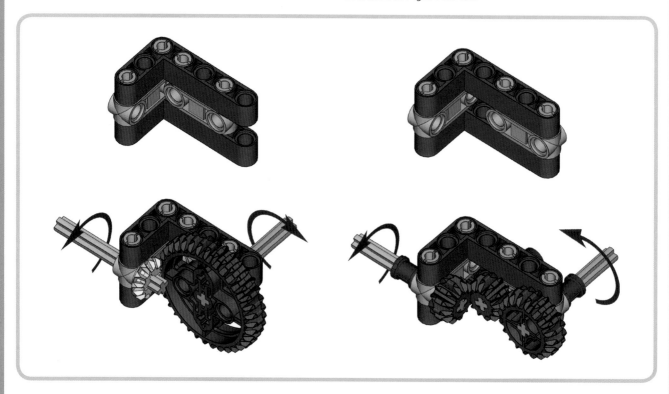

Figure 11-18: Using two L-shaped beams and two connector blocks to secure two perpendicular axles. The gear train on the right contains both a perpendicular connection and a connection between two parallel axles; the 12T gear in the middle acts as the idler gear.

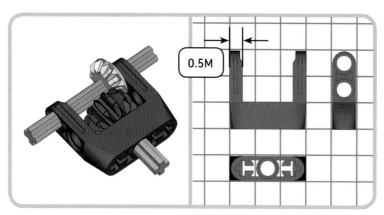

Figure 11-19: Creating a compact perpendicular connection

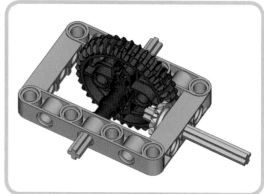

Figure 11-20: Connecting perpendicular axles using a frame

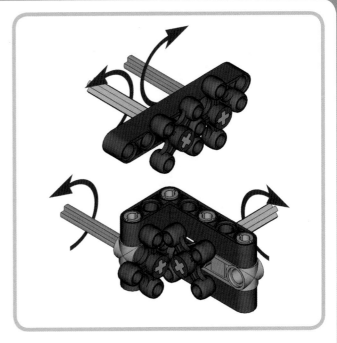

Figure 11-21: Using the knob wheel with parallel axles (top) and perpendicular axles (bottom)

using knob wheels

The *knob wheel* can be used to transfer motion between two parallel axles or between two perpendicular axles, as shown in Figure 11-21. The knob wheel can handle greater torques than bevel gears without slipping, making it suitable for use with highly loaded perpendicular axles. Due to its unique shape, a knob wheel can be used to drive only another knob wheel; it cannot drive spur, bevel, or double-bevel gears. The gear ratio between two knob wheels is always 1.

using worm gears

The *worm gear* can drive spur gears to achieve substantial speed reduction, as shown in Figure 11-22. In calculating the gear ratio, you may consider the worm as an input gear with just one tooth. When the output is a 24T gear, the resulting gear ratio is 24 ÷ 1 = 24, thereby reducing the output speed by a factor of 24. In principle, the torque increases by a factor of 24 as well, but because this gear configuration has more friction than gear trains with normal gears, some of the torque gain is lost.

Unlike with the other gear configurations you've seen so far, the motion transfer works only one way: You can rotate the worm gear to make the spur gear turn, but you cannot turn the spur gear to make the worm gear turn. This can be

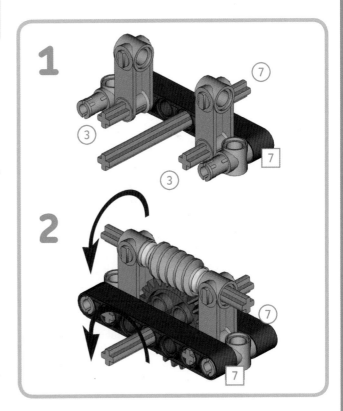

Figure 11-22: You can drive a 24T gear using a worm gear to reduce the output speed by a factor of 24. The specialized grey connector blocks place the worm gear at just the right distance while allowing the 24T gear to turn. You can use this geometry as a starting point for your own constructions with a worm gear.

an advantage in your design. For instance, if you use the worm gear to control a robotic arm, the arm won't move back down when you stop applying power to the motor. If you used normal gears, gravity acting on the arm would cause the gears to turn in the opposite direction and lower the arm.

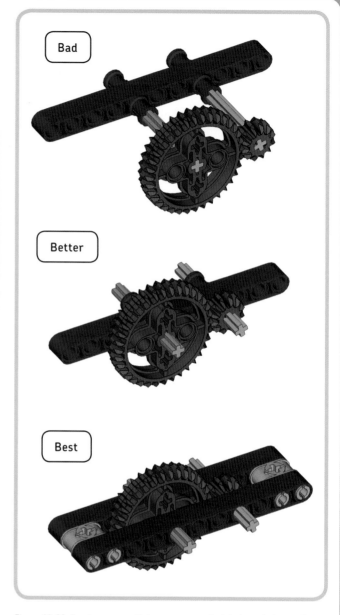

DISCOVERY #64:
WORM DRIVE!

Difficulty: ✹✹ **Time:** ◷◷
Can you create a gear train that reduces the output speed by a factor of 8?

HINT First, decrease the speed by a factor of 24, and then increase it by a factor of 3. Why does this have the same effect?

---

constructing sturdy gear trains

Once you've selected the gears for your gear train, you'll need to mount them on your robot. The possibilities for mounting axles with gears are different for every robot, but it is always important to mount the axles securely so that the axles do not bend or twist and so that the gear teeth do not slip.

bracing gears with beams

The force between the teeth of two gears pushes the axles on which they are mounted away from each other. If the distance between the gears increases too much because of this force, they no longer mesh properly, and their teeth slip. You can reduce the deflection of the axles by placing the gears next to a beam, as shown in Figure 11-23. Your gear train becomes even more robust if you brace the gears between two beams. The force between the gear teeth is still there, but the beams keep the axles with the gears in place so that the gears can't slip.

If bracing the gears with a second large beam is not possible in your design, you can add a short beam or a connector block (see Figure 11-24). The solution is not as robust, but it does prevent the gears from slipping.

Figure 11-23: Bracing gears with beams ensures that their teeth do not slip.

Figure 11-24: Bracing a gear with a short beam

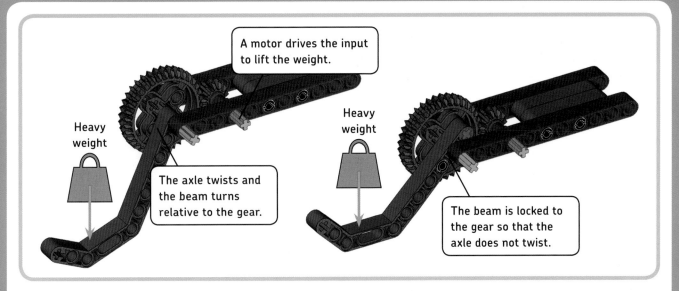

Figure 11-25: High torques can cause axles to twist (left). You can reduce the load on the output axle by attaching the angled beam that carries the heavy load directly to the mounting holes of the output gear (right).

preventing axles from twisting

When you increase torque using gears, it is possible to achieve torques that can twist axles, as shown on the left of Figure 11-25. You can avoid twisting the output axle by attaching the part of the mechanism that carries a heavy load directly to a 36T double-bevel gear instead of to the axle, as shown on the right.

   For the same reason, it's a good idea to connect beams directly to the holes on the motor shaft rather than connecting them only with an axle (see Figure 10-29 on page 118).

reversing direction of rotation

You can reverse an axle's direction of rotation by reversing the motor on the input, but you can also reverse the direction using gears, as shown in Figure 11-26. This is useful if a single motor drives two mechanisms on the same axle, but the mechanisms should turn in opposite directions.

building with gears and EV3 motors

You'll often use gears to transfer motion from a motor to a mechanism, such as a robotic arm. The Large Motor does not have many attachment points near the rotating shaft, but you can add those yourself using beams, as shown in Figure 11-27.

   Figure 11-28 shows how you can drive axles parallel (a) and perpendicular (b) to the Medium Motor shaft.

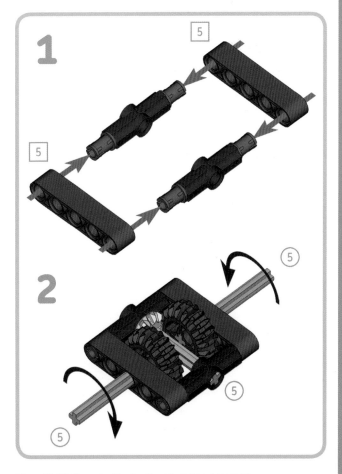

Figure 11-26: Reversing the direction of rotation using gears

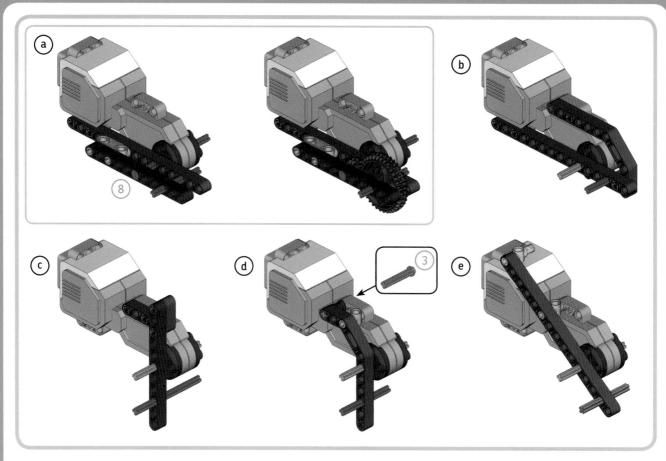

Figure 11-27: You can add beams to the Large Motor to create attachment points for axles and gears. The grey axles are connected to the motor shaft, while the orange axles show where you can connect the 36T gear to achieve a gear ratio of 3, as in example a. The beam in example e is placed at the 53.13 degree angle, as discussed in Chapter 10. The connector shown in green aligns with the unit grid.

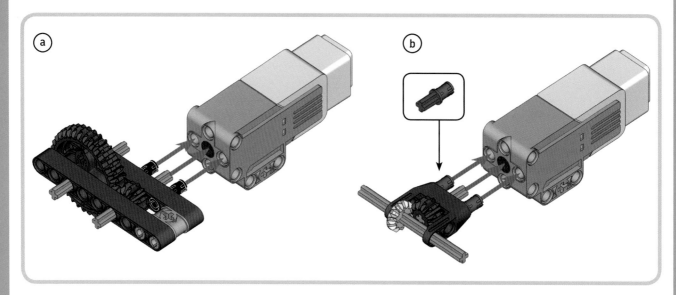

Figure 11-28: Connecting gears to the Medium Motor. The output axle is parallel to the motor shaft in example a and perpendicular in example b.

further exploration

In this chapter, you've learned how gears work and how you can use them to change the speed and torque provided by EV3 motors. You've also learned how the gear ratio, friction, and backlash affect the performance of the gear train. In addition, you've seen how to create sturdy gear trains with spur gears, bevel gears, and double-bevel gears. In the next part of the book, you'll put your building and programming skills to work as you create a race car and a robotic insect, but first, solve some of the following Design Discoveries to gain more experience with gears.

If you want to learn more about gearing principles and other building techniques, I highly recommend that you read *The Unofficial LEGO Technic Builder's Guide* by Paweł "Sariel" Kmieć (No Starch Press, 2012), which covers LEGO Technic elements in much more detail.

DESIGN DISCOVERY #15: DRAGSTER!

Building: ✹✹ Programming: ▭

Can you create a really fast drag-racing robot? Design a robot with four wheels and use Large Motors to drive two of them. (It won't be necessary to add steering functionality in this Design Discovery.) Use gears to speed up your robot. Which gear ratio makes your robot go the fastest?

HINT Add the Infrared Sensor at the front of your robot, and program the robot to stop if it sees an obstacle up close.

DESIGN DISCOVERY #16: SNAILBOT!

Building: ✹✹ Programming: ▭

What is the largest gear ratio you can achieve with the gears in the EV3 set? Find the ratio and use it to create the slowest robot of all time. (It should still move, though!)

HINT Include the worm gear in your gear train.

DESIGN DISCOVERY #17: CHIMNEY CLIMBER!

Building: ✹✹✹ Programming: ▭

Can you create a robot that can climb vertically between two walls, as if climbing up a chimney? To create the "chimney," place a bookcase about 30 cm (12 inches) from a wall, making sure that the bookcase is perfectly parallel to the wall. Be sure to place a pillow between the two walls, in case your robot falls unexpectedly.

How can a robot climb vertically? Can you use wheels to drive up the walls?

HINT Visit *http://robotsquare.com/* to see how I've created such a robot with the previous generation of LEGO MINDSTORMS. Can you do the same with EV3?

DESIGN DISCOVERY #18: TURNTABLE!

Building: ✹✹✹ **Programming:** ▭

Can you build an automated turntable? A *turntable* is a rotating platform that can carry and rotate heavy objects, such as trains or cars. For EV3 robots, such a platform can act as the base of a stationary machine, like a robotic arm or a robot that drops assorted LEGO bricks in different storage compartments. Use one motor to make the turntable turn clockwise and counterclockwise.

HINT Design a platform using beams, and add four wheels under it, positioned as shown in Figure 11-29. In which direction should each of the wheels turn? Do you need to drive all of the wheels, and why might the mechanism of Figure 11-26 be useful for this application?

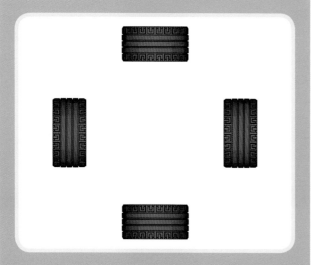

Figure 11-29: The wheel configuration of Design Discovery #18

DESIGN DISCOVERY #19: ROBOTIC ARM!

Building: ✹✹✹ **Programming:** ▭▭

Can you build a robotic arm that can grab and lift objects around it? Use one motor to enable the robot to turn in place, use another motor to lower and raise the robotic arm, and use a third motor to open and close the gripper. Create a program that lets you control each motor using remote control commands.

HINT Use the turntable of Design Discovery #18 as the robot's base.

PART IV

vehicle and animal robots

12

Formula EV3: a racing robot

Now that you've learned how to program the EV3 to control motors and sensors, you can begin making more sophisticated robots, such as autonomous vehicles, robotic animals, and complex machines. This chapter presents the *Formula EV3 Race Car*, shown in Figure 12-1.

Unlike the EXPLOR3R you built earlier, the race car uses three motors. Two Large Motors in the rear make the car drive, while the Medium Motor lets you steer the front wheels. Think

of the rear motors as the car's engine and the motor in the front as the car's steering wheel.

Once you've built the race car, you'll create several My Blocks to make it easy to program the car to drive and steer. Then you'll combine these blocks in one program that lets you control the car remotely and another program that makes the robot drive around autonomously and avoid obstacles. Finally, you'll be challenged to add more functionality to the design and to make it race faster using gears.

Figure 12-1: The Formula EV3 Race Car

building the Formula EV3 Race Car

Build the race car by following the instructions on the subsequent pages. Before you start building, select the pieces you'll need to complete the robot, shown in Figure 12-2.

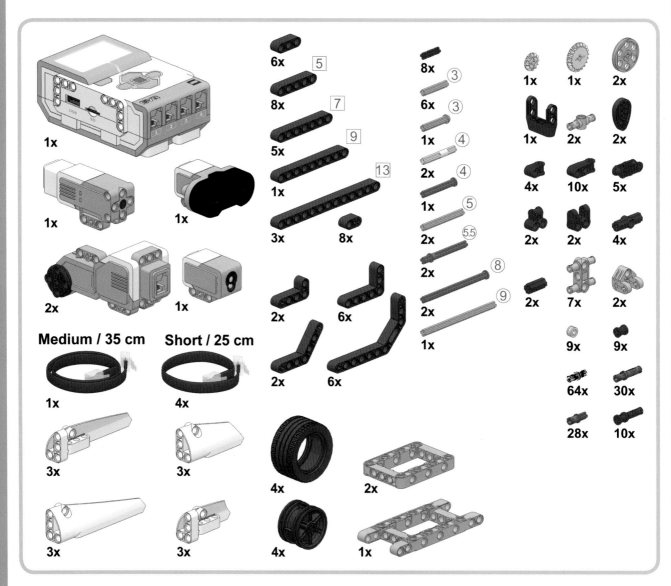

Figure 12-2: The pieces needed to build the Formula EV3 Race Car

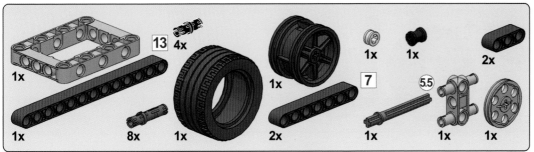

1

2

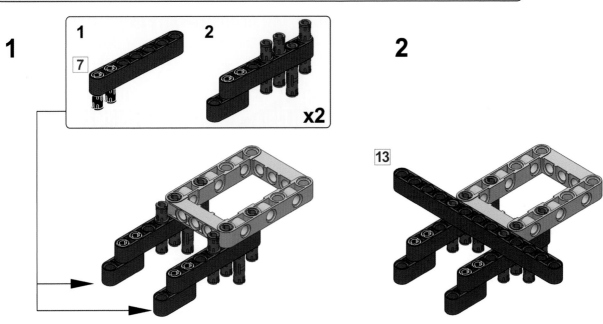

3

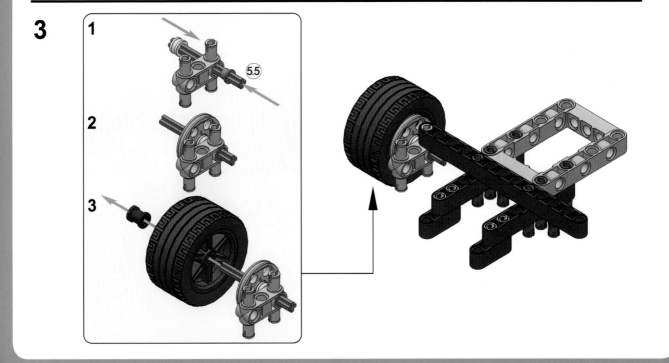

1x   13   1x   5.5

1x   1x   1x   1x   1x   2x

4

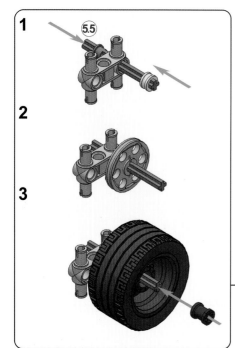

1

5.5

2

3

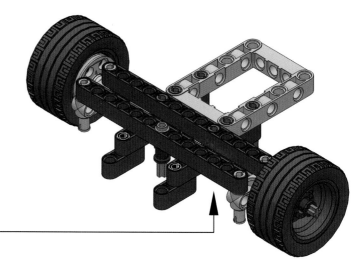

5

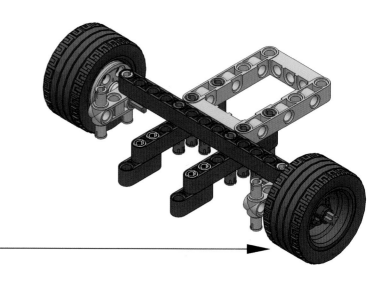

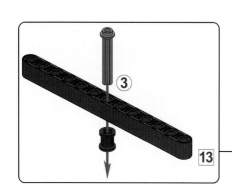

3

13

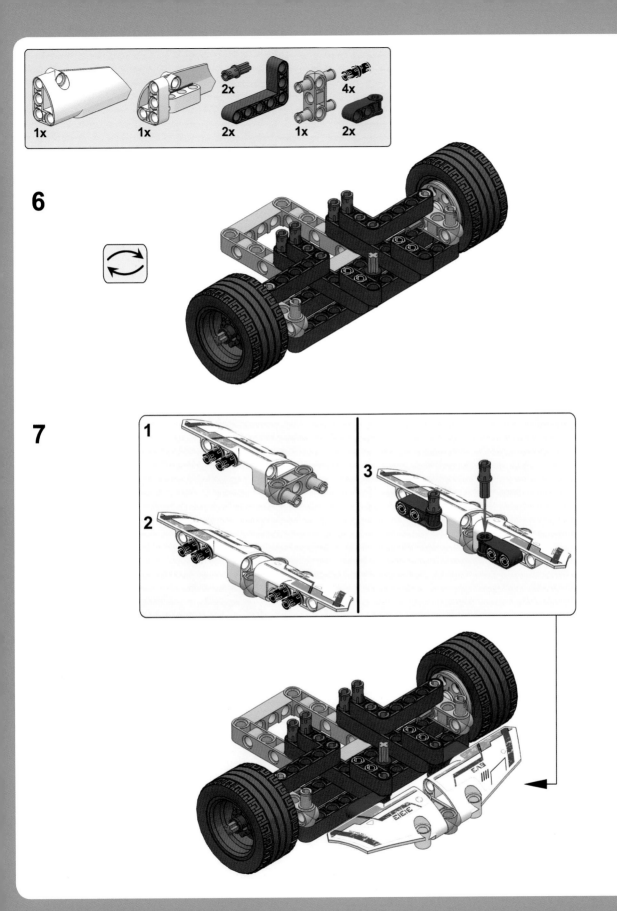

6

7

1

2

3

8

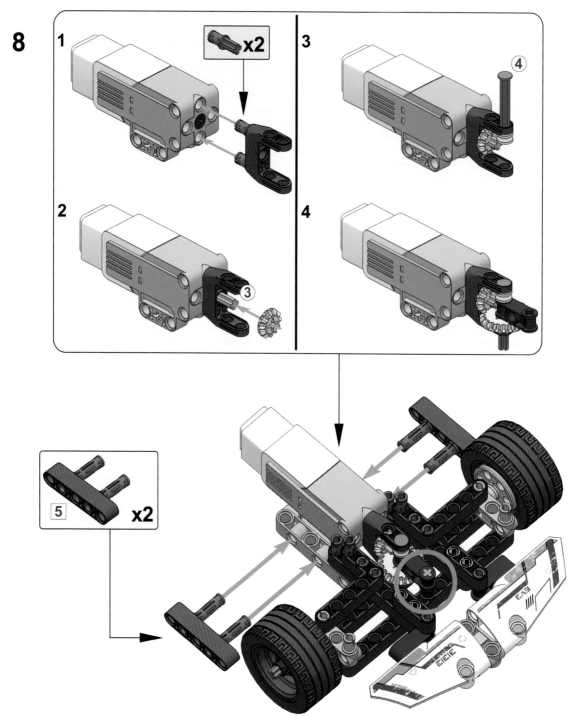

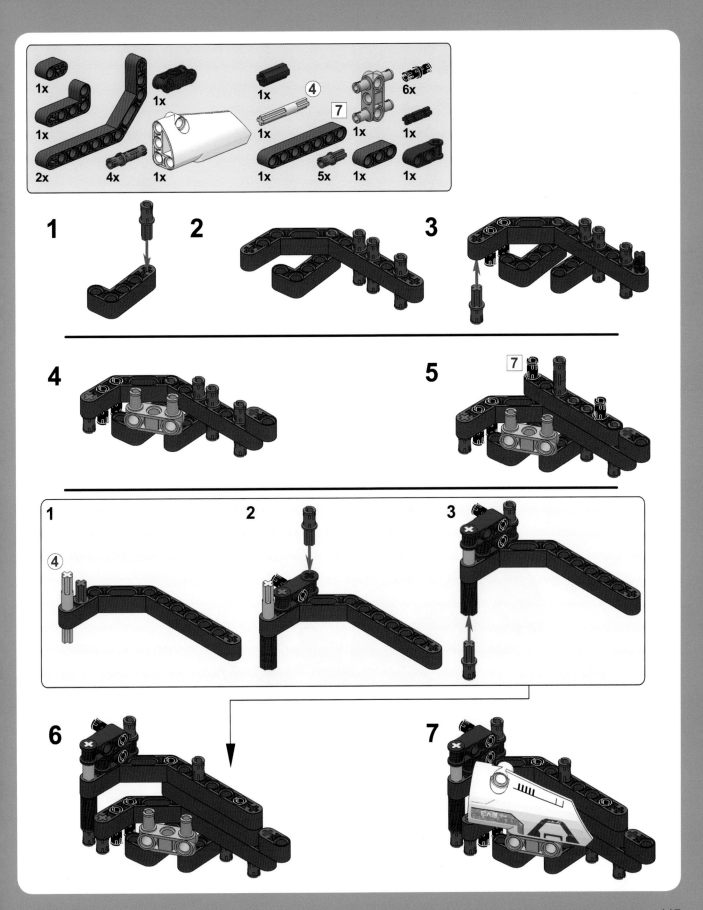

Short / 25 cm

1x 1x 1x 1x 1x

8

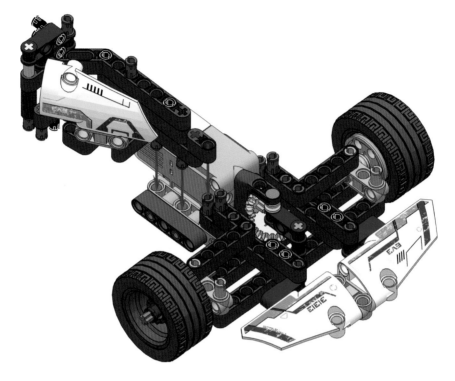

9

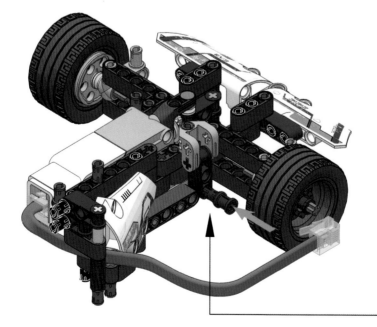

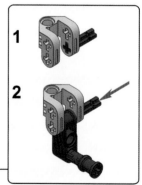

1

2

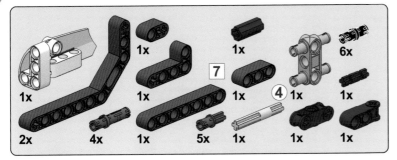

1

2

3

4

5

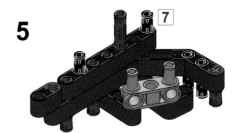

1   **2**   **3**

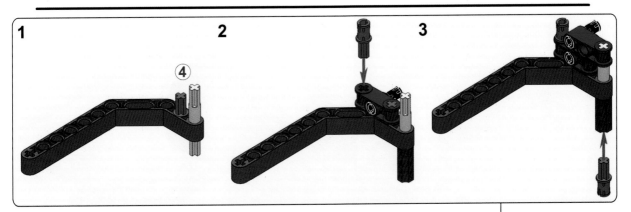

6

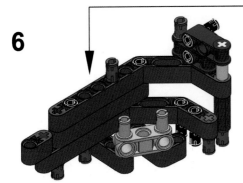

7

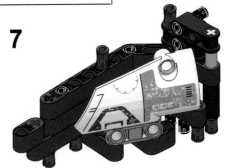

8

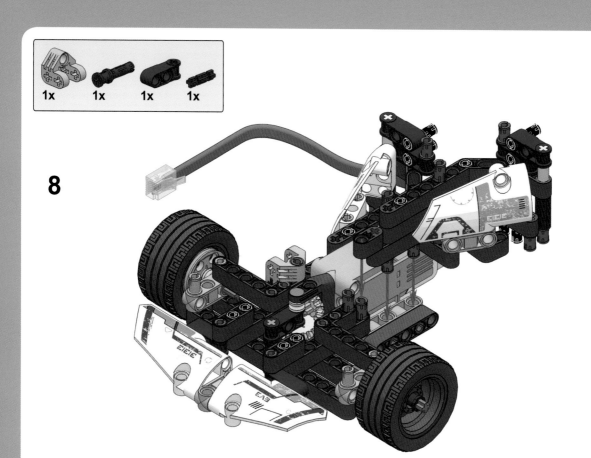

9

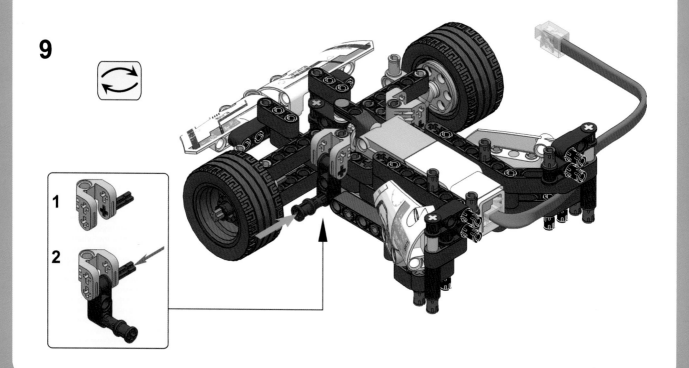

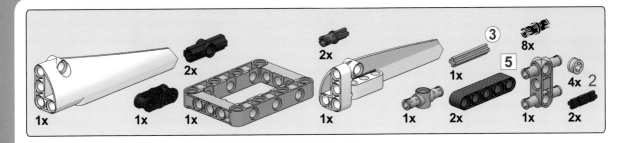

1

5

2

3

3

4

1

2

5

5

6

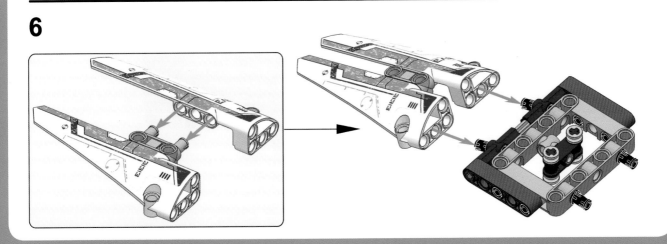

7

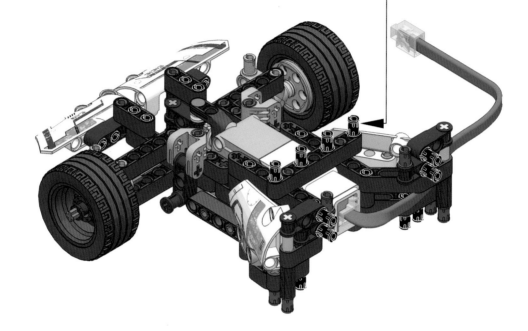

8

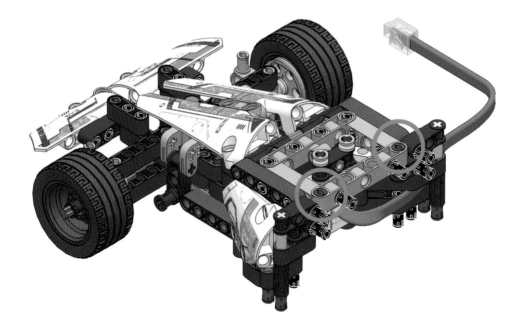

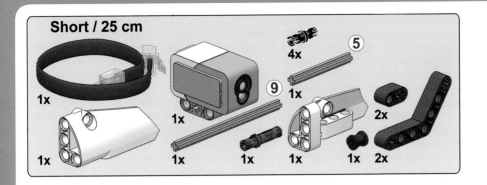

Short / 25 cm

1x · 1x · 1x · 1x · 4x ⑤ 1x ⑨ 1x · 1x · 1x · 1x · 2x · 2x

1

⑨

2

⑤

3

4

5

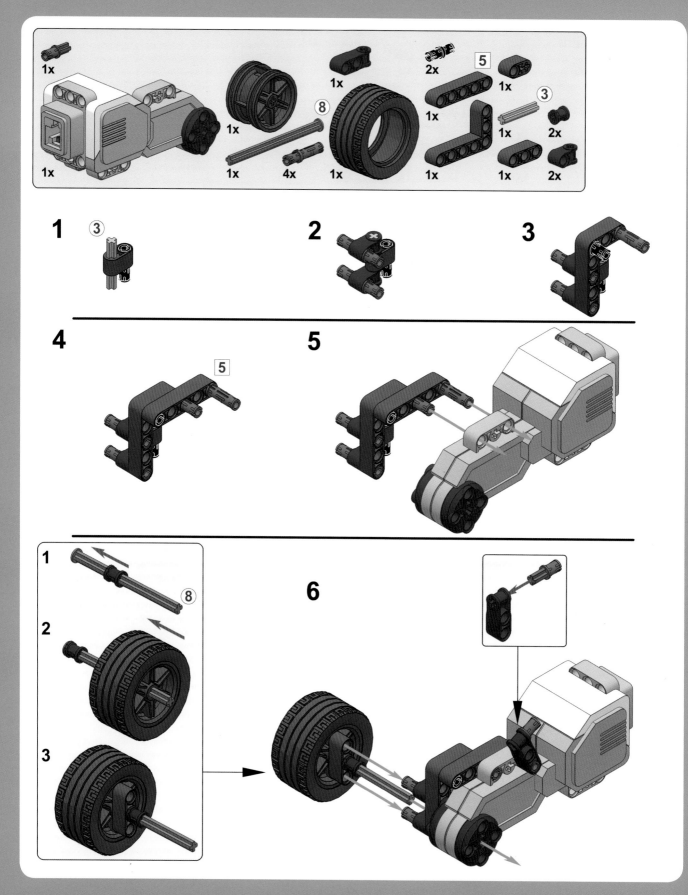

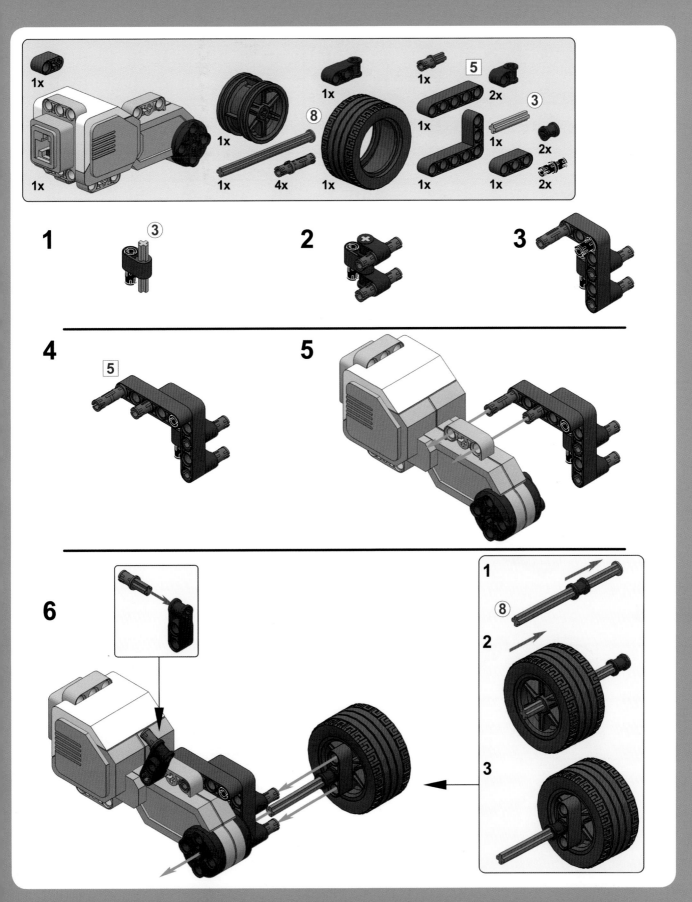

1

2

3

4

5

6

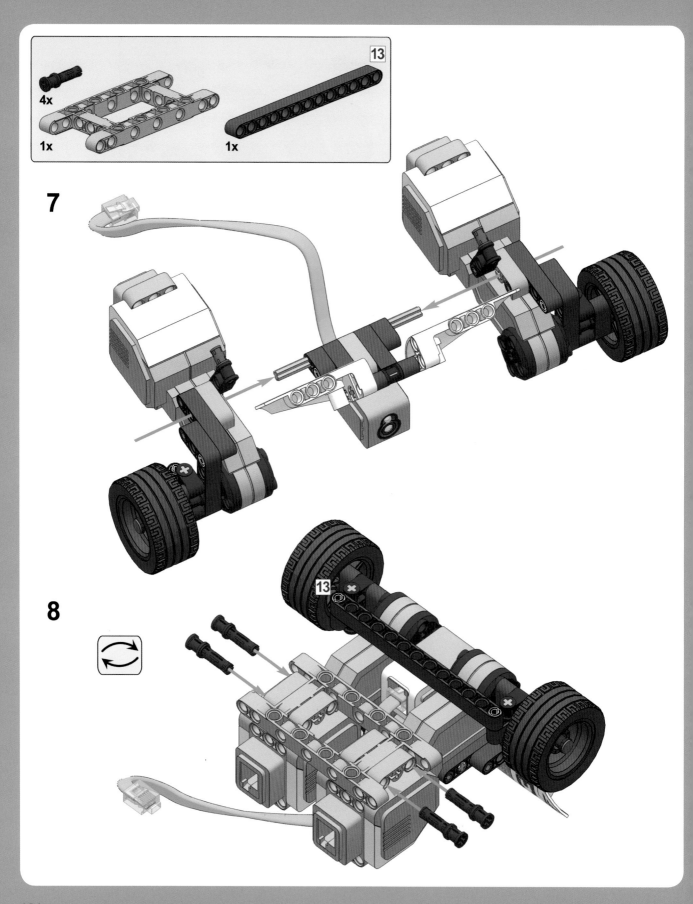

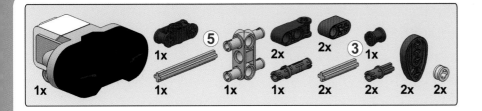

1

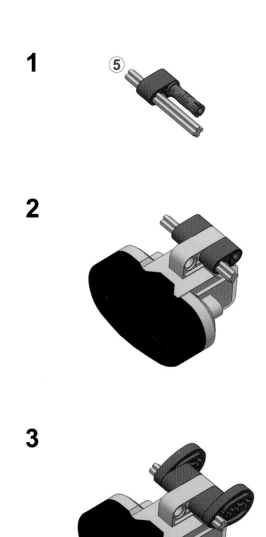

2

3

4

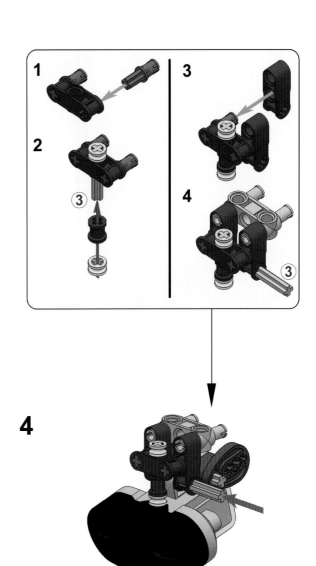

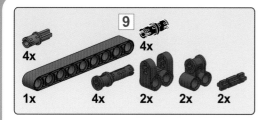

9
4x 1x 4x 2x 2x 2x
4x

1

2

1 2 3

x2

3

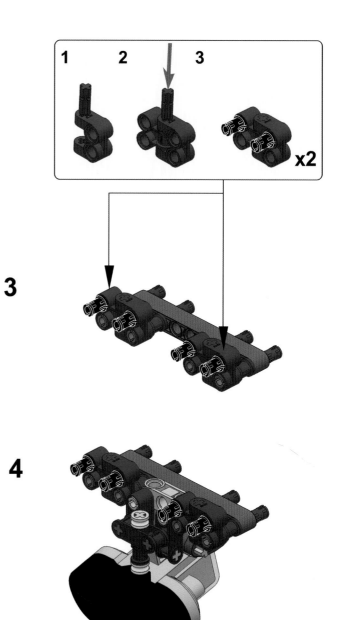

4

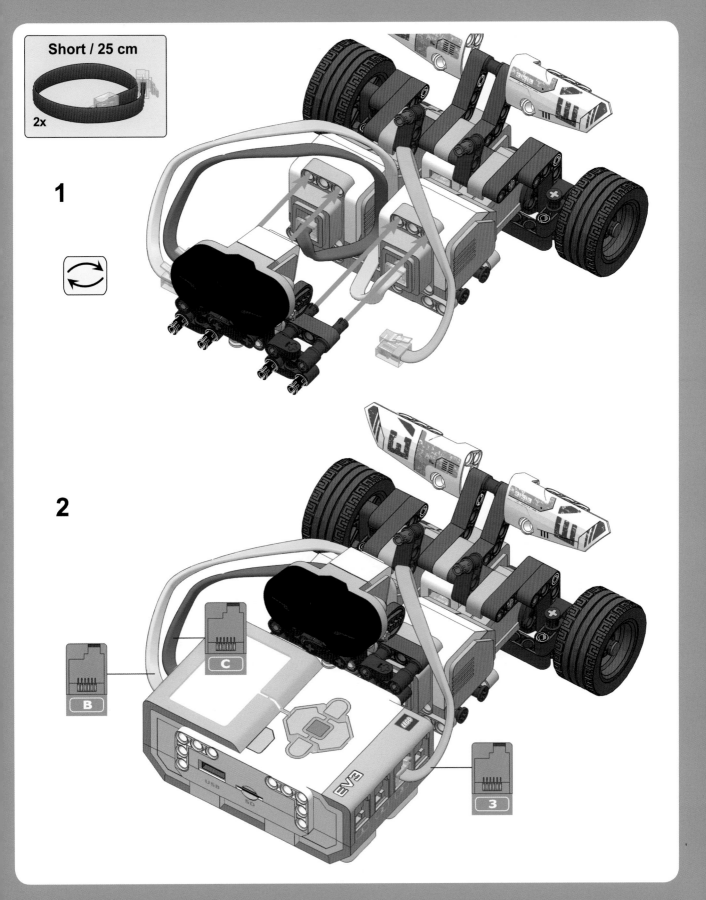

Short / 25 cm

2x

1

2

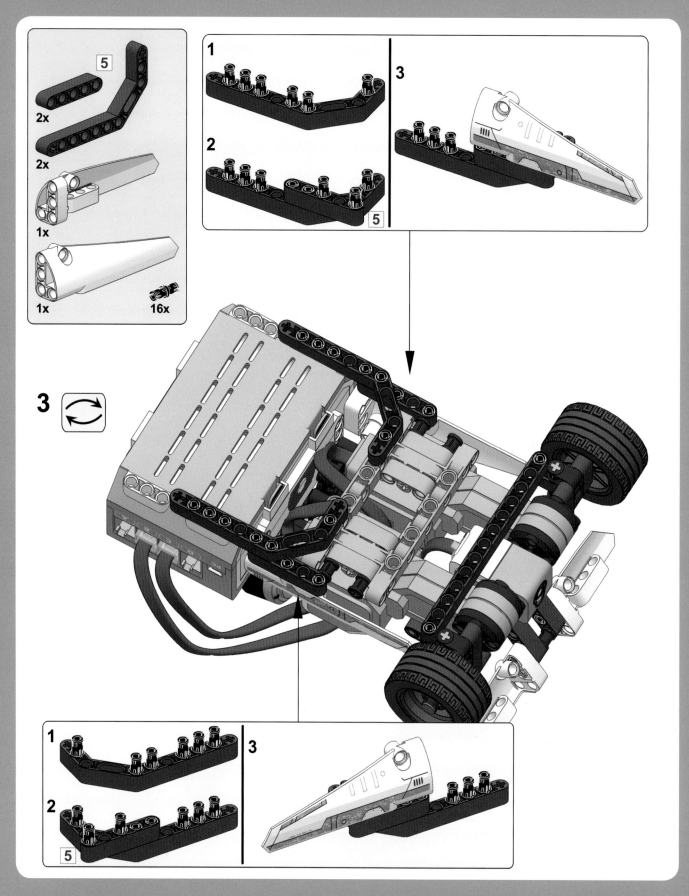

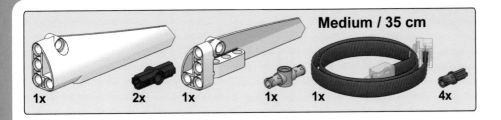

Medium / 35 cm

1x 2x 1x 1x 1x 4x

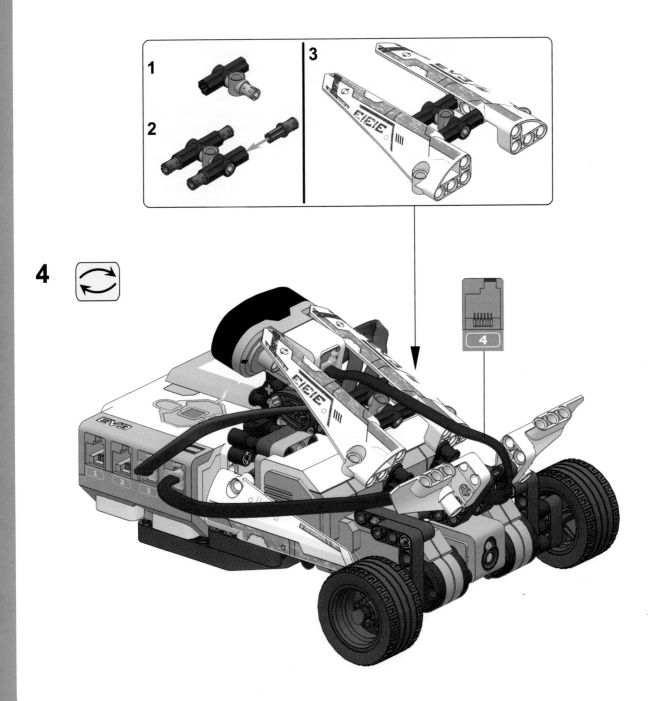

1

2

3

4

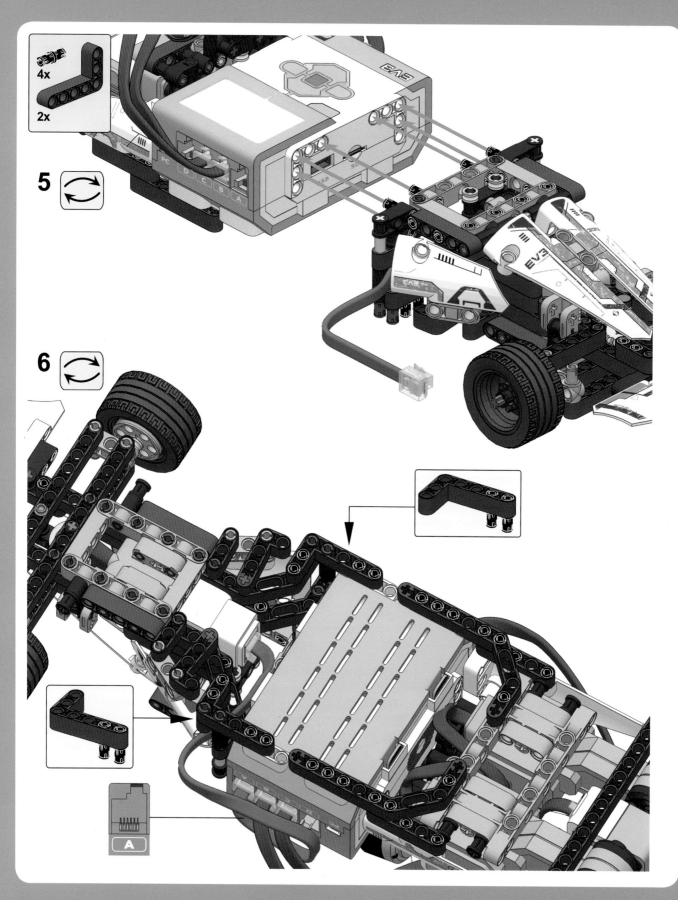

5

6

driving and steering

The Formula EV3 Race Car uses two separate mechanisms to drive and steer. To drive the car, you switch on the Large Motors in the rear. You can steer the car by turning the front wheels to the left or to the right with the Medium Motor in the front. When you combine driving and steering in one program, you can move the car in any direction.

creating my blocks for steering

In a moment, we'll create some custom My Blocks that will make it easy to program the Formula EV3 Race Car to steer in different directions. But first, let's have a look at how the steering mechanism works. Your program will use the Rotation Sensor inside the Medium Motor to accurately control the orientation of the front wheels, which determine the direction in which the robot will drive.

To see how this works, use your hands to set the front wheels in the centered position, as shown in Figure 12-3. Then, go to **Port View** on the EV3 brick, and observe the Rotation Sensor value (port A) as you manually steer the front wheels to the left and to the right. You'll find values around 60 degrees for the left, 0 degrees for the center, and –60 degrees for the right. Note that these angles indicate the position of the motor. The wheels turn left and right by a smaller amount because of the gears in the steering mechanism.

As shown in Figure 12-3, the Medium Motor should move toward the point where the sensor measures 60 degrees in order to make the robot steer left. To steer right, it should move toward the –60 degree measurement, and to drive straight ahead, it should move back to 0 degrees. You'll make three My Blocks, called *Left*, *Right*, and *Center*, to put the wheels in each of these positions.

These blocks will work only if the sensor value is 0 degrees when the wheels are centered. Because you don't want to center the front wheels manually each time you run the program, you'll make another My Block, called *Reset*, to do this for you. This My Block centers the wheels and sets the Rotation Sensor value to 0. You'll place it at the start of each program for this robot.

my block #1: reset

The front wheels can be in any position when the program starts, so you'll have to move them to a known orientation before you can center them. To accomplish this, you'll steer the wheels all the way to the left by having the Medium Motor rotate forward until it stalls.

The stalled motor position should be 78 degrees past the center, which means you can move the wheels all the way to the left, then use a Medium Motor block to turn 78 degrees backward to reach the center position. Once the front wheels are aligned properly, you set the Rotation Sensor value to 0. From that point on, Rotation Sensor values near 0 indicate that the front wheels are aligned in the center.

Begin by creating a new project called *FormulaEV3* for all of the programs for your race car. Then create the Reset My Block, as shown in Figure 12-4.

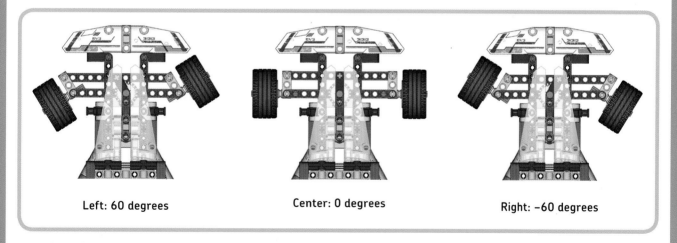

Left: 60 degrees   Center: 0 degrees   Right: –60 degrees

Figure 12-3: Rotation Sensor values for various positions of the front wheels. Make sure the wheels are in the center position before you start Port View in order to see these measurements.

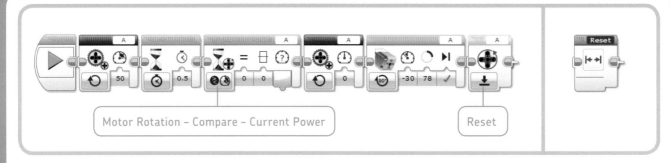

Figure 12-4: The Reset My Block moves the steering wheel to the center and resets the Rotation Sensor to 0. The completed My Block is shown on the right.

my block #2: left

The next My Block makes the front wheels steer to the left by turning the Medium Motor forward until the Rotation Sensor measures 60 degrees, but only if the wheels aren't steered left already.

You'll use a Switch block to determine whether the wheels are already in the left orientation. If they are, the motor is turned off; if not, the motor turns forward until it reaches 60 degrees. This works regardless of whether the wheels are currently centered or steered right because the motor turns until it is in the correct position, rather than turning a fixed number of degrees.

The Brake at End setting is set to True, locking the motor in this position until you have it make another movement. Place and configure the blocks shown in Figure 12-5, and turn them into a My Block called *Left*.

NOTE The Switch and Wait blocks in the Left, Right, and Center My Blocks are all configured in Motor Rotation – Compare – Degrees mode.

my block #3: right

The *Right* My Block does just the opposite: First, it determines whether the wheels are already in the right orientation. If so, the motor is switched off; if not, the motor turns backward until it reaches the −60 degree position. To accomplish this, the Medium Motor is switched on at negative speed (−30%), a Wait block pauses the program until the Rotation Sensor goes below −60 degrees, and then the motor is turned off.

Create the Right My Block, as shown in Figure 12-6.

my block #4: center

The front wheels are perfectly centered if the sensor value is exactly 0, but in this My Block, you'll consider any position between −5 and 5 degrees to be close enough to the center. (If you make this range any smaller, the motor tends to move past the center position each time it tries to reach it.)

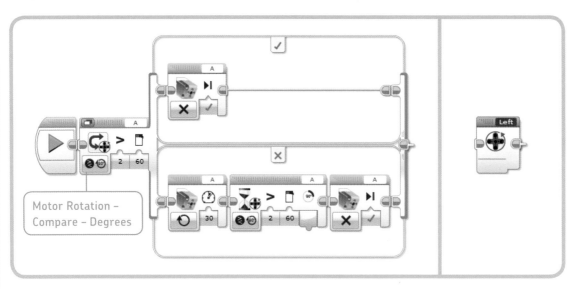

Figure 12-5: The Left My Block makes the front wheels steer to the left. The completed My Block is shown on the right.

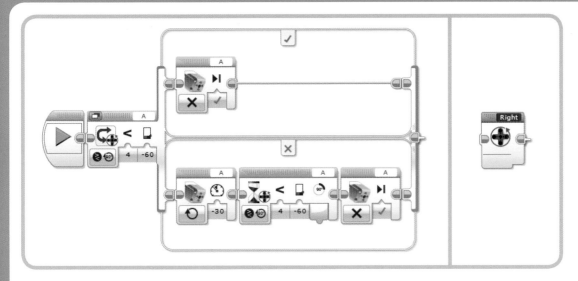

Figure 12-6: The Right My Block makes the front wheels steer to the right. The completed My Block is shown on the right.

If the motor is already in the center region when this My Block runs, the motor is switched off. If the wheels are to the left of the center (the sensor value is greater than 5 degrees), they turn right until they reach the center (less than 5 degrees). If the wheels are to the right (less than –5 degrees), they turn left until they reach the center (greater than –5 degrees). You use two Switch blocks to determine which of these positions the motor is in.

Create the *Center* My Block using Figure 12-7.

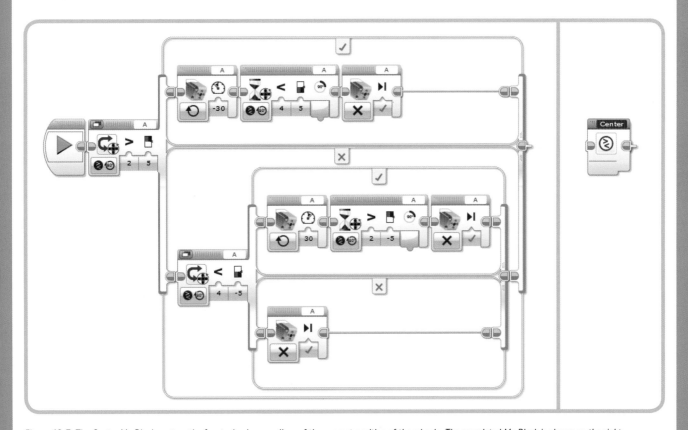

Figure 12-7: The Center My Block centers the front wheels regardless of the current position of the wheels. The completed My Block is shown on the right.

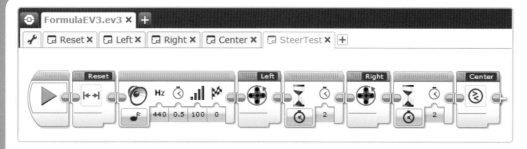

Figure 12-8: Use the SteerTest program to test your My Blocks. Run the program a few times, trying a different starting position for the front wheels each time.

testing the my blocks

Before you create a complete program, you should test the My Blocks to see whether they work as expected. Create the *SteerTest* program shown in Figure 12-8 and run it. The front wheels should automatically line up with the rear wheels when the Reset My Block runs. Next, they should turn to the left, to the right, and then back to the center.

creating a remote control program

Now that you've created My Blocks for the steering mechanism, it's easy to create a remote control program using the techniques you learned in Chapter 8. Your next program makes the car drive in all directions as you press the buttons on the infrared remote, as shown in Figure 12-9. For each of the button combinations, the robot runs one of the My Blocks to steer the front wheels and a Move Tank block to power the rear wheels.

You use Move Tank blocks so you can choose the speed of both motors separately. If the robot goes straight forward or backward, both wheels turn at 75% speed. When turning, the outermost wheel should go slightly faster than the inner wheel, so you'll make the faster wheel drive at 80% speed and the slower wheel at 70%.

A *negative* speed value, such as –75, makes the robot go *forward* due to the orientation of the motors in this robot. Positive values, such as 75, make the robot go backward. Now create the *RemoteControl* program using Figure 12-10.

NOTE If the robot does not drive in a straight line when driving forward, adjust the degrees value in the Medium Motor block in the Reset My Block shown in Figure 12-4. Try a value slightly larger than 78 degrees if you see a deviation to the left; try a smaller value if the robot deviates to the right.

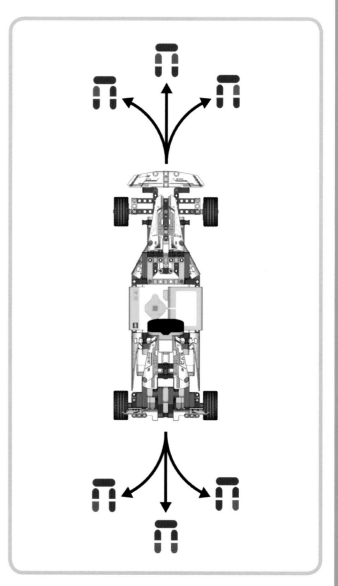

Figure 12-9: The button combinations to make the Formula EV3 Race Car drive in any direction.

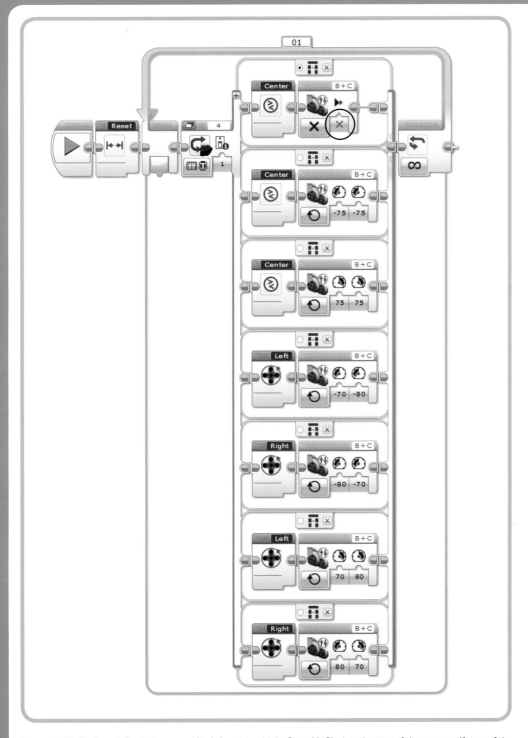

Figure 12-10: The RemoteControl *program. Don't forget to add the Reset My Block at the start of the program. If none of the buttons on the remote is pressed (default case), the robot centers the front wheels and then switches off the rear wheels.*

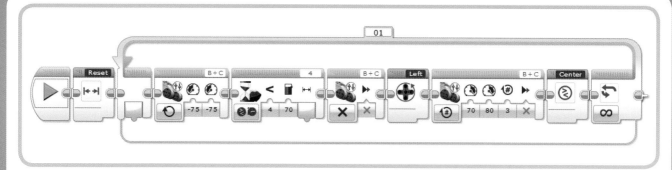

Figure 12-11: The AutonomousDrive *program*

racing autonomously

Now you'll create a program to make the race car drive around your room autonomously, avoiding obstacles with the Infrared Sensor. The robot begins by driving straight forward until the sensor detects a proximity value below 70%. Then, the robot reverses and steers left to move away from the obstacle. Finally, the robot centers its wheels so that it can drive straight forward in a new direction. Create the *AutonomousDrive* program, as shown in Figure 12-11.

further exploration

In this chapter, you had a chance to build and program a robot from instructions. This is, of course, a lot of fun, but it's even more fun to create your own robot designs. For example, you could expand the design with gears or larger wheels to make the car faster, or you could turn the robot into a completely different vehicle, such as a passenger car or an off-road vehicle.

Don't worry if you don't succeed on the first try; you'll gain more and more building experience as you continue to try new designs. Try out the Design Discoveries at the end of this chapter to get started, and be sure to use the techniques you learned in Chapters 10 and 11.

DISCOVERY #65: OVERSHOOT EXPERIMENT!

Difficulty: ▱   **Time:** ◷
In the Center My Block, you consider any position between −5 and 5 to be close enough to the center. To see why the robot should use such a large range of values, try changing the settings of the Wait and Switch blocks to reduce the center range boundaries to −1 and 1. What happens when you run the *RemoteControl* program?

DISCOVERY #66: NIGHT RACING!

Difficulty: ▱   **Time:** ◷
So far you haven't used the Color Sensor that's built into the rear of the robot. Can you use this sensor to have the robot drive only when the lights in the room are off? Does the robot still detect obstacles when it's dark?

HINT Begin by modifying the *AutonomousDrive* program to measure the ambient light intensity. You'll need a Switch block to determine whether the lights are on or off. What is the threshold value, and how do you place the other blocks in the switch?

DISCOVERY #67:
WIRED GAS PEDAL!

Difficulty: ▭▭ **Time:** ◷◷

Can you create a program to control the race car's speed with the Touch Sensor and to control the steering with the remote? Create two parallel sequences in your program: One sequence should control the orientation of the front wheels using the remote control; the other sequence should control the speed of the rear wheels with the Touch Sensor.

When you're ready, add engine sounds to your program using Sound blocks configured to play the *Speeding*, *Speed idle*, and *Speed down* sounds.

HINT Connect the Touch Sensor to input port 1 on the EV3 brick using a long cable. Remember that the Touch Sensor can detect only whether the red button is pressed or not; it can't detect any intermediate position.

DISCOVERY #68:
BLINKING REAR LIGHT!

Difficulty: ▭▭ **Time:** ◷◷

Formula 1 race cars have a bright red light in the rear that blinks in poor weather conditions to make them visible to other drivers. Can you make the light on the Color Sensor switch between blue and red every second to mimic the blinking pattern? There's no standard block that changes the light color, so you'll have to make one My Block to switch the light to blue and another to switch the light to red.

HINT Use a Switch block to measure ambient light intensity, but don't place any blocks in the switch. The robot won't do anything with the measurement, but it will change the light to blue when doing an ambient light measurement.

DISCOVERY #69:
CRASH DETECTION!

Difficulty: ▭▭▭ **Time:** ◷◷◷

When you ran the *AutonomousDrive* program, you probably noticed that the Infrared Sensor is good at detecting the presence of walls and other large objects, but it doesn't always detect small objects, like a chair leg. Can you make the robot avoid obstacles in other ways?

Use Unregulated Motor blocks to drive, and use the Rotation Sensors in the rear motors to detect a sudden drop in rotational speed. Make the robot reverse and drive away when it either sees an object with the Infrared Sensor or detects one by running into it.

HINT Use the techniques you learned in Discovery #53 on page 102.

DESIGN DISCOVERY #20: RACING FASTER!

Building: ✹✹ **Programming:** 🔲

Can you improve the Formula EV3 Race Car design to make the car go faster? Use 36T and 12T gears to speed up the rear wheels by a factor of 3, as shown in Figure 12-12. You can also make the car go faster by using bigger wheels from other LEGO Technic sets, but you may need to adjust the front and rear wings to make space.

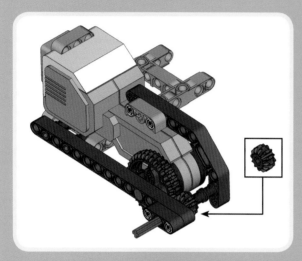

Figure 12-12: You can make the race car go faster using gears found in your EV3 set. (The wheel and the rest of the vehicle are not shown for better visibility.)

DESIGN DISCOVERY #21: CAR UPGRADE!

Building: ✹✹✹ **Programming:** 🔲

Can you create your own vehicle based on the robot you built in this chapter? Take the Formula EV3 Race Car apart, except for the front-wheel steering mechanism (page 146). By choosing a new placement for the EV3 brick and the rear motors, you can create a completely different vehicle.

To create a passenger car, for example, place the rear wheels closer to the front wheels and place the EV3 brick on top of the motors. You can also build an off-road vehicle by positioning the motors at an angle to create more ground clearance. Test your creation using the *RemoteControl* program you made in this chapter.

13

ANTY: the robotic ant

By now, you've gotten the hang of building models that move on wheels. Another fun, but slightly more challenging, kind of model you can build is an animal robot that uses legs to move instead of wheels. In this chapter, you'll build and program ANTY (see Figure 13-1). ANTY is a six-legged, insect-like creature that walks around and responds to its environment by expressing different types of behavior based on what its sensors "see."

The Infrared Sensor serves as ANTY's eyes, allowing the robot to detect objects in its surroundings and to find food. The

Color Sensor in the robot's tail enables ANTY to detect changes in its environment. Different surroundings—that is, different colored objects—make the robot behave differently. Green objects give the robot a sense of safety, meaning it can take a nap. Red objects indicate danger, making the robot shake aggressively to scare off enemies. Blue objects scare ANTY, prompting it to run away. Finally, yellow objects make the robot hungry, causing it to start looking for food.

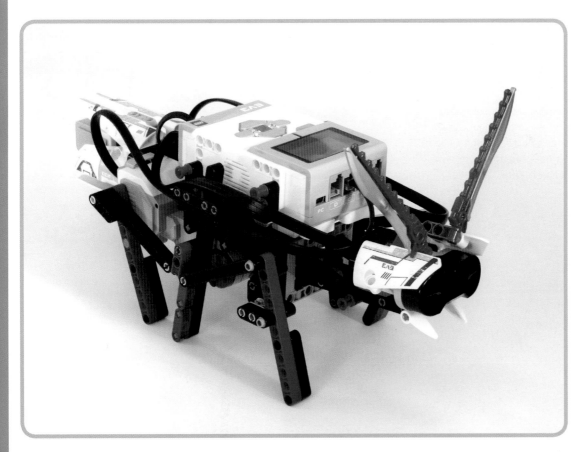

Figure 13-1: ANTY

understanding the walking mechanism

ANTY uses two motor assemblies to walk, each of which controls a set of three legs. Turning one motor forward makes the three legs connected to it step forward, as shown in Figure 13-2. Turning both motors forward simultaneously makes the whole robot walk forward.

But this works only if the legs on the left side are in the *opposite* position of the legs on the right side. For example, the left legs should be in position 2 when the right legs are in position 4 to ensure that there are always at least three legs in contact with the ground. This will happen when one motor is exactly 180 degrees ahead of the other. (The green dot in position 4 is 180 degrees ahead of the green dot in position 2.)

The robot uses the Touch Sensor to determine the *absolute position* of each motor so that the motors can be placed in opposite positions when the program begins. The robot knows that a set of legs is in position 1 when the Touch Sensor is pressed by a cam element, as shown in Figure 13-2.

After you've built the robot, you'll make a My Block that places the legs on either side in the required position, after which the robot can start walking using Move Tank blocks. As long as both motors turn at the same speed, the motors will remain 180 degrees apart.

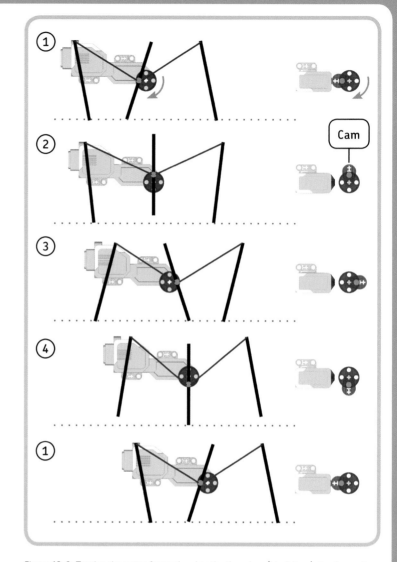

Figure 13-2: Turning the motor forward makes the three legs (black lines) step forward. The motor moves the middle leg by rotating the pivot (green dot) in a circle. In turn, the middle leg makes the outer legs move using a set of beams (blue lines). After one full rotation, the legs are back in position 1, and the robot has moved forward.

building ANTY

Now that you've learned a bit about the robot's functionality, you're ready to build it. Follow the directions on the next pages, but first select the pieces you need (see Figure 13-3).

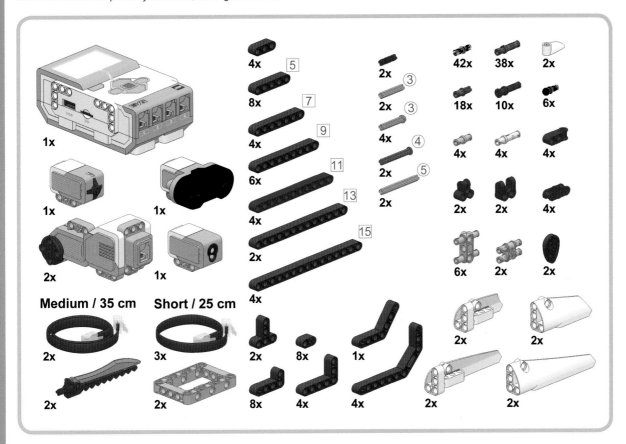

Figure 13-3: The pieces needed to build ANTY

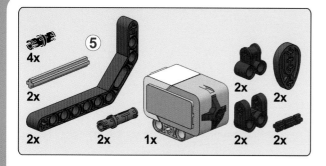

1

2

⑤

3

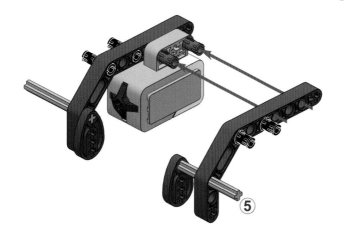

⑤

4

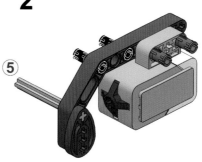

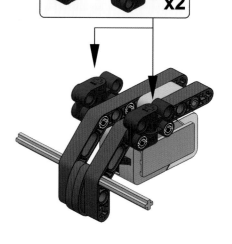

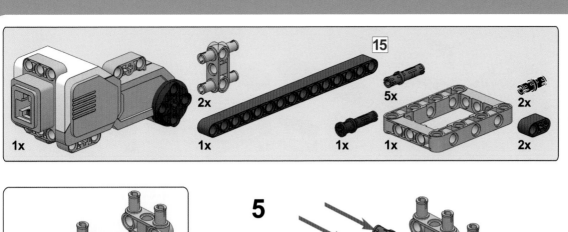

15

2x
1x
1x
5x
1x
1x
2x
2x

5

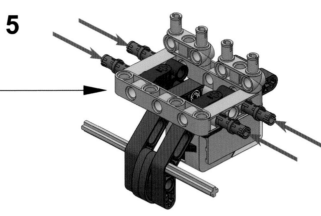

1

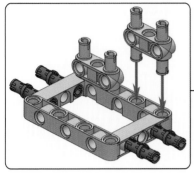

6

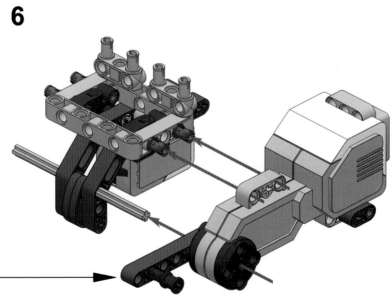

2

15

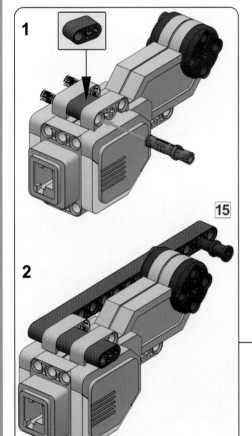

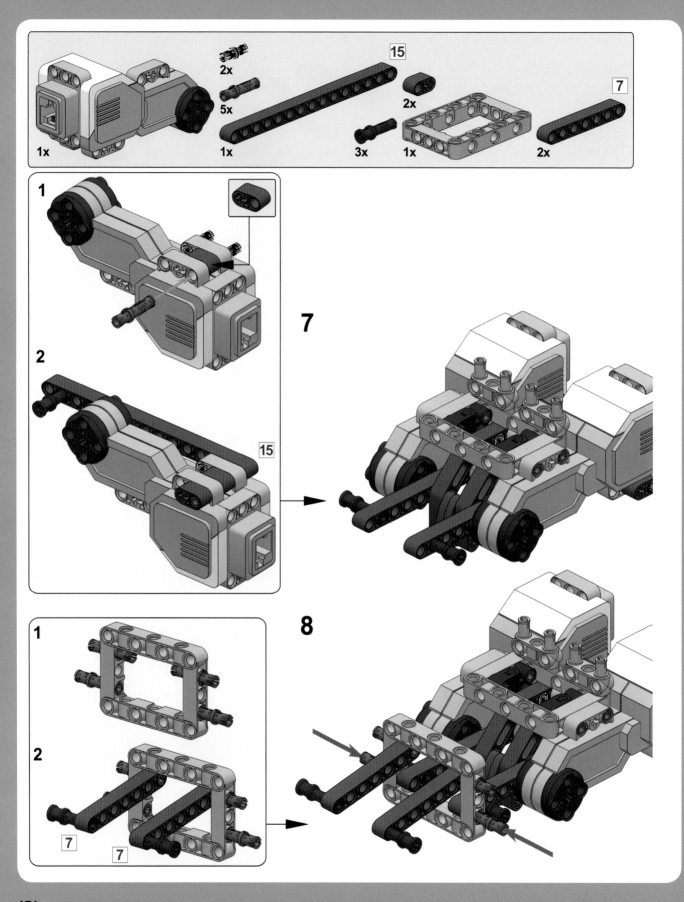

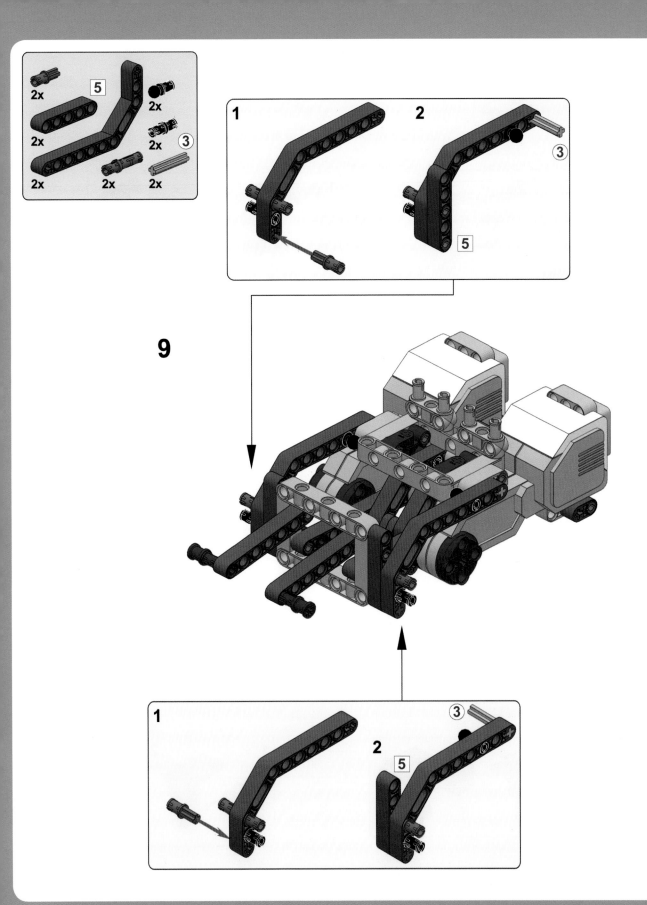

9

ANTY: THE ROBOTIC ANT 177

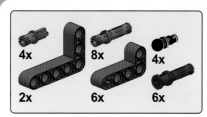

10

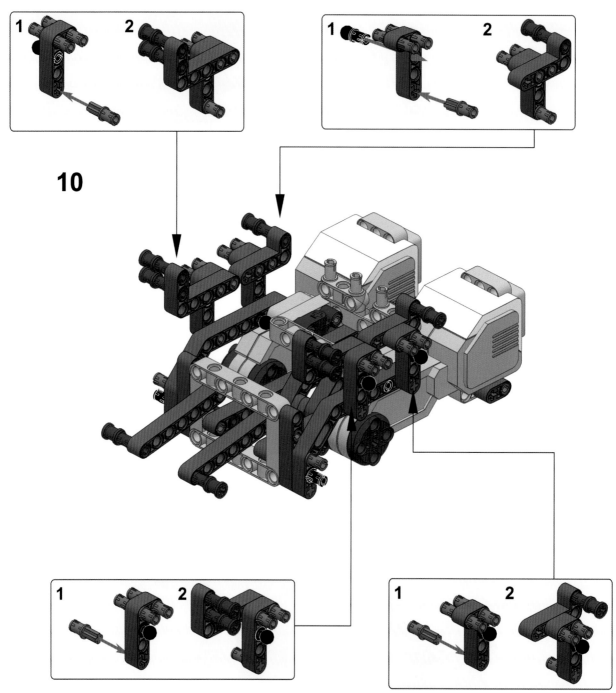

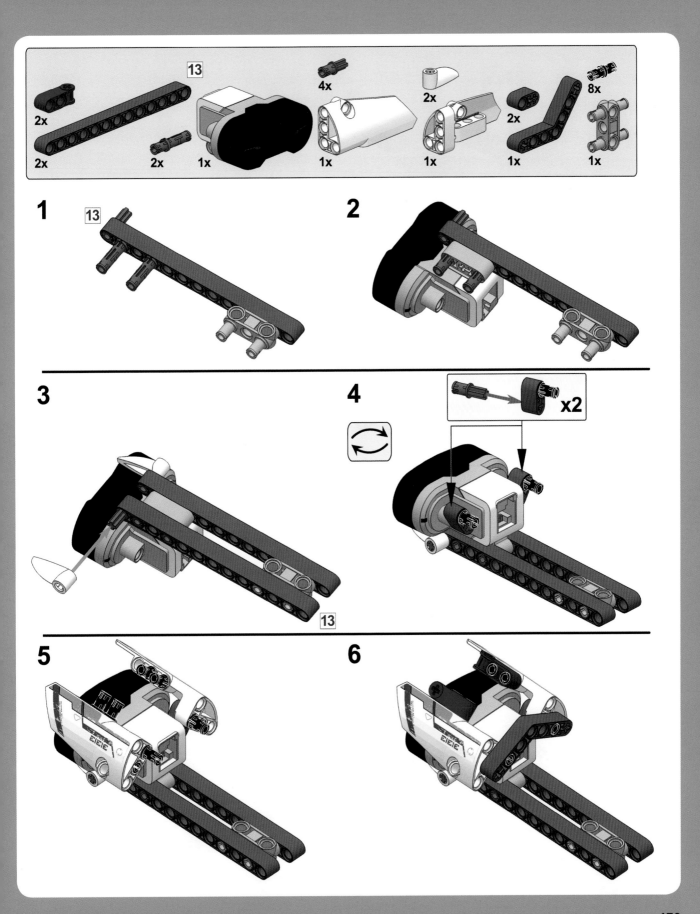

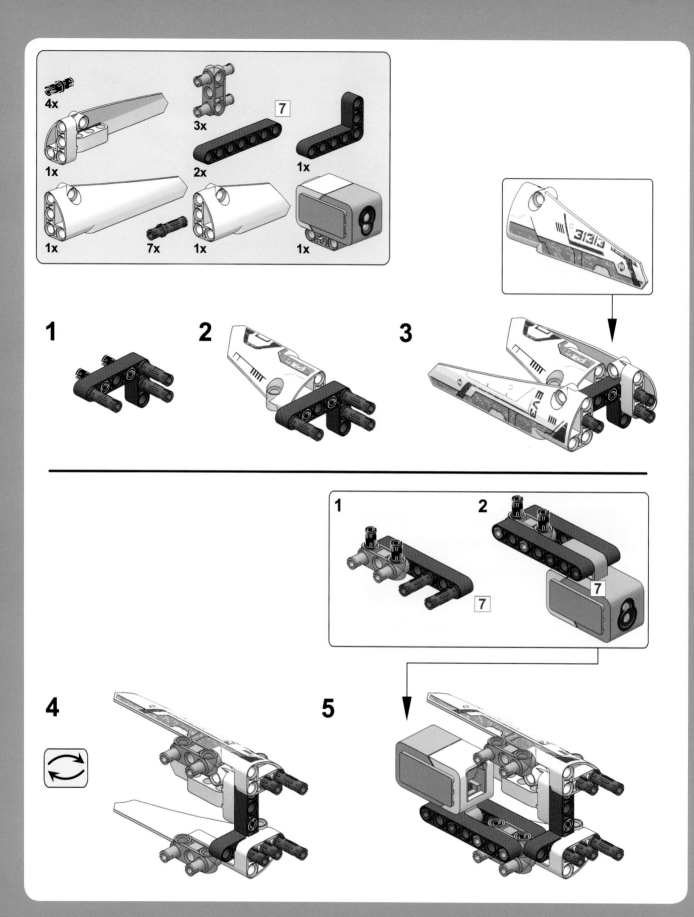

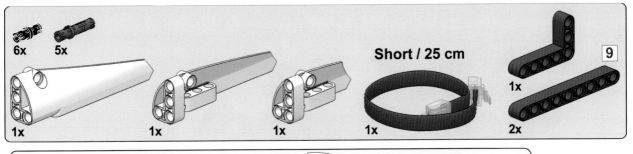

Short / 25 cm

9

6x 5x

1x 1x 1x 1x 1x 2x

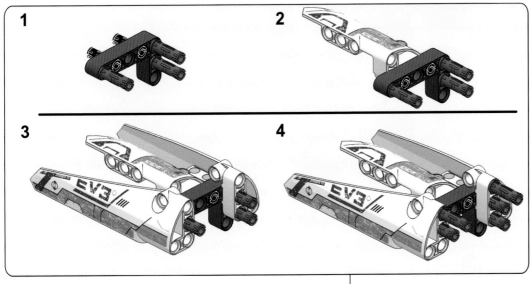

1

2

3

4

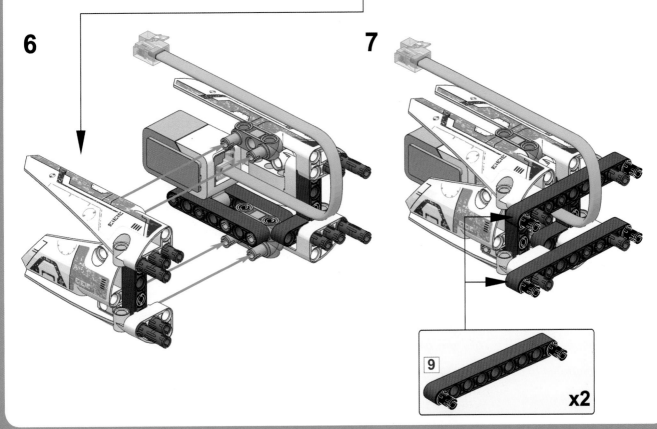

6

7

9

x2

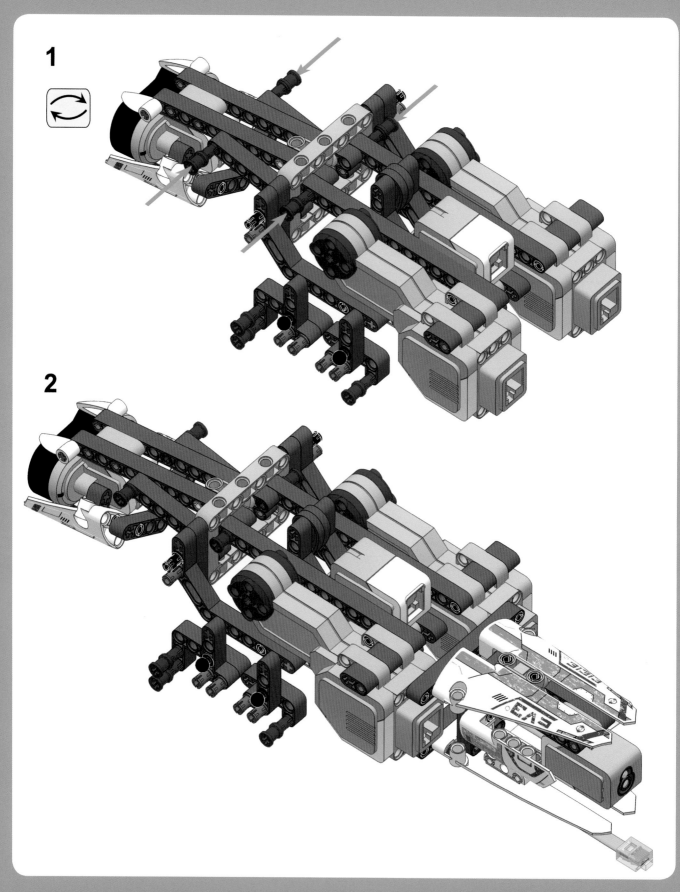

1

2

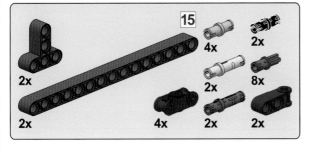

3

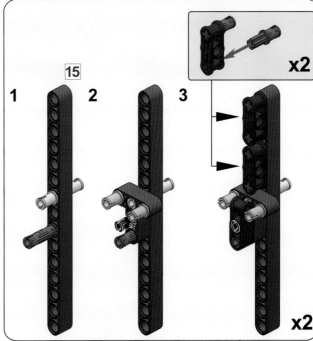

Make sure that the cam elements (circled in green) are aligned as shown *before* you attach the legs. If necessary, manually rotate the motors until they are in the correct position.

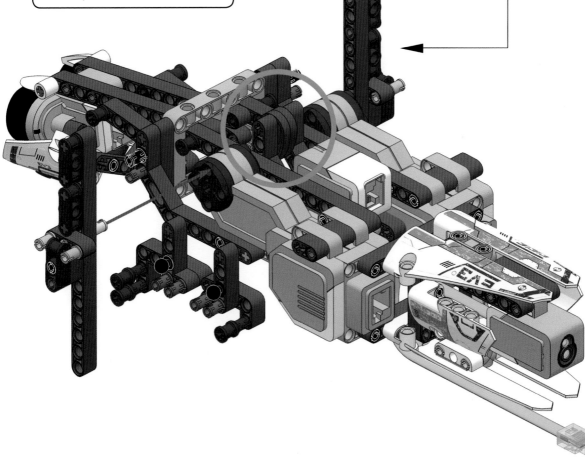

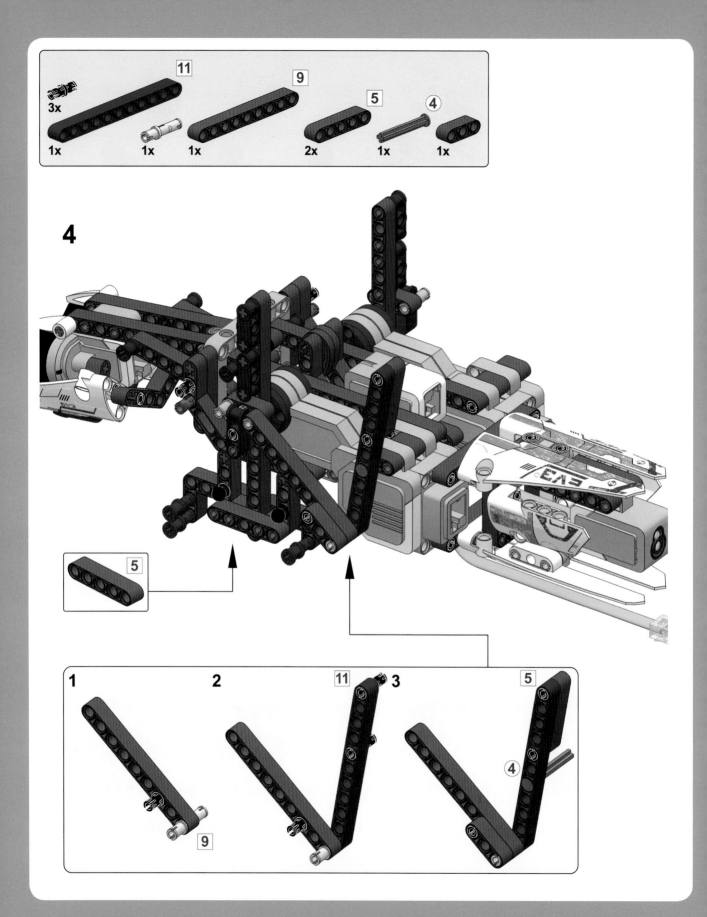

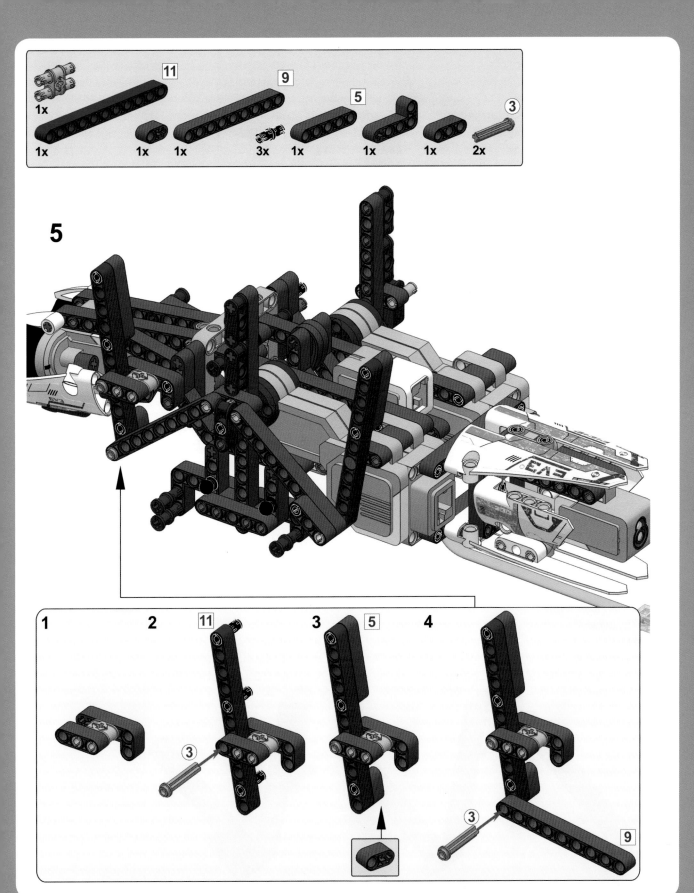

5

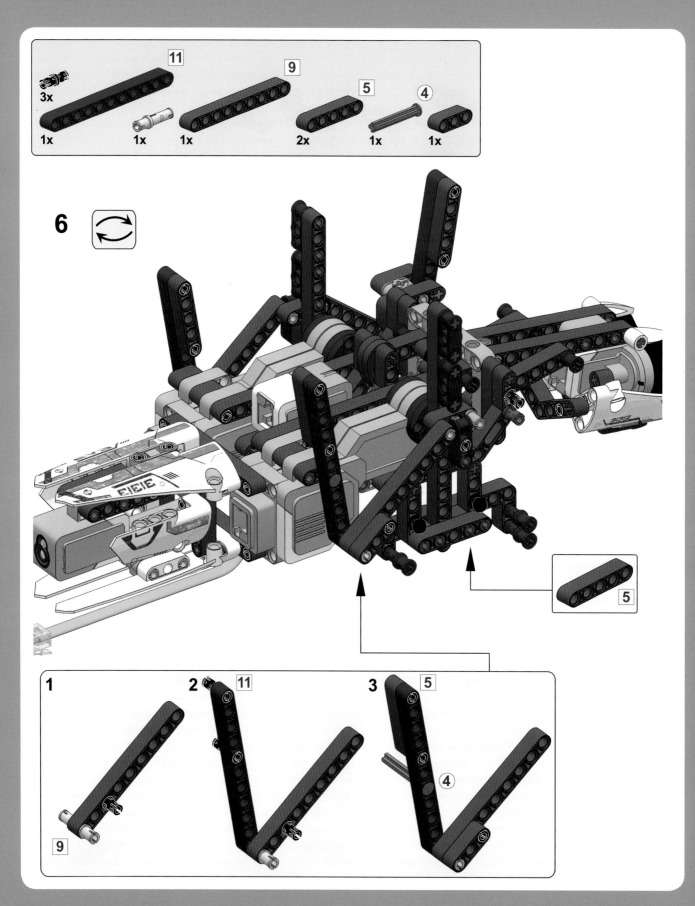

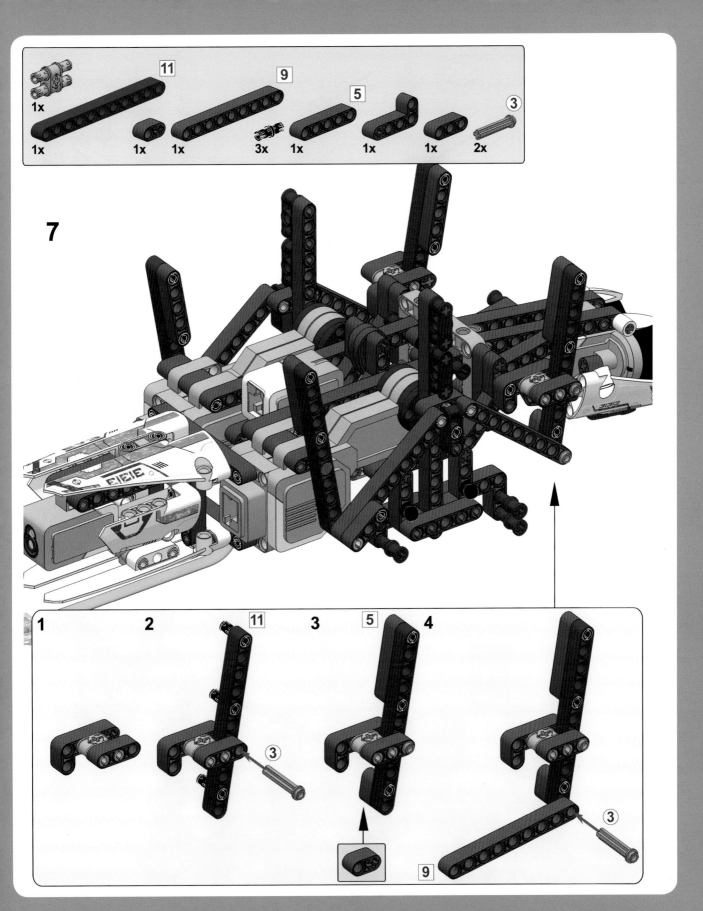

7

1x 1x 1x 3x 1x 1x 1x 2x

11 9 5 3

1 2 11 3 5 4

3 3

9

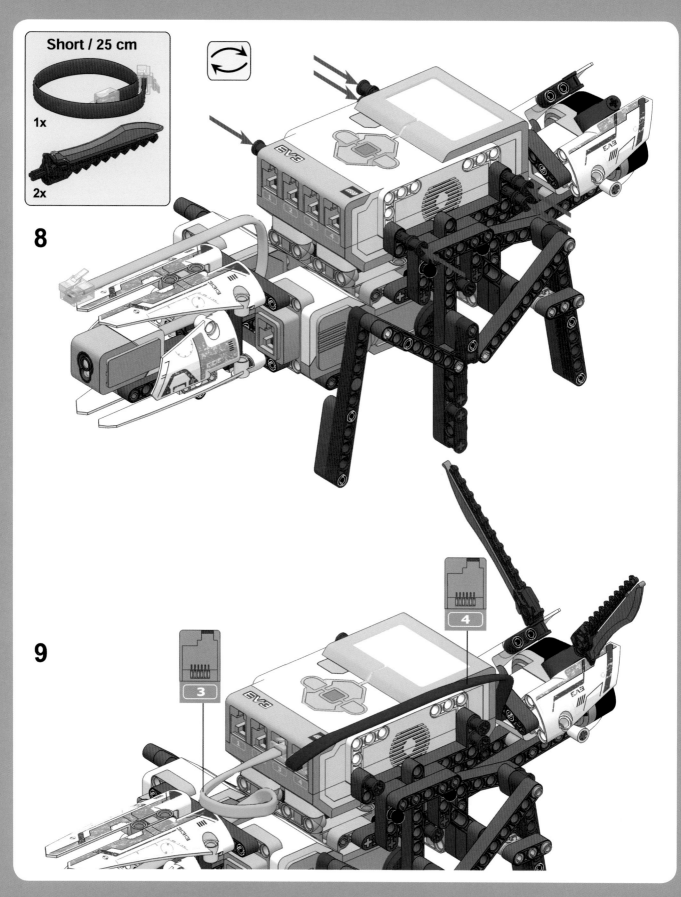

Short / 25 cm

1x

2x

8

9

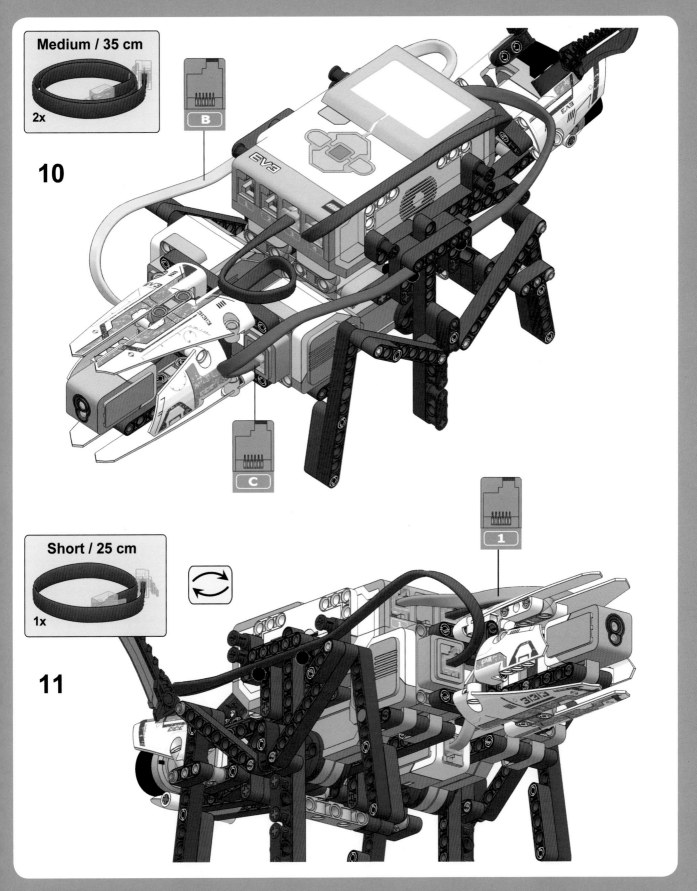

Medium / 35 cm
2x

10

B

C

Short / 25 cm
1x

11

1

making ANTY walk

You learned earlier that you make ANTY walk forward by having both motors rotate forward simultaneously, but only after the legs on either side are placed in opposite positions (180 degrees apart). You'll create a My Block to place the right legs in position 1 and the left legs in position 3 (see Figure 13-2), using the fact that a motor presses the Touch Sensor with a cam element when it's in position 1. You'll use this block at the start of each program for ANTY.

Once the block completes running, Move Tank blocks will make the robot walk by turning both motors *at the same speed* to keep them in opposite positions.

creating the opposite my block

Because both motors use the same Touch Sensor to detect their absolute position, the robot doesn't know whether the sensor is being pressed by the left motor (B), the right motor (C), or both. However, you can still determine the position of each motor by moving only one motor at a time, following these steps in order:

1. Turn both motors simultaneously until the Touch Sensor is released. (The robot can do this by repeatedly moving both motors forward by a small amount until the sensor is released.)

2. Rotate the left motor forward until the Touch Sensor is bumped. The left legs are now just beyond position 1.

3. Rotate the left motor 180 degrees forward. The left legs are now just beyond position 3.

4. Rotate the right motor forward until the Touch Sensor is bumped. The right legs are now just beyond position 1, exactly opposite the left legs.

Create a new project called *ANTY*. Place and configure the blocks that implement these steps, as shown in Figure 13-4. Then, turn them into a My Block called *Opposite* and run the block to test it.

If the My Block does not place the legs in opposite positions, even if your program looks just like the one given, you may have incorrectly attached the cam elements. Review page 183 and fix your robot if necessary. (You may not need to take the whole robot apart: It's easier to leave the cam elements in place. Just disconnect the middle legs from the motors and connect them to match the instructions.)

avoiding obstacles

Once the legs are in place, you make the robot walk forward using Move Tank blocks configured to rotate both motors forward at the same speed. The robot can turn left or right by rotating one of the motors backward. You can use these techniques to create an obstacle avoidance program.

Rather than using Move blocks in On mode, as when programming the EXPLOR3R, this program has the motors make four *complete turns* using On For Rotations mode. This movement is repeated until the Infrared Sensor's proximity measurement drops below 50%. Then, the robot turns left by reversing the left motor and turning the right motor forward three rotations.

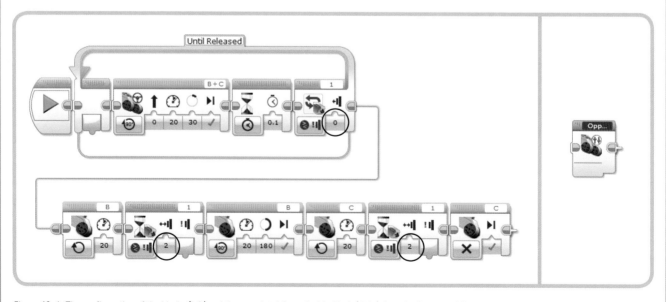

Figure 13-4: The configuration of the blocks (left) and the completed Opposite My Block (right). I used a Sequence Wire to split the program in two for better visibility, but you won't need to do this in your program.

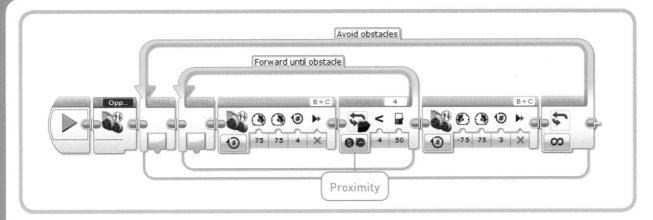

Figure 13-5: The ObstacleAvoid *program*

Create the *ObstacleAvoid* program using the instructions in Figure 13-5. Note that it's necessary to rotate both motors at the same speed and for a fixed number of complete turns to ensure that the legs always end up in the original opposite position, ready to take more steps. If you program a movement that makes your motors end in a different position, you'll need to run the Opposite My Block again to reset the legs.

NOTE ANTY works best on smooth surfaces, such as a tile or wood floor. If the robot has difficulty walking forward, try setting the power settings in the Move Tank blocks to 40 instead of 75.

programming behavior

Next, you'll create the program that makes your robot express different kinds of behavior based on changes in its surroundings. We'll start with the most complicated behavior—searching for food—so we can test this part of the program separately. When you're ready, you'll combine it with the robot's other behaviors.

finding food

We'll mimic the search for food by making the robot search for the infrared beacon. You'll use the techniques you learned in Chapter 8. ANTY looks for the beacon, walks toward it, and stops once it's found it. To accomplish this, the robot takes two steps to the left or to the right, depending on which side it sees the beacon, and then it takes two steps forward. It repeats this behavior until it sees the beacon up close, indicating that it has reached its destination.

Create the *FindingFood* program, as shown Figure 13-6. Once you've verified that your program works, turn the Loop block with its contents into a My Block called *Find* so that you can use it in your next program. (Don't include the Opposite block in the Find My Block—you'll place it elsewhere in the final program.)

NOTE Remember to toggle the top button of the infrared beacon (Button ID 9) so that the green indicator light remains on. Also, remember to hold the beacon at the same height as the robot's eyes.

sensing the environment

The final program makes ANTY express different kinds of behavior based on the color it sees with the Color Sensor in its tail. For example, ANTY will take a quick nap when you hold a green object near the sensor, because the color green indicates the safety of a grass field.

Let's begin by creating the basic structure of the *ColorBehavior* program. The robot should repeatedly check which color it sees and do something different in response to each color. You program the robot to do this with a Switch block in Color Sensor – Measure – Color mode that you place in a Loop block, as shown in Figure 13-7.

The switch has five cases: No Color, Green, Yellow, Blue, and Red. The default case is No Color, so the blocks on this tab will run when black, white, or brown is detected. The next step is to add blocks to each tab of the switch.

no color: sitting still

In the absence of green, yellow, blue, or red, ANTY simply sits still while making a chirping sound. Place a Sound block on the No Color tab of the switch, as shown in Figure 13-8. Note that the *Insect chirp* sound file plays two chirp sounds, so the sensor takes a measurement after every second chirp.

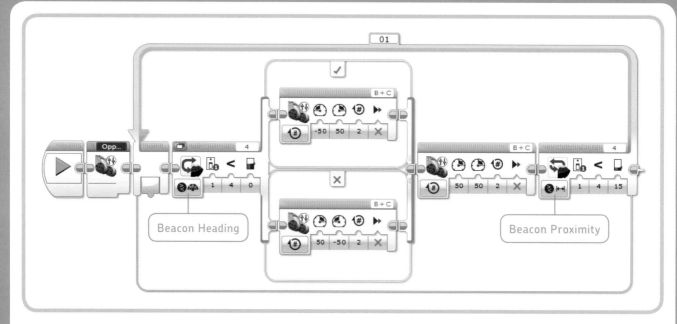

Figure 13-6: The FindingFood *program. The Loop block will become the Find My Block.*

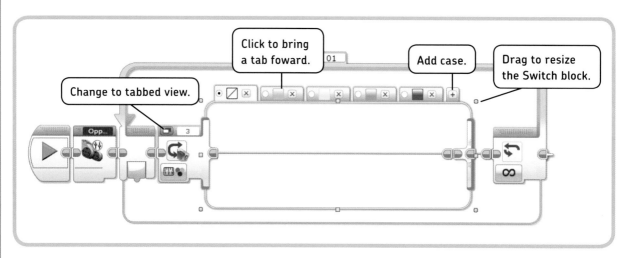

Figure 13-7: The basic structure of the ColorBehavior *program. You'll add blocks to each of the tabs.*

green: safety

When ANTY senses the green grass of a large field, it knows it's safe to take a quick nap. Add the blocks that make the robot do this to the Green tab, as shown in Figure 13-9.

yellow: food

Seeing a yellow object makes ANTY hungry, causing it to start looking for food (the beacon) using the Find block, as shown in Figure 13-10. See "Finding Food" on page 191 if you haven't made the My Block yet.

blue: predators

Blue objects indicate the presence of predators. ANTY's best option is to simply run away, using Move Tank blocks, as shown in Figure 13-11. The blocks here make the robot take five steps forward and then two steps to the left.

red: aggression

Red causes ANTY to become aggressive and shake violently in order to scare off enemies, as shown in Figure 13-12.

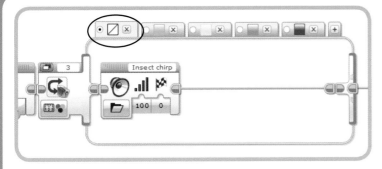

Figure 13-8: ANTY makes a chirping sound if it doesn't detect green, yellow, blue, or red.

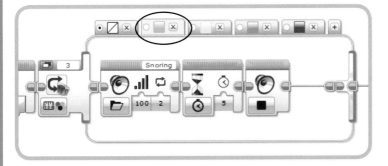

Figure 13-9: ANTY takes a quick nap when it sees a green object.

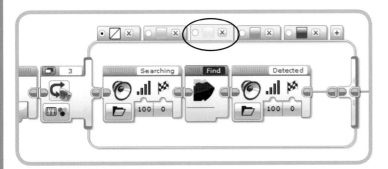

Figure 13-10: ANTY locates the infrared beacon and walks toward it when its Color Sensor sees yellow.

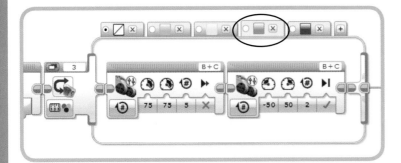

Figure 13-11: ANTY runs away when it sees something blue. Note that the motors turn for a fixed number of rotations so that the legs end up in the original opposite position, ready to run away again.

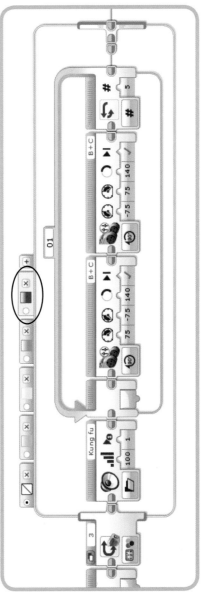

Figure 13-12: ANTY shakes aggressively when it sees a red object.

The program is now complete, so run it and test each of ANTY's behaviors by presenting objects of different colors to the Color Sensor.

further exploration

You can use the EV3 to build not only vehicles and machines but also robotic animals and creatures. In this chapter, you built ANTY, a six-legged walker. You learned how to make it walk with two Large Motors combined with the Touch Sensor to detect the absolute position of each motor.

In addition to building a walking robot, you created a program that controls your robot's behavior, so that ANTY would perform different actions based on sensor input to imitate the behavior of a real animal.

Now try some of the Discoveries to explore this robot design further. What cool creatures will you make?

DISCOVERY #70: REMOTE CONTROL!

Difficulty: 🔲 **Time:** ⏱⏱

Can you create a remote control program for ANTY? Make the robot walk in any direction with the infrared remote. Use the techniques you learned in Chapter 8.

TIP Make the program run the Opposite My Block when you press the button at the top of the remote. Press this button to reposition ANTY's legs whenever they aren't positioned correctly.

DISCOVERY #71: NIGHT CREATURE!

Difficulty: 🔲🔲 **Time:** ⏱⏱

Modify the *ObstacleAvoid* program to adjust ANTY's walking pace to the time of day. Make your robot walk around normally at night, make it creep slowly during dusk and dawn, and make it sit completely still during the day. Use the Ambient Light Intensity mode to estimate the time of day.

TIP Test your program in a dark room to simulate nighttime, switch on just one lamp to mimic dawn, and switch on all the lights to imitate daylight. How do you calculate the threshold values in your program?

DISCOVERY #72: HUNGRY ROBOTS!

Difficulty: 🔲🔲🔲 **Time:** ⏱⏱

Can you create a program that makes the robot walk around and avoid objects until it becomes "hungry"? When it's hungry, it should walk toward the infrared beacon to find food. After repeating this behavior three times, the robot should take a nap.

HINT Determine how hungry ANTY is by measuring how many rotations its motors have turned since it last had food (50 steps is a good threshold value). Once the robot has found food, reset the value of the Rotation Sensors to 0 in order to restore its "energy" level.

DESIGN DISCOVERY #22: ROBOTIC SPIDER!

Building: ✳✳✳ **Programming:** 🔲

Can you create a robotic spider? Remove ANTY's tail and rearrange the Infrared Sensor and EV3 brick to create a shape that looks more like a spider. You'll also need to expand the design by two legs because spiders have eight. When you're ready, add decorative elements to the legs and the robot's body to make the spider look more realistic (and scarier).

TIP If you're not sure how to make the two extra legs move, just add stationary elements that look like legs and have the robot walk on its six working legs. If you do this, make sure that the two new legs don't touch the ground or the other legs while walking.

DESIGN DISCOVERY #23: ANTENNAE!

Building: ✹✹ **Programming:** ▭▭

Insects are able to "feel" their environments using antennae, usually found on their heads. Can you create antennae for ANTY that let it detect obstacles and move away when it runs into one? The Touch Sensor is already in use, so can you think of a way to detect *contact with an object* using the other sensors?

HINT Design the antennae in such a way that they press the buttons of the infrared remote when ANTY comes across an obstacle (see Figure 13-13), and use the Infrared Sensor to detect a remote control button press.

Figure 13-13: You can use the remote control with the Infrared Sensor as a contact sensor. Pressing an antenna (blue arrow) causes a button on the remote to be pressed. Use the Infrared Sensor to detect which button is pressed or whether they're both pressed.

DESIGN DISCOVERY #24: CREEPY CLAWS!

Building: ✹✹✹ **Programming:** ▭▭

Can you expand ANTY by giving it a set of claws so that it can grab objects and drag them to its nest? Use the Medium Motor to make the claws open and close. For an extra challenge, make ANTY search for the infrared beacon, walk toward it, and clamp it between its claws.

TIP Remove or modify the robot's head to create space for the Medium Motor.

creating advanced programs

14

using data wires

In this fifth part of the book, you'll learn how to use data wires (this chapter), Data Operations blocks (Chapter 15), and variables (Chapter 16) to create more advanced programs for your robots. Chapter 17 demonstrates how you can combine these techniques to create a larger program that lets you play an Etch-A-Sketch-like game on the EV3 brick. You'll do all this with a robot called *SK3TCHBOT*, shown in Figure 14-1.

In earlier chapters, you configured programming blocks by entering the desired settings of each block manually. One of the fundamental concepts in this chapter is that blocks can configure each other by sending information from one block to another with a *data wire*. For example, one block can measure the Infrared Sensor's proximity value and send it to a Large Motor block. The Motor block can use the received value to set the motor's speed. As a consequence, the motor moves slowly for low sensor values (27% proximity results in 27% speed) and quickly for high sensor values (85% proximity results in 85% speed).

This chapter teaches you how to make programs that use data wires. You may find this a bit difficult at first, but as you go through the sample programs and the Discoveries, you'll master all of these programming techniques!

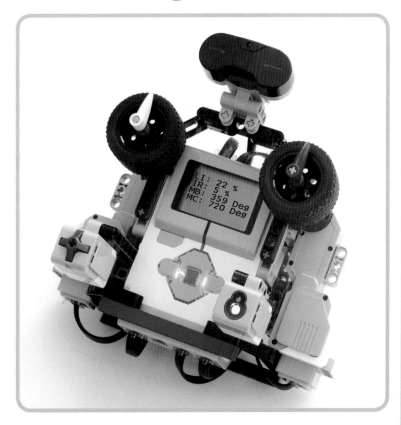

Figure 14-1: You'll use SK3TCHBOT to learn many new programming techniques and to play an Etch-A-Sketch-like game on the EV3 screen.

building
SK3TCHBOT

You'll use SK3TCHBOT to test your programs with data wires and variables. The robot consists of the EV3 brick, three sensors, and two Large Motors, which you'll use as inputs and outputs for your programs so you can see data wires in action. In Chapter 17, you'll learn how to create a program that turns SK3TCHBOT into an Etch-A-Sketch-like device,

letting you make drawings on the screen by turning the input dials attached to the motors.

Build SK3TCHBOT by following the instructions on the next pages, but first select the required pieces, as shown in Figure 14-2.

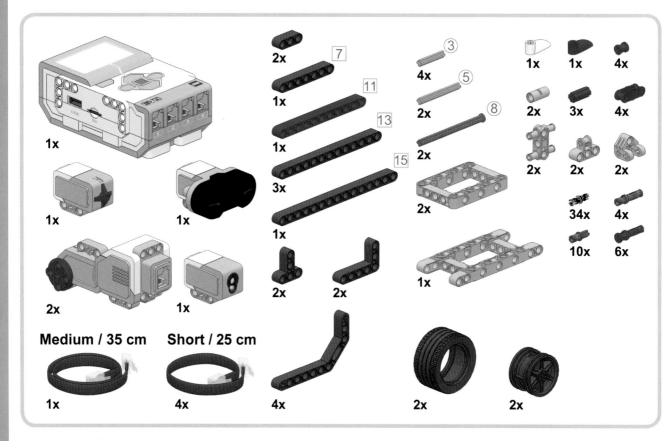

Figure 14-2: The pieces needed to build SK3TCHBOT

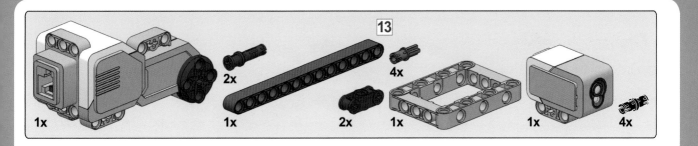

1

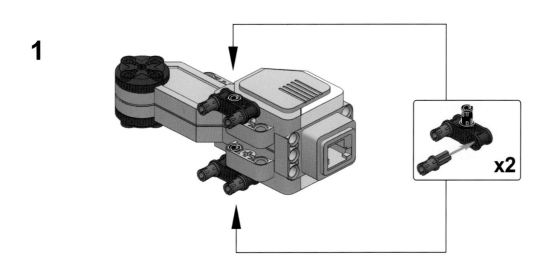

x2

2

13

3

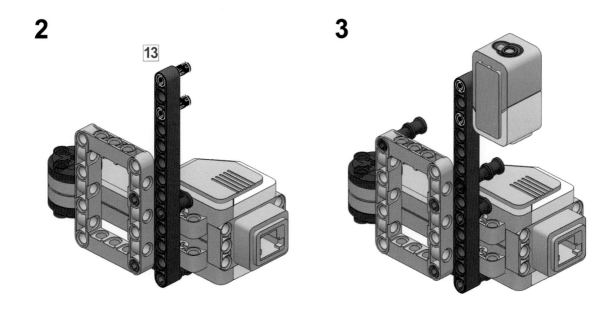

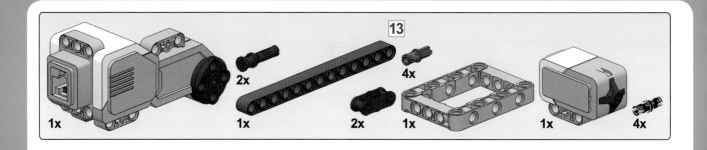

1

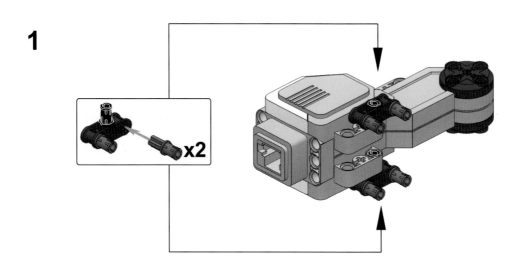

x2

2

13

3

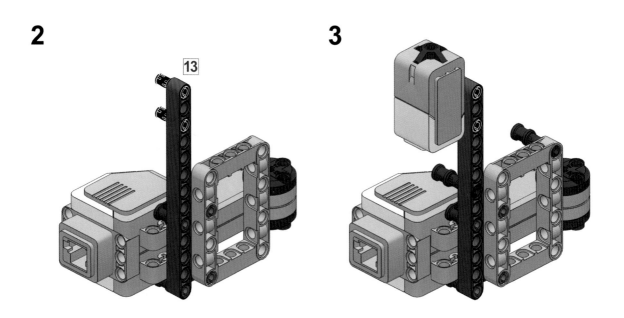

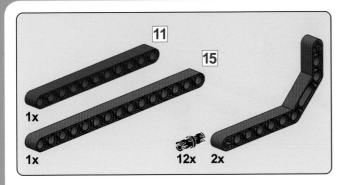

1x
1x
12x 2x

4

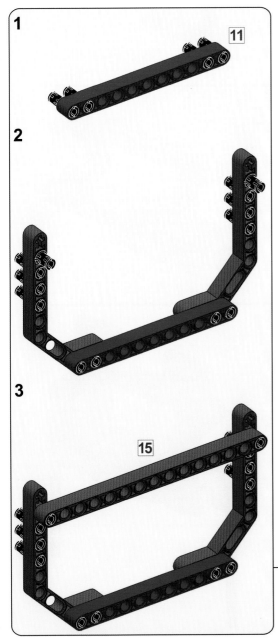

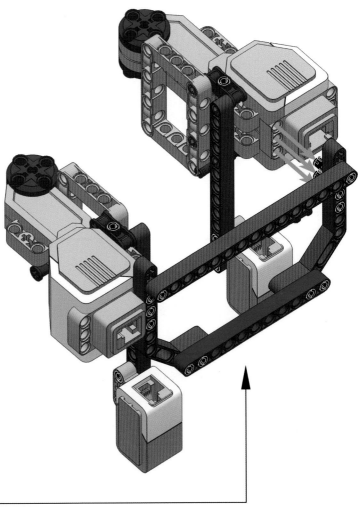

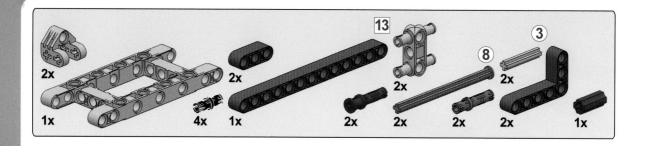

1

2

3

4

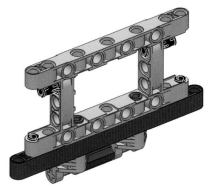

5

6

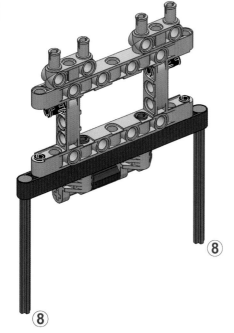

7

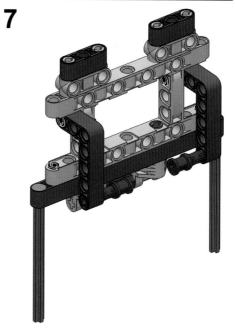

8

9

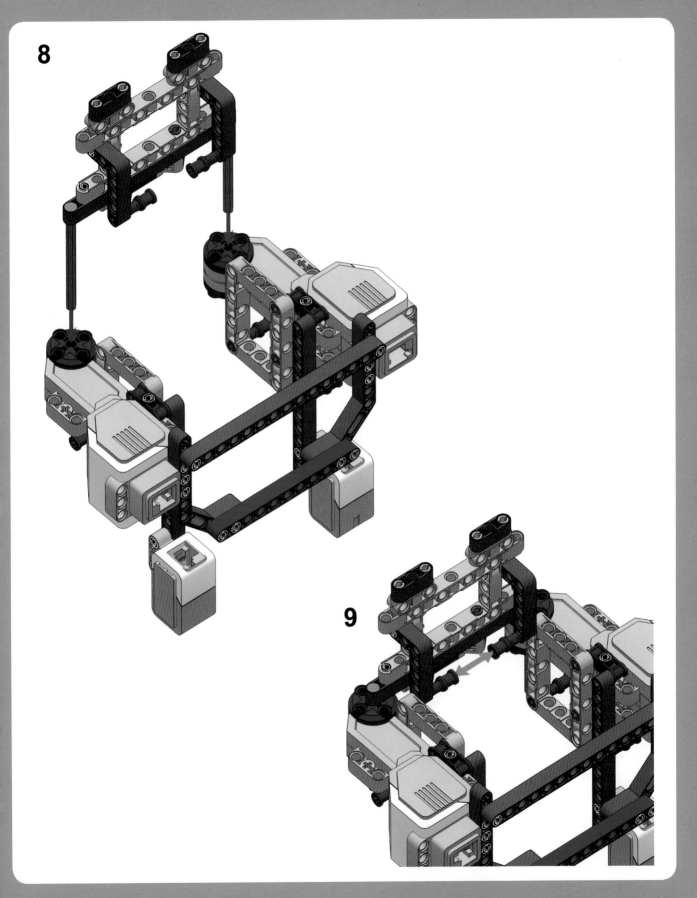

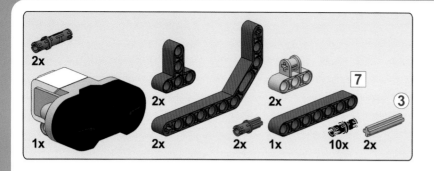

1

2

3

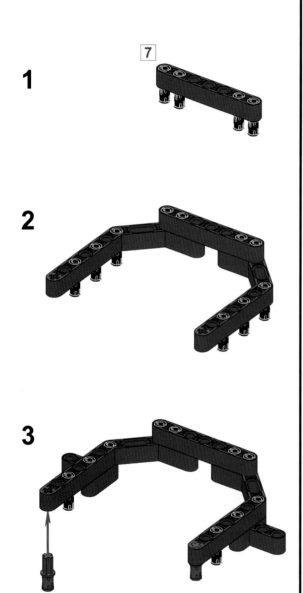

1

2

3

4

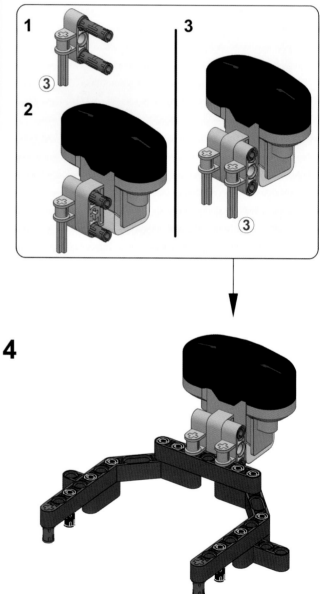

5

6

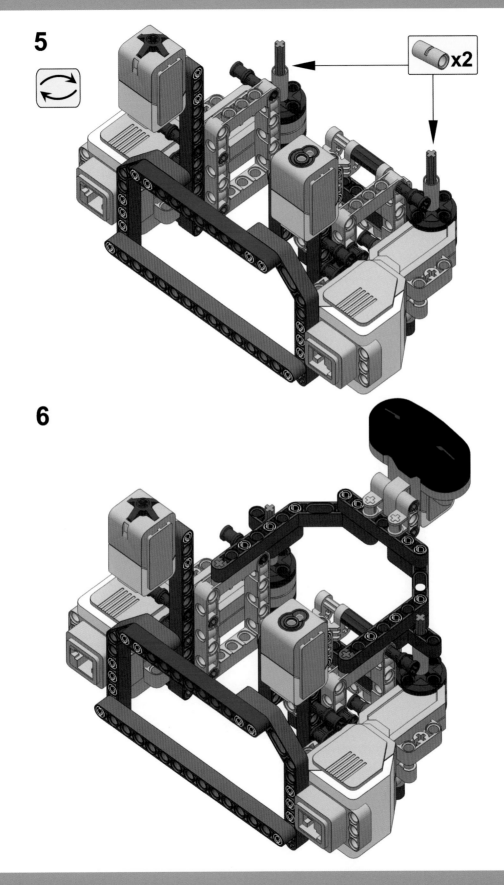

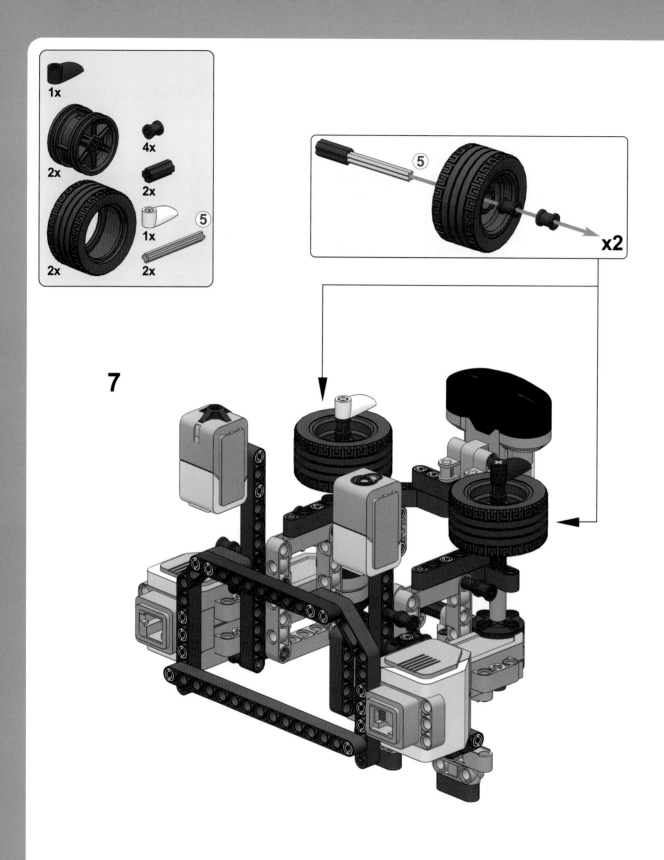

7

Medium / 35 cm

1x

Short / 25 cm

4x

8

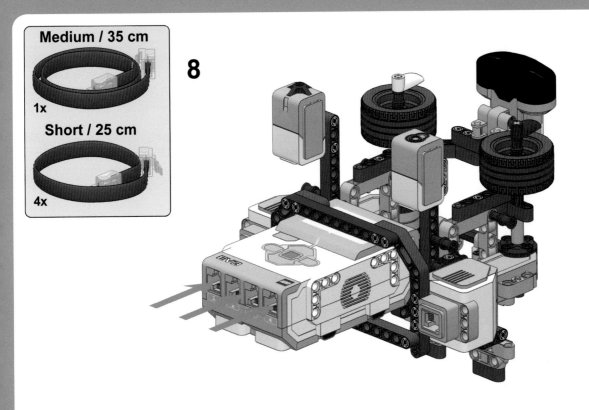

9

Push the red pins into the EV3 brick as shown. Then connect the motors and sensors to the indicated ports on the EV3 brick.

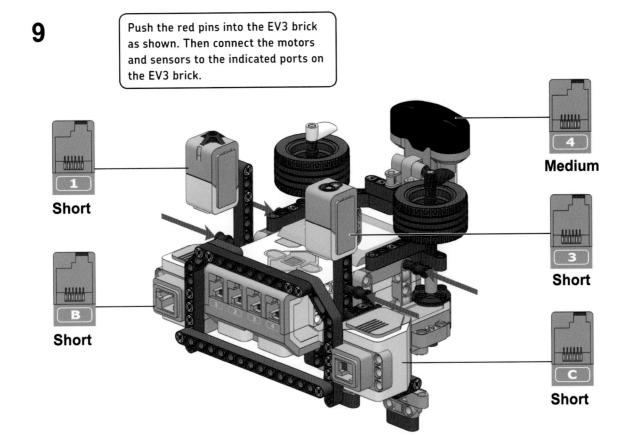

1
Short

B
Short

4
Medium

3
Short

C
Short

getting started with data wires

To see how data wires work, you'll create a small program that makes SK3TCHBOT play a sound and then rotate the white dial for 3 seconds. The motor's Power setting, and therefore its speed, will respond to what the Infrared Sensor sees: If the sensor's proximity value is 27%, the motor's speed will be 27%; if the sensor reads 85%, the motor's speed will be 85%; and so on.

You use the *Infrared Sensor block* to read the sensor. (You'll learn more about Sensor blocks later.) Create a new project called *SK3TCHBOT-Wire* with a program called *FirstWire*, as shown in Figures 14-3 and 14-4, and download it to your robot.

You should notice that the white dial turns at a different speed each time you run the program. For example, if you hold your hand close to the sensor and run the program, the wheel should spin slowly, but it should spin faster if you run the program again with your hand farther away.

Congratulations! You've just created your first program with data wires. Let's have a look at how each block works. The first Sound block simply plays a sound. Once the sound finishes playing, the Infrared Sensor takes one measurement (let's say it reads a proximity value of 27%). The yellow data wire carries the sensor measurement to the Large Motor block, which then makes motor B turn for 3 seconds. The Power setting of the Large Motor block depends on the value carried by the data wire; it's 27% in this case. Figure 14-5 shows an overview of what happens.

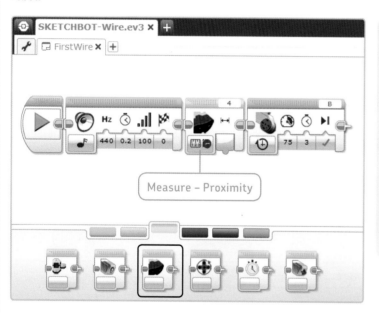

Figure 14-3: Step 1: Place all the necessary blocks on the canvas, and configure them as shown. You'll find the Infrared Sensor block on the yellow tab of the Programming Palette. Be sure to select **Measure – Proximity** mode.

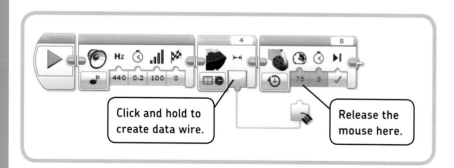

Figure 14-4: Step 2: Create and connect the yellow data wire as shown. Then run the finished program.

DISCOVERY #73: SOUND IN THE DISTANCE!

Difficulty: ▭ **Time:** ◷
Remove the Large Motor block from the *FirstWire* program, and replace it with a Sound block configured to say "Hello." Connect the data wire from the Infrared Sensor block to the Volume input of the Sound block. This should make the robot say "Hello" softly when it sees you up close or loudly when it sees you from afar.

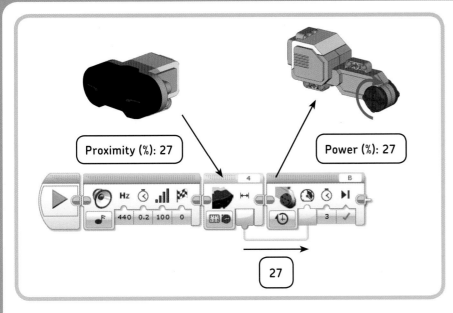

Proximity (%): 27

Power (%): 27

27

Figure 14-5: The Infrared Sensor block reads the sensor and sends the measured value through a data wire to the Large Motor block, which uses this value to set the motor's speed.

working with data wires

As you've seen, you use a data wire to carry information between blocks. The information is always sent from an *output plug* to an *input plug*, as shown in Figure 14-6. In the example program, the data wire carries the sensor value from the Sensor block's Proximity output plug to the Motor block's Power input plug.

Note that the data wire hides the value that you originally entered in the Power setting (75). The program ignores this value and uses the value it gets from the data wire instead. On the other hand, because you didn't connect data wires to the other input plugs of the Motor block (*Seconds* and the *Brake at End*), they operate normally. For example, you entered 3 in the Seconds setting, making the motor rotate for 3 seconds.

Now let's have a look at some other properties of data wires.

seeing the value in a data wire

You can see the value carried by a data wire by placing your mouse on the wire while the program runs, as shown in Figure 14-7. This can help you understand exactly what's going on in your program.

You'll see data wire values only if the robot is connected to the computer and you launch the program using the Download and Run button in the EV3 software; you won't see them if you start the program manually using the EV3 buttons.

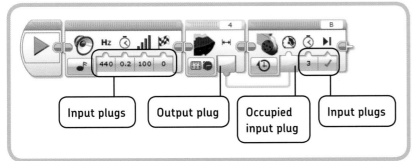

Input plugs   Output plug   Occupied input plug   Input plugs

Figure 14-6: The data wire carries information from an output plug (on the Infrared Sensor block) to an input plug (on the Large Motor block).

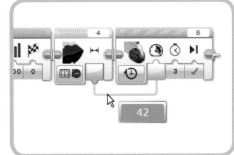

42

Figure 14-7: Hover your mouse over a data wire to see its value.

deleting a data wire

Delete a data wire by disconnecting the right end, as shown in Figure 14-8.

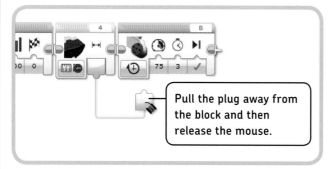

Pull the plug away from the block and then release the mouse.

Figure 14-8: Deleting a data wire

data wires across a program

When configuring programs with data wires, you don't have to connect a block to the one right next to it; you can connect blocks when there are other blocks in between, as shown in Figure 14-9. The *WirePause* program reads the sensor, pauses for 5 seconds, and makes the motor move. The motor speed is based on the sensor measurement taken at the start of the program.

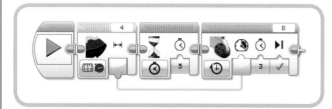

Figure 14-9: The WirePause *program takes one measurement when the Sensor block runs (say, 34%). After 5 seconds, the motor begins turning at 34% speed, even if the sensor value has changed in the meantime.*

The reverse isn't possible, as shown in Figure 14-10. You can't make the Motor block use the data wire because the wire doesn't contain a value until the measurement is made. In other words, the block with the input plug (the Motor block, in this case) must come after the block with the output plug (the Sensor block).

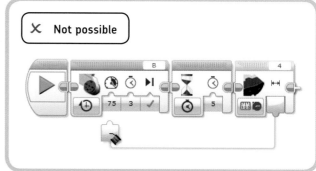

X Not possible

Figure 14-10: You can't connect a data wire from the right to the left. The EV3 software doesn't allow you to make this connection.

using multiple data wires

You can use an output plug as a starting point for multiple data wires, as shown in Figure 14-11. The *MultiWire* program sends the proximity measurement to the Power setting of the motor on port B (white dial) and to the Degrees setting of the motor on port C (red dial). For example, if the sensor measurement is 34%, motor B rotates at 34% speed while motor C rotates 34 degrees at 75% speed.

On the other hand, you can't connect multiple data wires to a single input plug. Once an input is *occupied* (see Figure 14-6), you can't connect another wire to it. (If it were possible to connect two or more wires to the same input, the block wouldn't know which of the values to use.)

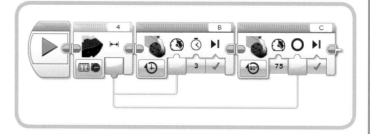

Figure 14-11: Two data wires with the same starting point in the MultiWire *program. The sensor performs one proximity measurement and passes the sensor value to both Motor blocks.*

repeating blocks with data wires

In the *FirstWire* program, the motor moved for 3 seconds at a constant speed based on one sensor value taken just before the motor started moving. You'll now expand this program to make the motor adapt its speed to the sensor value continuously. To do so, place the Sensor block and the Motor block in a Loop block, and set the mode of the Large Motor block to **On**, as shown in Figure 14-12.

When you run the *RepeatWire* program, the motor's speed should change gradually as you slowly move your hand away from the sensor. If you suddenly place your hand up close to the sensor, the sensor value drops and the motor quickly stops.

The speed changes continuously because the blocks in the loop take hardly any time to run. The Sensor block takes a measurement, and the Motor block switches on the motor at the desired speed. Then, the program goes back to run the blocks in the loop again, instantly taking a new measurement to adjust the speed, and so on.

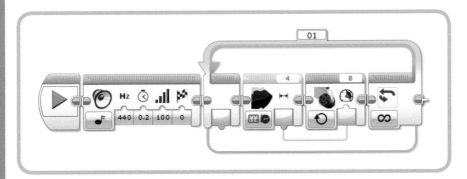

Figure 14-12: The RepeatWire *program continuously adapts the motor's speed to the proximity sensor* measurement.

DISCOVERY #74: BAR GRAPH!

Difficulty: ⬚ **Time:** ◔

The incomplete program shown in Figure 14-13 plots a bar graph on the EV3 screen, but a data wire is missing. How do you connect the data wire to make the bar's length represent the proximity measurement?

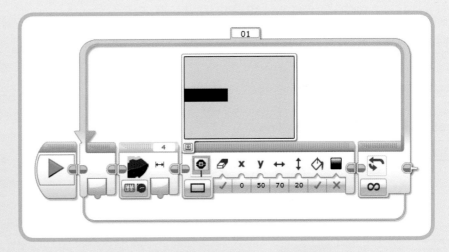

Figure 14-13: The incomplete program for Discovery #74

data wire types

So far you've used data wires to transfer numerical values only, but there are three basic types of information that data wires can carry: *Numeric*, *Logic*, and *Text*. Each type has its own color and plug shape (round/yellow, triangular/green, and square/orange), as shown in Table 14-1. The plug shape matches the shape of the input it connects to like a puzzle piece. This match helps indicate what data wire type can be connected to a particular input.

numeric data wire

The *Numeric data wire* (yellow) carries numeric information that may include whole numbers (such as 0, 15, or 1427), numbers with decimals (such as 0.1 or 73.14), and negative numbers (such as –14 or –31.47). The Infrared Sensor's proximity is an example of such a numeric value (it ranges between 0 and 100).

logic data wire

The *Logic data wire* (green) can carry only two values: *true* or *false*. These wires are often used to define settings of a block that can have only two values, such as the Clear Screen setting of the Display block. Similarly, because the Touch Sensor has only two possible values (pressed and released), it uses a Logic data wire to send the sensor state. The Touch Sensor block in Measure – State mode gives you a Logic data wire carrying true if the sensor is pressed and false if it's not pressed.

Create the *LogicClear* program to see a Logic data wire in action, as shown in Figure 14-14. The program begins by displaying an image of two angry eyes on the EV3 screen. Two seconds later, the robot determines the Touch Sensor value and passes it to the Clear Screen setting of the Display block with a Logic data wire. If the sensor is pressed (true), the screen is cleared before displaying the word *MINDSTORMS*. If it's not pressed (false), the screen isn't cleared and the word is simply drawn on top of the image.

text data wire

The *Text data wire* (orange) carries text to, for example, a Display block so that it appears on the EV3 screen. Text can be a word or phrase, like *Hello*, *I'm a robot*, or *5 apples*. We'll get back to this data wire type in Chapter 15.

table 14-1: basic data wire types

Type		Example values
Numeric		–5 0 3.75 75
Logic		True False
Text		Hello I'm a robot 5 apples

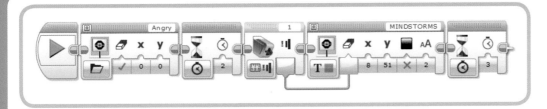

Figure 14-14: The LogicClear program. You'll find the Touch Sensor block between the Sensor blocks (the yellow tab of the Programming Palette).

numeric array and logic array

In addition to the Numeric, Logic, and Text data wire types, the EV3 software has the *Numeric array* and the *Logic array* data wire types. A Numeric array is basically a list of Numeric values used to send multiple numbers across a single wire. For example, you could use an array to send the five most recent sensor values to a custom My Block that displays all of them on the EV3 screen.

Similarly, a Logic array is a list of Logic values. We won't use arrays here, but you can experiment with an example program that uses arrays at *http://ev3.robotsquare.com/* after you've finished reading this book.

type conversion

Generally, you should connect Numeric data wires to inputs that accept numeric values (rounded plug), Logic data wires to logic inputs (triangular plug), and Text data wires to text inputs (square plug).

However, the EV3 software allows three more connections, as shown in Table 14-2. You can remember which connections work by looking at the shape of the plugs. In some cases, the shapes don't match exactly, but they still fit. For example, the triangular plug fits in a rounded socket, which means you can control a Numeric input with a Logic data wire. On the other hand, a rounded plug doesn't fit in a triangular socket, which means you can't control a Logic input with a Numeric data wire.

In the three situations shown in Table 14-2, the information carried by the wire is *converted* to the proper type to make your program work. Let's make two programs to see what this means.

table 14-2: data wire type conversions

From			To	Effect
Logic			Numeric	*True* becomes 1 *False* becomes 0
Logic			Text	*True* becomes 1 *False* becomes 0
Numeric			Text	The number is turned into a format understood by the Display block so that it can be displayed.

converting logic values to numeric values

The *ConvertWire* program (Figure 14-15) demonstrates the conversion of a Logic value to a Numeric value. The Wait block requires a numeric value for the Seconds setting, but it receives a Logic value from the Touch Sensor block instead. This works because the software converts the Logic value to a Numeric value: True becomes 1; False becomes 0. Consequently, you'll hear a 1-second pause between the beeps if you press the sensor, but you'll hear no pause (0 seconds) if the sensor isn't pressed.

displaying numbers on the EV3 screen

When using a Display block to display text on the EV3 screen, you can either type something in the Text field or select *Wired*, as shown in Figure 14-16. Selecting Wired creates an extra input plug, allowing you to supply the text with a data wire.

The block expects you to connect a Text data wire to the Text input, but you can also connect a Numeric data wire.

Computers like the EV3 brick store numbers and text lines in two different ways. Therefore, you normally can't send a Numeric value to a Text input. Fortunately, the software converts the number to a text format that the Display block understands. This conversion makes it possible to display numbers on the EV3 screen, as demonstrated by the *DisplayNumeric* program (see Figure 14-17).

The program continuously updates the sensor value and shows it on the screen. Displaying values is a useful technique to test your programs. Of course, you already know how to monitor sensor values with Port View, but with this data wire technique, you can also perform calculations on sensor values and display the results in real time, for example, as you'll see in the next chapter.

In addition to Text and Numeric data wires, a Text input plug can accept Logic data wires. Again, the EV3 brick stores Logic values in a different way, but the program converts them to a Text format when necessary. *True* causes the Display block to display 1, while *false* makes the block display 0. Replace the Infrared Sensor block in the *DisplayNumeric* program with a Touch Sensor block to try this out.

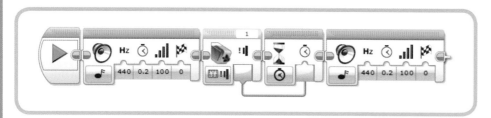

Figure 14-15: The ConvertWire *program*

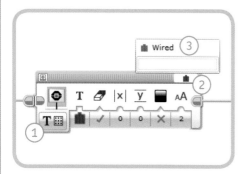

Figure 14-16: Pick a Display block from the palette, choose **Text – Grid** mode (1), click the Text field (2), and choose Wired (3) to reveal the Text input plug.

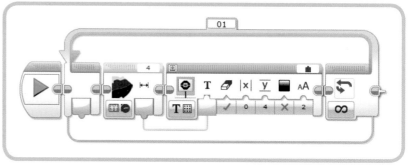

Figure 14-17: The DisplayNumeric *program*

using sensor blocks

Figure 14-18: From left to right: Brick Buttons block, Color Sensor block, Infrared Sensor block, Motor Rotation block, Timer block, and Touch Sensor block

In Part II, you learned to work with sensors by creating programs with the Wait, Loop, and Switch blocks. The final way to read sensors is with *Sensor blocks*. These blocks are useful if you want to retrieve a sensor value and transfer it to another block with a data wire, as you saw with the *FirstWire* program.

There's a Sensor block for each sensor on the Sensor tab of the Programming Palette, as shown in Figure 14-18. Each block can be used in Measure mode or Compare mode.

measure mode

A Sensor block in *Measure mode* takes one measurement and passes the measured value on to another block with a data wire. You choose the type of measurement by selecting one of the sensor's operation modes. For example, you passed

the proximity sensor value to a Motor block using an Infrared Sensor block in Measure – Proximity mode.

Everything you have learned about sensor operation modes and sensor values so far holds true for Sensor blocks, too. Table 14-3 provides a summary of the sensor values for each sensor operation mode. Use this chart as a reference when making your own programs, whether you use Wait, Loop, Switch, or Sensor blocks.

table 14-3: sensor values for each sensor operation mode

Sensor block	Operation mode	Min.	Max.	Meaning	Page	Notes
Brick Buttons	Brick Buttons	0	5	0 = None, 1 = Left, 2 = Center, 3 = Right, 4 = Up, 5 = Down	97	Detects only one button at a time.
Color Sensor	Color	0	7	0 = No color detected, 1 = Black, 2 = Blue, 3 = Green, 4 = Yellow, 5 = Red, 6 = White, 7 = Brown	77	
	Reflected Light Intensity	0	100	0 = Lowest reflectivity 100 = Highest reflectivity	81	
	Ambient Light Intensity	0	100	0 = Darkness 100 = Very bright light	85	
Infrared Sensor	Proximity	0	100	0 = Very close 100 = Very far	89	
	Beacon (Proximity)	1	100	1 = Very close 100 = Very far	93	The value is *undefined* if no beacon signal is detected. See Figure 14-26.
	Beacon (Heading)	–25	25	–25 = Left 0 = Middle 25 = Right	93	The value is also 0 if no beacon signal is detected or if the signal direction cannot be resolved.
	Remote	0	11	The number represents a combination of pressed buttons on the remote.	92	

(continued)

table 14-3 (continued)

Sensor block	Operation mode	Min.	Max.	Meaning	Page	Notes
Motor Rotation	Degrees	N/A	N/A	Number of degrees that the motor has turned since the start of the program.	98	The value can be reset to 0 with Reset mode.
	Rotations	N/A	N/A	Number of rotations that the motor has turned since the start of the program.	98	The value is a decimal number, such as 1.5 for one and a half rotations. The value can be reset to 0 with Reset mode.
	Current Power	–100	100	The number represents the rotational speed of the motor. 100% speed is equivalent to 170 rpm for the Large Motor; 100% speed is equivalent to 267 rpm for the Medium Motor.	99	This mode measures rotational speed; it does not measure current or power consumption. The measured value is independent of the EV3 brick's battery level.
Timer	Time	0	N/A	Time elapsed in seconds since the start of the program.	235	The value is a decimal number, such as 1.5 for one and a half seconds. The value can be reset to 0 with Reset mode.
Touch Sensor	State	False	True	False = Released True = Pressed	66	The value is carried out with a Logic data wire.

compare mode

Just as in Measure mode, a *Sensor block in Compare mode* takes one measurement and outputs the sensor value with a data wire. In addition, it compares the measured value to a threshold value, and it outputs the result with a Logic data wire. The *Compare Result* plug carries out *true* if the condition (for example, "The proximity value is greater than 40%") is true; it carries out *false* if the condition is false. You specify the condition by entering a Threshold Value and choosing a Compare Type in the block's settings, as you've done for Wait, Loop, and Switch blocks.

You'll now create a program to see how this works. The *SensorCompare* program (see Figure 14-19) contains an Infrared Sensor block in Compare – Proximity mode to measure the sensor value and check whether it's greater than 40%. The proximity value controls the speed of motor B, which runs for 5 seconds. The Compare Result output controls the Pulse setting of the Brick Status Light block. If the comparison is true (the sensor value is indeed greater than 40%), the Pulse setting will be true, making the status light blink. If it's false, the light just stays on.

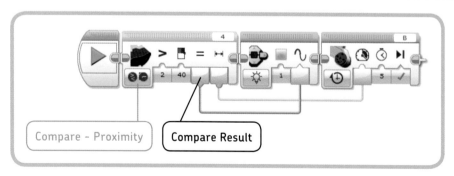

Figure 14-19: The SensorCompare program

compare mode and beacon values

Beacon Proximity and Beacon Heading are combined into a single mode (called Beacon) if you use an Infrared Sensor block in Measure mode (see Figure 14-23). They appear as separate modes if you choose Compare mode, but otherwise there is no difference. You'll see an example of Beacon mode later in this chapter.

compare mode and the touch sensor

A Touch Sensor block in Compare mode can tell you whether the sensor has been bumped (pressed and released) since the last time you used any block to interact with the Touch Sensor. That is, if you bump the sensor and check the sensor state later, the block will tell you that the sensor has been bumped: The (Numeric) Measured Value plug outputs 2. But this is a bit ambiguous—if you check whether the sensor is released after you bump it, the block outputs false (because it was bumped), even though the sensor is actually released.

Usually you'll just want to know whether the Touch Sensor is pressed or released when the block runs, so it's better to use the Touch Sensor block in Measure mode. In this mode, the block simply outputs true if the sensor is currently pressed and false if the sensor is currently released, regardless of what happened earlier during the program.

DISCOVERY #77: SENSOR THROTTLE!

Difficulty: ▭ **Time:** ◷
Can you make a program to control the *speed* of the white dial (motor B) using the *position* of the red dial (motor C)? Turn the red dial manually to test your program.

HINT Use the *RepeatWire* program (see Figure 14-12) as a starting point, and use a Motor Rotation block in Measure – Degrees mode.

---

data wire value range

When making programs with data wires, it's important to consider what happens when a data wire value is outside the *allowed range* of values. For example, the brick status light accepts three values that set the color to green (0), orange (1), or red (2), but what happens if you control the color using a data wire that has a value of 4? To find out, create the *ColorRange* program shown in Figure 14-20; refer to Table 14-3 to see what the data wire value will be for each of the EV3 buttons.

DISCOVERY #78: MY PORT VIEW!

Difficulty: ▭▭ **Time:** ◷◷
Expand the *DisplayNumeric* program (see Figure 14-17) to display the Color Sensor's reflected light intensity, the Touch Sensor value, and the positions of the Rotation Sensors on the EV3 screen. The values should be updated four times a second.

When you're ready, turn the blocks inside the loop into a My Block called *MyPortView*. You can use it anytime in your programs for SK3TCHBOT to see information about each sensor.

HINT Place a Wait block in the loop to pause the program for 0.25 seconds.

DISCOVERY #79: COMPARE SIZE!

Difficulty: ▭▭ **Time:** ◷
Can you show a circle in the middle of the EV3 screen and control its size and color using the proximity measurement? Use the *Radius* setting on the Display block to control the circle's size; use the *Fill* setting to display a filled circle if the proximity is greater than 30% and an empty circle otherwise. Place the Sensor block and the Display block in a Loop so that the circle is continuously redrawn.

HINT Use an Infrared Sensor block in Compare – Proximity mode.

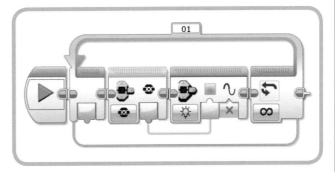

Figure 14-20: The ColorRange program

You should find that the light is green if no buttons are pressed (0), orange if the Left button is pressed (1), and red if the Center button is pressed (2). All of the other buttons (3, 4, and 5) also result in a red light.

You can find the allowed values for each input plug by going to **Help ▶ Show EV3 Help**, but the documentation doesn't tell you what happens if you go beyond this range. As a rule of thumb, just remember that the EV3 software uses the *nearest allowed value*, but you'll have to create a simple experiment like the *ColorRange* program to be sure whether this generalization applies to your program. (The rule works in this case: The value 2, or red, is the nearest allowed value when the data wire value is 3, 4, or 5.)

advanced features of flow blocks

Now that you know how data wires work, you're ready to explore the features of Wait, Loop, and Switch blocks that require the use of data wires. You'll also learn to use the Loop Interrupt block.

data wires and the wait block

Recall that a Wait block pauses the program until a sensor reaches a certain trigger or threshold value. The *Measured Value* output gives you the measurement that made the block stop waiting. For example, the *WireWait* program (see Figure 14-21) waits for the Color Sensor to see either blue (2), green (3), yellow (4), or red (5), after which it displays the last measurement on the screen.

data wires and the loop block

Loop blocks have two features that require data wires: Loop Index and Logic mode. You'll try out each feature with a sample program.

using the loop index

The *Loop Index* output plug (see Figure 14-22) gives you the number of times the Loop block has finished running the blocks inside it. The index begins at 0, incrementing by 1 each time the blocks in the loop run.

You'll now make the *Accelerate* program, which uses the Loop Index as an input for the motor speed. When you start the program, the index is 0, which makes the motor run at 0 speed (it stands still) for 0.2 seconds. When the Loop block returns to the beginning, the Loop Index increases to 1, and it repeats the Motor block, this time setting the speed to 1. When it repeats, the motor speed is 2, and so on.

The Loop block is configured to run 101 times so that the speed is 100 when it runs for the last time. You'll then hear a sound and the program ends. You could configure the loop to run, say, 150 times, but the motor will stop accelerating when the index exceeds 100 because 100 is the maximum speed.

ending a loop in logic mode

You learned earlier that you can make the Loop block stop repeating after a certain number of repetitions, after a certain number of seconds, or when a sensor reaches a trigger value. In **Logic** mode, you can make a loop stop repeating using a Logic data wire.

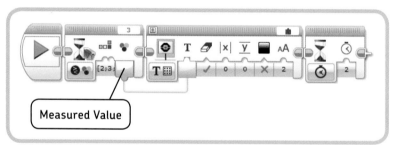

Figure 14-21: The WireWait program

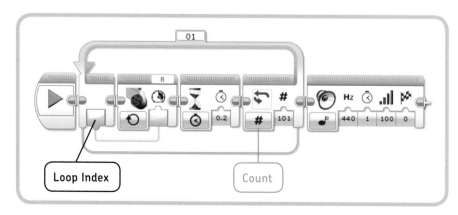

Figure 14-22: The Accelerate program

The Loop block checks the data wire value each time it's done running the blocks in the loop. If the data wire value is false, the blocks run again. If it's true, the loop ends. In other words, the blocks repeat until the data wire value is *true*.

The *LogicLoop* program (see Figure 14-23) demonstrates this technique, using a Loop block in Logic mode and an Infrared Sensor block in Measure – Beacon mode. The Sensor block gives you the beacon heading and beacon proximity, but here you'll use only the *Detected* output plug, which tells you whether the sensor successfully detects the beacon's signal. If a signal is detected, the data wire carries out *true*, making the loop end; if no signal is detected, the data wire carries out *false* and the loop runs again. In other words, the program waits until the sensor picks up a signal, and then it plays a sound.

When you run the program, you should find that the sensor can successfully detect a signal up to around 3 meters (10 feet) away.

DISCOVERY #80: IR ACCELERATION!

Difficulty: 🔲 **Time:** ⏱

Can you make motor B accelerate *until* the Infrared Sensor successfully picks up a signal from the beacon?

HINT Combine the *Accelerate* program (Figure 14-22) and the *LogicLoop* program (Figure 14-23) into a single program.

data wires and the switch block

As you'll recall from Chapter 6, you use Switch blocks to have your robot make decisions. The robot uses a sensor value to determine whether a given condition (such as "The proximity value is greater than 30%") is true. If it's true, the blocks at the top branch of the Switch block run (✔); if it's false, the blocks at the bottom run (✖), as demonstrated by the *SwitchReminder* program in Figure 14-24.

logic mode

Rather than using a sensor value to make a decision, you can use a Logic data wire to control the Switch block by choosing **Logic** mode, as shown in Figure 14-25. If the data wire value is true, the blocks at the top of the switch run; if it's false, the blocks at the bottom run.

The *LogicSwitch1* program (see Figure 14-25) continuously checks whether the Infrared Sensor successfully detects the beacon. If so (true), the robot displays "Success!" on the EV3 screen and motor B moves; if not (false), it shows "Error!" and the motor stops. (The sensor stops receiving a signal about a second after you release the button, so it takes about a second for the error message to appear.)

numeric mode

If you set the Switch block to **Numeric** mode and you connect a Numeric data wire to it, you can choose specific actions to run for each value using the techniques you learned for Switch blocks with more than two cases, as discussed in Chapter 7 (see Figure 7-10 on page 81). For example, you can make the robot say "Hello" if the data wire value is 3, "Good morning" if it's 10, and "No" for all other values (the default case). You'll try this out in the next chapter.

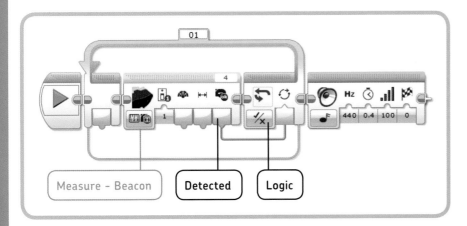

Figure 14-23: The LogicLoop *program plays a sound once the Infrared Sensor successfully detects a signal from the infrared beacon.*

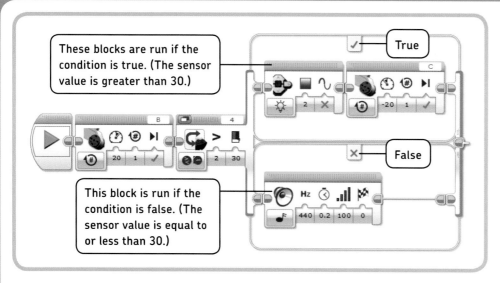

These blocks are run if the condition is true. (The sensor value is greater than 30.)

True

False

This block is run if the condition is false. (The sensor value is equal to or less than 30.)

Figure 14-24: The SwitchReminder *program*

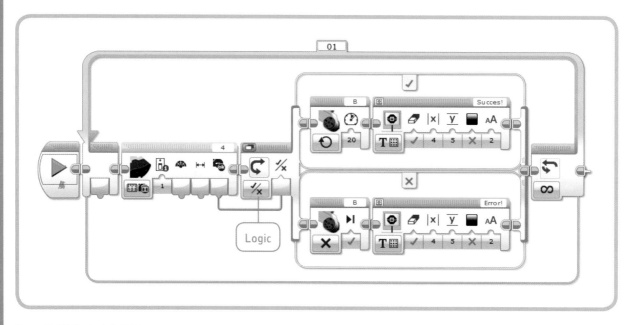

Figure 14-25: The LogicSwitch1 *program*

connecting data wires to blocks inside switch blocks

In some cases, it can be useful to connect data wires from outside a Switch block to blocks inside the switch. For example, you can modify your previous program to control the motor speed with the beacon proximity value without having to use another Infrared Sensor block.

To do so, toggle the Switch block to Tabbed View and connect the data wire as shown in Figure 14-26. The finished

LogicSwitch2 program controls the motor's speed with the beacon proximity if a signal is detected; it stops the motor otherwise.

As shown in Table 14-3, the beacon proximity value is *undefined* if no signal is detected. If the beacon proximity value is connected to a Motor block while no signal is detected, the Motor block receives no value, and it moves unpredictably. You can avoid this potential problem by making sure to use the value only when it contains a proper sensor measurement—that is, only if the Detected output is true.

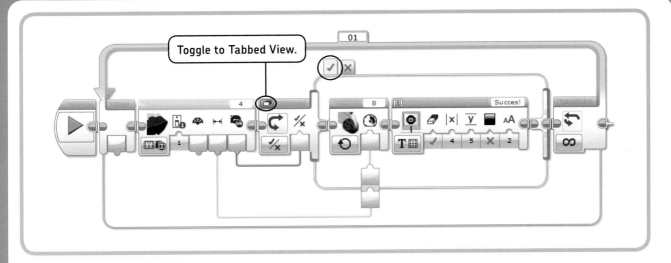

Figure 14-26: The LogicSwitch2 program. Start with the LogicSwitch1 program, switch to Tabbed View, bring forward the True tab, and connect the Numeric data wire as shown. (A pair of input and output plugs across the edge of the switch should appear automatically as you try to connect the wire.) The blocks on the False tab remain unchanged.

That's why you need to use the Switch block in the *LogicSwitch2* program. The sensor value is used to control the motor speed only when a signal is detected. Otherwise, the blocks on the False tab run and a Large Motor block in Off mode makes the motor stop.

In Beacon Heading mode, on the other hand, the output value will be 0 if either the beacon is directly in front of the sensor or no signal is received at all. You can distinguish between these two cases using the same technique as in our last program. For example, you could display the Beacon Heading value on the screen if the signal is detected and make the screen read "Error" otherwise.

NOTE You can connect data wires to blocks inside a Switch block only when it's in Tabbed View. Changing back to Flat View will remove the data wires. You can connect data wires to blocks inside a Loop block in the same way.

the loop interrupt block

The final way to make a Loop block stop repeating is to use the *Loop Interrupt block*. A Loop block normally checks a sensor condition or Logic value each time it finishes running the blocks inside it, but the Loop Interrupt block makes a specified loop end *immediately*.

You choose which loop you want to interrupt by selecting the Loop name from a list, as shown in Figure 14-27. You can use the Loop Interrupt block to end a loop from the inside or to end a loop that's running on a parallel sequence. When a loop is interrupted, the program continues with the blocks that are placed after the loop.

breaking a loop from the inside

The Loop Interrupt block can be useful to break out of a Loop block at any point in the loop, rather than having to wait for all of the blocks in the loop to finish. Consider the *BreakFromInside* program in Figure 14-27, which repeatedly turns motor B for one rotation and then says "LEGO." If the Infrared Sensor's proximity value is less than 50% after the robot says "LEGO," the loop ends normally, and you'll hear "MINDSTORMS" right after.

Thanks to the Loop Interrupt block, you can also end the loop by keeping the Touch Sensor pressed just after motor B turns. When you do, the program skips to the block after the loop, and you'll hear only "MINDSTORMS."

Create the program as shown in Figure 14-27, and run it a few times to determine which sensor should be triggered (and when) to make the loop end.

breaking a loop from the outside

You can also interrupt a Loop block from a sequence running in parallel. When you do this, the loop ends and the program begins to run the blocks after the loop. At the same time, the program attempts to finish the block that was running when you ran the Loop Interrupt block.

For example, the *BreakFromOutside* program (see Figure 14-28) has a Loop block that repeatedly turns motor B for one rotation. On a parallel sequence, a Loop Interrupt block is run once the Infrared Sensor is triggered. The loop then ends, and the Sound block immediately plays a sound.

If you trigger the sensor when the motor is halfway through making its rotation, it will finish its rotation while the sound plays. So momentarily, the Large Motor block and the Sound block will run simultaneously.

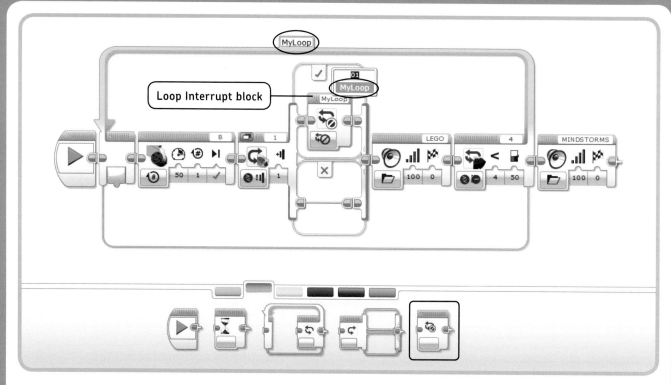

Figure 14-27: The BreakFromInside *program contains a loop called* MyLoop*, which ends if the Touch Sensor is pressed after the motor moves or if the Infrared Sensor is triggered after the robot says "LEGO."*

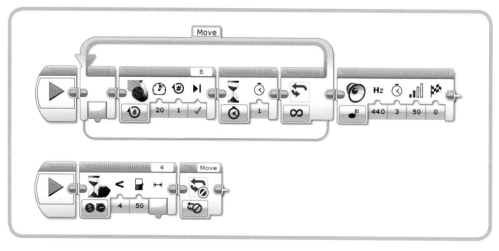

Figure 14-28: The BreakFromOutside *program interrupts a loop called* Move *when the Infrared Sensor is triggered.*

Now change the Duration setting of the Sound block to 0.1 seconds and run the program again. If you trigger the sensor when the motor is halfway through its rotation, the program ends almost immediately, stopping the motor before it completes its rotation.

I recommend that you use this technique with caution. Interrupting a loop that runs in parallel can make the robot

behave unpredictably. (What happens, for instance, if you have motor B make another movement just after the loop? Will the previous block continue, or will the new block run?)

Note that interrupting the loop *from the inside* doesn't introduce ambiguity as to what the program might do: The *BreakFromInside* program interrupts the loop, but it doesn't interrupt another block that's running.

DISCOVERY #81:
INTERRUPTING
INTERRUPTS!

Difficulty: ⬜⬜ **Time:** 🕐

Can you create a program like the *BreakFromInside* program that continuously moves the motor and plays a sound until the Touch Sensor is pressed *and* the Color Sensor sees something green? Triggering these sensors simultaneously should end the loop after the motor moves. (You'll learn about another way to make a loop end when multiple sensors are triggered in the next chapter, but for now, just use the Loop Interrupt block.)

HINT Add a Switch block to the *BreakFromInside* program (see Figure 14-27).

further exploration

In this chapter, you learned to use data wires to carry information from one block to another. In addition, you've learned to read sensor values with Sensor blocks and to work with the advanced features of Wait, Loop, and Switch blocks.

Most of the programs you've made so far with data wires are fairly small, and data wires may not seem very useful yet. However, data wires are essential to programming advanced robots like the ones you'll build in Part VI. The following Discoveries will let you practice the programming skills you gained in this chapter.

DISCOVERY #82:
SENSOR PRACTICE!

Difficulty: ⬜⬜ **Time:** 🕐🕐

Create a program that makes the white dial turn at a speed based on the Infrared Sensor's proximity, but only while you press the Touch Sensor and touch the Color Sensor simultaneously. If either of these sensors isn't triggered, the motor shouldn't move.

HINT You've learned a lot about data wires and Sensor blocks in this chapter, but sometimes you'll still need to use Wait, Loop, or Switch blocks to work with sensors.

DISCOVERY #83:
POWER VS. SPEED!

Difficulty: ⬜⬜ **Time:** 🕐

Create a program that makes motor B turn at *30% speed* (51 rpm) using a Large Motor block, and motor C turn at *30% power* using an Unregulated Motor block. Then, continuously display the speed of both motors using Motor Rotation blocks in Current Power mode. (Recall that this mode gives you the motor *speed*, as discussed in Chapter 9).

Now observe what happens as you try to slow down the motors with your hands. You should see the speed of motor C drop quickly: 30% power isn't enough to overcome the friction applied by your hand. However, motor B keeps turning at almost 30% speed because the block makes the EV3 apply additional power to the motor when it is being slowed down.

DISCOVERY #84: REAL DIRECTION!

Difficulty: ⬜⬜ **Time:** 🕐

Can you make a program that displays the Beacon Heading value if a beacon signal is detected, and can you make it display "Error!" if no signal is detected?

HINT Use the *LogicSwitch2* program (see Figure 14-26) as a starting point.

DISCOVERY #85: SK3TCHBOT IS WATCHING YOU!

Difficulty: ⬜⬜ **Time:** 🕐

Create a program that counts the number of people walking by your robot. Place your robot in such a way that the Infrared Sensor can sense people in front of it. Place two Wait blocks inside a Loop block, and configure the first block to wait until the sensor sees someone passing by; use the second one to wait until the person is out of sight. Ultimately, the loop should go around once each time it senses someone going past. Now when you display the Loop Index on the EV3 screen, you're displaying how many people have walked by. To ensure that your program works properly, place a Sound block inside the loop so that you'll hear a sound each time someone passes by.

DESIGN DISCOVERY #25: BIONIC HAND!

Building: ✶✶✶ **Programming:** ⬜⬜

Can you design a robotic claw that you can attach to your arm? Use sensors and the EV3 buttons to control the movements of the claw. Add the Infrared Sensor and program the robot so that the arm will warn you when the sensor sees that you're approaching a wall. Use the programming techniques you learned in this chapter to display sensor values on the screen or to make sounds based on sensor measurements.

DISCOVERY #86: OSCILLOSCOPE!

Difficulty: ⬜⬜⬜ **Time:** 🕐🕐🕐

Can you turn your EV3 brick into a measurement device that records sensor measurements on the screen, as shown in Figure 14-29? Display proximity measurements as small circles whose *x* coordinate is determined by the Loop Index and whose *y* coordinate is determined by the proximity measurement. As the loop progresses, you should see new measurements appear one by one to form a plot like the one shown in the figure.

Because the screen has 178 pixels in the *x* direction, the Loop block should run 178 times to display 178 circles, with a 0.05-second pause between each measurement. Then, the program should erase the screen and start over.

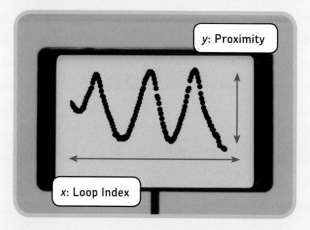

Figure 14-29: The plot of Discovery #86. This particular pattern of measurements could result from repeatedly moving an object back and forth in front of the sensor.

HINT To have a Display block empty the screen after 178 measurements, make it display a rectangle that fills the screen with *Color* set to white (true).

15

using data operations blocks and my blocks with data wires

Now that you know how data wires work, you can do some really interesting things with more of the programming blocks. For example, you can make the EV3 combine and process sensor values so that they can be used as input values for other actions. With the tricks you learn in this chapter, you'll be able to program your robot to choose a random action or do something only when two sensors are triggered at the same time, rather than performing strictly preprogrammed actions.

You'll learn in this chapter how to use the Math block to allow your robots to make calculations you can use in your programs. For example, the robot can calculate the distance it should travel based on a sensor reading. I'll also introduce you to a few new programming blocks, such as the Random, Compare, and Logic blocks, and show you how to make My Blocks with inputs and outputs.

These techniques and programming blocks are essential to the advanced robot programs you'll create in Part VI as well as for advanced programs that you'll create for your own robots. As in Chapter 14, you'll use SK3TCHBOT to test the new programs in this chapter.

The Discoveries in this chapter might be a bit challenging at first, but they will help you master essential programming skills, which will let you create much more interesting programs and build some really smart robots!

using data operations blocks

The Programming Palette contains a series of blocks that you have not used yet: *Data Operations blocks* (see Figure 15-1). Data Operations blocks include the Math block, the Random block, the Compare block, and the Logic Operations block. Each block has its own function, but they all process values carried by data wires and generate new values based on the input values. This section will explain how to use these blocks in your programs.

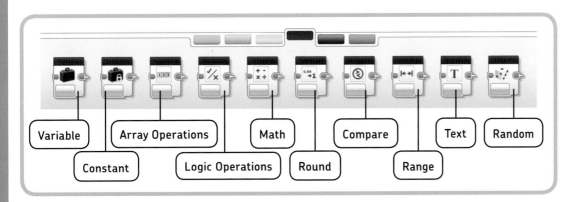

Figure 15-1: Data Operations blocks

the math block

The *Math block* (see Figure 15-2) allows the EV3 to do arithmetic operations, such as addition, subtraction, multiplication, and division. You enter two numbers in the *a* and *b* fields on the block, and you use the Mode Selector to choose which operation, such as division, should be applied to them (in the case of division, *a* will be divided by *b*). A data wire outputs the result. Instead of entering numbers for *a* and *b* yourself, you can supply them with a data wire.

the math block in action

The program in Figure 15-3 shows the Math block performing multiplication on the Color Sensor's measurement to control the speed of motor B. The Color Sensor value (a number between 0 and 7) isn't suitable to control the speed because it would make the motor turn very slowly (7% speed at best). The Math block multiplies the sensor value by a factor of 10 and sends the result (a number between 0 and 70) to the Motor block's Power setting. As a result, the motor moves at 10% speed for black, 20% speed for blue, and so on, up to 70% speed for brown.

Create a new project called *SK3TCHBOT-Data* with a program called *MathSpeed*, as shown in Figure 15-3.

DISCOVERY #87: 100% MATH!

Difficulty: 🔲 **Time:** ⏱

The Math block in the *MathSpeed* program makes it possible to use the detected color to control the motor's speed, but the speed only goes up to 70%. Can you modify the program to use the full range of the motor speed values? Add a Display block to your program to verify that the Math block's output reaches 100% when brown is detected.

HINT It's possible to enter decimal numbers in the *a* and *b* fields of the Math block.

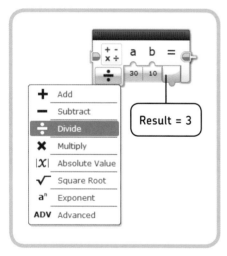

Figure 15-2: The Math block. The output value of this block is 30 ÷ 10 = 3.

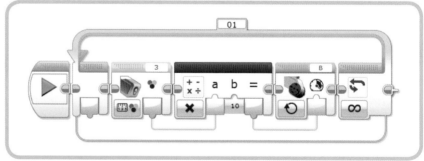

Figure 15-3: The MathSpeed *program*

using advanced mode

Some calculations require more than one arithmetic operation. For example, you might want to subtract two numbers and multiply the result by a third number. You could use two Math blocks to do this (one for subtraction and one for multiplication), but you can perform both calculations using just one Math block in *Advanced* mode, as shown in Figure 15-4.

In the *Equation* setting, you enter the calculation that the Math block should do just as you would enter it on a pocket calculator. For example, entering (7-3)*1.5 will result in an output of 6. (Just as on a calculator, you use parentheses to ensure that the subtraction is done before the multiplication.)

You can also use symbols in the equation (*a*, *b*, *c*, or *d*) and type a value for each symbol in the *a*, *b*, *c*, and *d* settings. For example, entering (b-c)*a as the equation and entering 7 for *b*, 3 for *c*, and 1.5 for *a* gives the same output result: 6. Finally, you can supply values for each symbol using a Numeric data wire.

To see the Advanced mode in action, you'll create a program that makes motor C (red dial) follow the movements of motor B (white dial), which you turn manually. Because motor C is mounted upside down in your robot, the dial will actually turn in the opposite direction, but it will turn by the same amount. To accomplish this, you'll set the speed of motor C as follows:

Speed of motor C = (Degr. of motor B – Degr. of motor C) × 1.5

To see why this formula works, consider what happens when you turn motor B forward by 70 degrees while motor C is at 60 degrees. The speed of motor C will be set to (70 – 60) × 1.5 = 15, making motor C turn forward to catch up with B. If C is ahead of B, the result is negative and motor C turns backward. The greater the difference between the motor positions, the faster motor C moves. Once both motors are in the same position, the result is 0, making the motor stop.

Create the *PositionControl* program shown in Figure 15-5. Turn the white dial manually, and watch the red dial follow the same movement in the opposite direction.

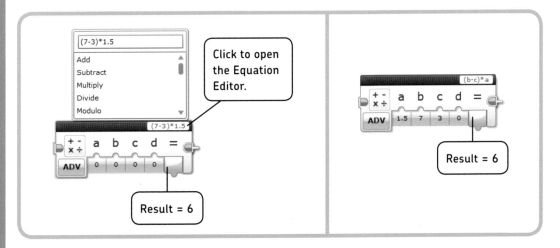

Figure 15-4: The Math block in Advanced mode. You can enter a formula with numbers (left), with variables (right), or a combination of both. You can enter operations such as * for multiplication and / for division manually, or you can select operations from the list.

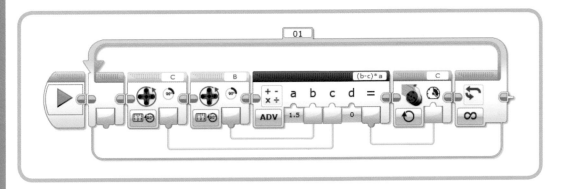

Figure 15-5: The PositionControl program. The Degrees output of motor B is connected to input b on the Math block; the Degrees output of motor C is connected to input c.

practicing with the math block

Because the Math block is an essential component of many programs that use data wires, it's a good idea to practice with it before you continue. These Discoveries will get you started.

DISCOVERY #88: ADDED VALUE!

Difficulty: ⬚ **Time:** ⏱

Can you create a program that continuously displays both an Infrared Sensor's proximity measurement and a Color Sensor's reflected light intensity measurement on the screen as well as the sum of the two?

DISCOVERY #89: INFRARED SPEED!

Difficulty: ⬚⬚ **Time:** ⏱⏱

Create a program similar to the *MathSpeed* program to control the speed *and* direction of motor B using the Infrared Sensor's proximity measurement. The motor should turn at 50% speed for 100% proximity, stand still for 50% proximity, and rotate at –50% speed (reverse) for 0% proximity.

HINT Use this formula: Speed = Proximity – 50. How do you configure a Math block to perform this operation?

DISCOVERY #90: DOUBLE INFRARED SPEED!

Difficulty: ⬚ **Time:** ⏱

Can you expand the program from Discovery #89 to make the motor speed range between –100 and 100?

HINT Use this formula: Speed = (Proximity – 50) × 2.

DISCOVERY #91: GAIN CONTROL!

Difficulty: ⬚⬚ **Time:** ⏱

What is the effect of the 1.5 value in the *PositionControl* program? Experiment by changing the value to a low number (0.1) or a high number (5) and observing how fast motor C can follow your movements.

DISCOVERY #92: DIRECTION CONTROL!

Difficulty: ⬚⬚⬚ **Time:** ⏱⏱

Can you modify the *PositionControl* program to have the red dial turn in the same direction as the white dial?

HINT Multiply the Degrees value of motor C by –1 so that forward movement is measured as backward movement, and vice versa. Likewise, multiply the resulting speed by –1 before passing it to the Large Motor block to reverse the direction.

the random block

The *Random block* allows you to generate a random value to use in your program. In *Logic* mode, the output is a Logic data wire carrying either true or false. Using the block in this mode is like tossing a coin, with the result being either true (heads) or false (tails). For a coin flip, the probability of getting true is 50%, resulting in true about half of the time. The Random block enables you to specify the likelihood of getting true using the Probability of True (%) setting. For example, entering 33 should result in true roughly one-third of the time, while the output would be false two-thirds of the time.

In *Numeric* mode, the output is a Numeric data wire carrying a random integer number in the range specified by the Lower Bound and Upper Bound settings. For example, setting 1 as the lower bound and 6 as the upper bound should result in 1, 6, or any whole number in between, with each number equally likely to occur, as when rolling a die.

The Random block is useful when you want your robot to do something unexpected. For example, you could use a Random block to randomly say "Left" or "Right" or to make the white dial turn at a random motor speed, as shown in the *RandomMotor* program in Figure 15-6.

The Switch block in the *RandomMotor* program runs either the block at the top (true) or the blocks at the bottom (false). You can have the program randomly choose between more than two cases by setting the Random block and the Switch block to Numeric mode. This allows you to specify actions for each possible random value, as demonstrated by the *RandomCase* program in Figure 15-7.

DISCOVERY #93: RANDOM FREQUENCY!

Difficulty: ☐ **Time:** ⏱

Can you make the EV3 play a random tone for half a second each time you press the Touch Sensor? Generate a random value and wire it into the Frequency plug of a Sound block. Finally, expand the program to also display the frequency on the EV3 screen.

HINT Which values does the Frequency input on the Sound block accept?

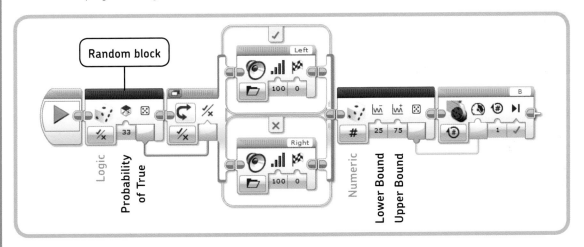

Figure 15-6: The RandomMotor *program. If you run the program many times, you should find that the robot says "Left" roughly one-third of the time. After the sound, motor B turns for one rotation at a random speed between 25% and 75%.*

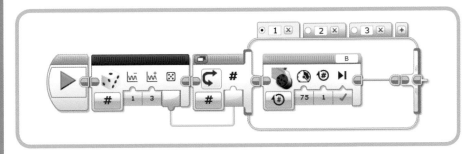

Figure 15-7: The RandomCase *program randomly chooses to run the blocks on the first, second, or third tab. (Only the block on the first tab is shown; just add any blocks you like to the other tabs.)*

the compare block

The *Compare block* checks to see whether a Numeric value is Equal To (=), Not Equal To (≠), Greater Than (>), Greater Than Or Equal To (≥), Less Than (<), or Less Than or Equal To (≤) another Numeric value. You can enter the values you'd like to compare in the block's settings (*a* and *b*), or you can supply the values to the block with data wires.

The Compare block outputs one Logic data wire (true or false) based on the result of comparing value *a* to value *b*. For instance, if you set the mode to Equal To (=), the block outputs true if value *a* is equal to *b*.

The *CompareValues* program in Figure 15-8 shows the Compare block in action. In this program, the robot waits until the proximity measurement drops below 80. The measurement that made the Wait block stop waiting is transferred to the Compare block, which determines whether it is less than 40. If so (true), SK3TCHBOT says "Down"; if not (false), it says "Up."

As a result, the robot says "Down" if you quickly place your hand in front of the sensor, approaching it from the side, but it says "Up" if you slowly approach from the front.

DISCOVERY #94: RANDOM MOTOR AND SPEED!

Difficulty: ⬜⬜⬜ **Time:** 🕐

Create a program that generates a random number between 10 and 100 to control the speed of either motor B or motor C. If the random value is less than 50, motor B should turn for one rotation with its Power setting controlled by the random value; if not, motor C should turn for one rotation at the random speed.

HINT You'll need a Random block, a Compare block, a Switch block (choose *Tabbed View*), and two Large Motor blocks.

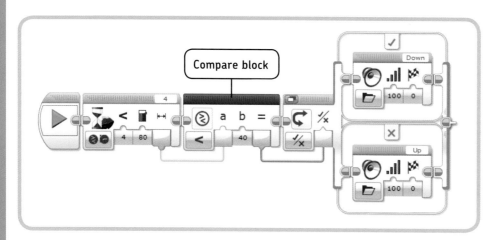

Figure 15-8: The CompareValues *program*

the logic operations block

The *Logic Operations block* compares two Logic data wire values and outputs the result with a Logic data wire. In *And* mode, it checks to see whether both input values (*a* and *b*) are true. If so, the Result output is true. If one or both inputs are false, the output is false.

You can use the Logic Operations block in this mode to create a program that draws a filled circle on the screen if both the Touch Sensor is pressed *and* the proximity measurement is less than 50%. If either or both conditions don't occur, the display shows an empty circle because the Logic Operations block's output is false. Figure 15-9 shows the *LogicAnd* program.

logic operations

When configuring the Logic Operations block, you can select one of four modes: And, Or, XOR, and Not. Each option will make the block compare the logic input values differently. The option you choose will depend on what you want your program to do. Table 15-1 lists the available modes as well as the input values that make the output value true.

table 15-1: the modes of the logic block and their output values

Mode		Output value is true when...
And	(A·B)	Both inputs are true
Or	(A·B)	One or both inputs are true
XOR	(A·B)	One input is true and the other is false
Not	(A)	The input is false

In *Or* mode, the block's output is true if either or both input values are true, as demonstrated by the *LogicOr* program shown in Figure 15-10. The program repeatedly checks whether the Touch Sensor or the Infrared Sensor is triggered. If one or both are triggered, the Result output is true, causing the loop to end. Once the loop ends, the Sound block plays a tone. This technique is useful because it allows you to make your program wait until at least one of several sensors is triggered.

Now change the mode of the Logic Operations block to *XOR*. When you run the program, you should hear the sound when either the Touch Sensor or the Infrared Sensor is triggered, but the sound won't play if both sensors are triggered at the same time.

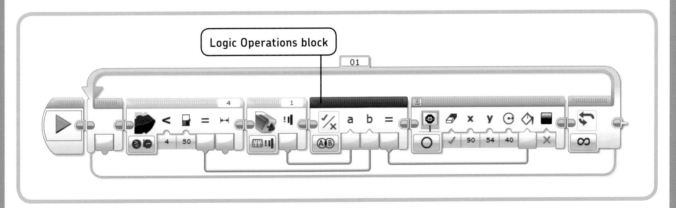

Figure 15-9: The LogicAnd *program*

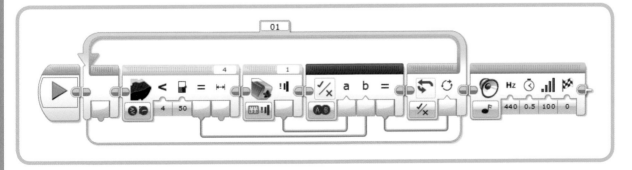

Figure 15-10: The LogicOr *program*

DISCOVERY #95: LOGIC SENSORS!

Difficulty: ⬜⬜ Time: ⏱

The *LogicOr* program plays a sound when the Touch Sensor or the Infrared Sensor is triggered, but it doesn't tell you which sensor caused the loop to end. Can you expand the program to say "Touch" if the Touch Sensor ended the loop and "Detected" if the Infrared Sensor was involved?

HINT Place a Switch block in *Logic* mode just after the Loop, and use the output of the existing Touch Sensor block to control the switch.

DISCOVERY #96: WAIT FOR THREE SENSORS!

Difficulty: ⬜⬜ Time: ⏱

Can you make a program that plays a sound when one of three sensors is triggered? Make the program wait for the Touch Sensor to be pressed, the proximity measurement to drop below 50%, or the reflected light intensity to go above 15%.

HINT Start with the *LogicOr* program, but add a Color Sensor block and another Logic Operations block. How do you connect the data wires?

not mode

When you select *Not* mode, the Logic Operations block has only one input. This mode just *inverts* the input signal: If the input value (*a*) is true, the output will be false; if the input value is false, the output will be true.

To see this mode in action, modify the *LogicOr* program you just made by removing the Infrared Sensor block, and change the mode of the Logic Operations block to Not. The output of the Logic Operations block will be false if the Touch Sensor is pressed and true if it's released so that the loop runs until you release the sensor.

the range block

The *Range block* determines whether a Numeric data wire carries a value within a range specified by a Lower Bound and Upper Bound. The boundary values themselves are considered to be *inside* the range.

When you select Inside mode, the block's Result output is true if the Test Value is in the range; the output is false if it's outside the range. When you select Outside mode, the opposite occurs: The Result output is true if the Test Value is outside the specified range; the output is false if it's inside the range.

The *SensorRange* program (see Figure 15-11) uses a Range block (Inside mode) and a Loop block (Logic mode) to wait for the Infrared Sensor's proximity value to attain a value of 40, 60, or somewhere in between.

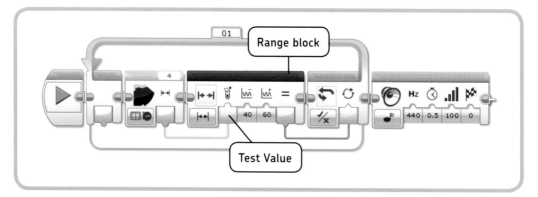

Figure 15-11: The SensorRange *program*

the round block

The *Round block* can turn a decimal input value into an integer by rounding it. You can select a mode to Round Up (1.2 or 1.8 becomes 2), Round Down (1.2 or 1.8 becomes 1), or round to the Nearest value (1.2 becomes 1, while 1.5 or 1.8 becomes 2). You can also choose Truncate mode to remove decimals from the number without rounding (1.877 becomes 1.8 if you choose to keep only one decimal).

The *RoundTime* program (see Figure 15-12) displays the time elapsed since the start of the program on the EV3 screen by rounding the decimal timer value (such as 3.508) down to a whole number of seconds (3).

We haven't used the *Timer block* in a program before, but its use is straightforward. The block measures time much like a stopwatch, starting at 0 when the program begins. A Timer block in Measure mode gives the current time as a decimal value, such as 1.500 for one and a half seconds. You can reset the timer to 0 by using the block in Reset mode.

You can use up to eight different timers in your program. For example, you can use timer 1 to measure how long the program has been running and timer 2 to measure how much time has passed since you pressed the Touch Sensor by resetting it each time you press the sensor. To specify which timer value you want to read or reset, you choose from one of eight *Timer IDs*. You used Timer ID 1 in the *RoundTime* program.

the text block

The *Text block* combines up to three Text data wire inputs into a single text line. For example, if the input text lines are "EV3 ", "is ", and "fun", the *Result* output value will be "EV3 is fun". You have to add a space after both "EV3" and "is", or the output will simply be "EV3isfun". If an input is empty, it will be ignored.

You can use the Text block to combine text and a number for display on the EV3 screen. For example, you can expand the previous program to say "Time: 41 s" rather than just displaying a number. To do so, you merge the word "Time: ", the numeric time value, and " s" and display the result on screen, as shown in the *TextTime* program in Figure 15-13.

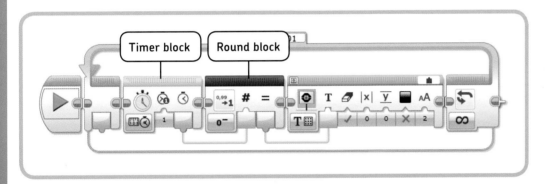

Figure 15-12: The RoundTime *program*

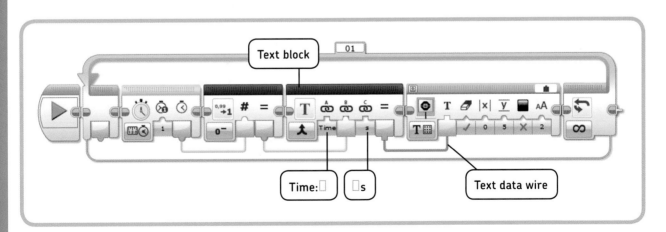

Figure 15-13: The TextTime *program. Don't forget the spaces in "Time: " and " s" (indicated here with blue boxes). After 41 seconds, the display should say "Time: 41 s".*

creating my blocks with data wires

Up to now, you've seen how to use data wires to send information between existing programming blocks, like Sensor blocks, Data Operations blocks, and Action blocks. You can also use data wires with your custom-made My Blocks. Doing so makes it possible to create My Blocks with *parameters* (inputs and outputs). For example, you can create a My Block to display the value passed to it by a data wire.

You learned how to create basic My Blocks without inputs or outputs in Chapter 5. In this section, you'll learn to create My Blocks with input and output plugs. You'll also learn about some strategies for creating useful My Blocks.

a my block with input

To get started, you'll create a My Block with two input plugs called *DisplayNumber*, as shown in Figure 15-14. The purpose of this block is to combine the information from the Text input called *Label* and the Numeric input called *Number* and display the result on the EV3 screen.

This block will make it easy to display a number on the screen with a label to describe it. For example, entering IR as the Label and 15 as the Number will make the block display IR: 15.

Create the My Block by following these steps:

1. Create a new program called *NumberTest*. You'll need it to test the My Block when it's ready.

2. Place and configure a Text block and a Display block on the canvas, as shown in Figure 15-15. You'll use the Text block to combine the Label input, a colon followed by a space, and the Number input into one text line, and you'll use a Display block to show it on screen. Then, select both blocks by dragging a selection around them with your mouse.

3. Open the My Block Builder by going to **Tools ▸ My Block Builder**.

4. Enter *DisplayNumber* as the My Block's name, enter a description, and choose an icon for your block, as shown in Figure 15-16.

5. Now, we'll configure the block's parameters. Add two inputs by clicking **Add Parameter** twice, as shown in Figure 15-17.

6. Open the Parameter Setup tab, select the first parameter, and configure it as a Text input called *Label*, as shown in Figure 15-18. Also, choose a descriptive icon, such as the T icon, for this input on the Parameter Icons tab.

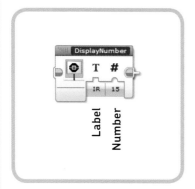

Figure 15-14: The DisplayNumber My Block. In this configuration, it displays IR: 15.

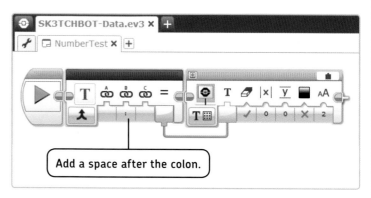

Figure 15-15: Configure a Text block and a Display block as shown and then select both blocks.

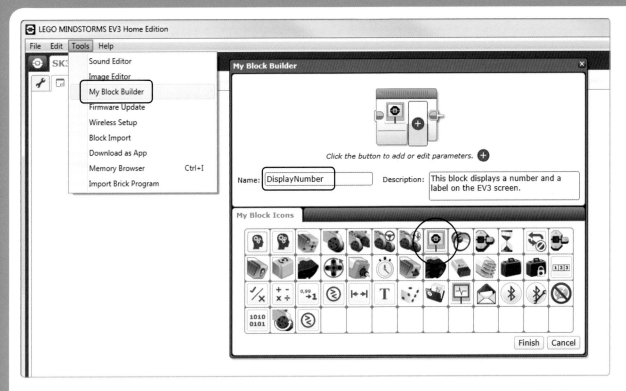

Figure 15-16: Open the My Block Builder, enter a name and a description, and choose an icon for the My Block. Because this block will display something on the EV3 screen, choose the icon with the EV3 screen.

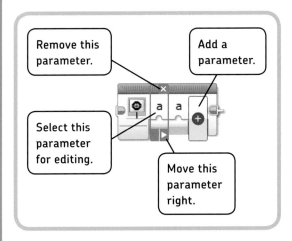

Figure 15-17: Adding, removing, and changing the order of parameters

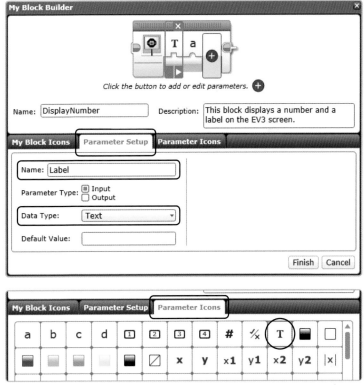

Figure 15-18: Configuring the Label input parameter

7. Now, select the second parameter, and configure it as a Numeric input called *Number*, as
shown in Figure 15-19. The Default Value is set
to 0, which means that the Number setting will
be 0 by default when you pick it from the Programming Palette later. Choose the standard
input plug in the Parameter Style option. This
creates a basic Numeric input plug that accepts
any value (like the inputs on a Math block),
rather than one with a slider that accepts only
a range of values (like the Power and Steering
settings on the Move Steering block).

8. Click **Finish**. You should now see the contents
of the DisplayNumber My Block on its own tab
inside your project, as shown in Figure 15-20.
You should also see the Label and Number
plugs, although the wires aren't connected yet.
The plugs carry the values *passed to the My
Block from the main program*. That is, if you
enter IR and 15, as shown in Figure 15-14,
Label carries IR and Number carries 15.

9. Complete the My Block by connecting the Label
and Number values to the Text block, as shown
in Figure 15-20, and save your project. (Recall
that the Text block combines the information
from its inputs into one text line. The Display
block shows it on the screen.)

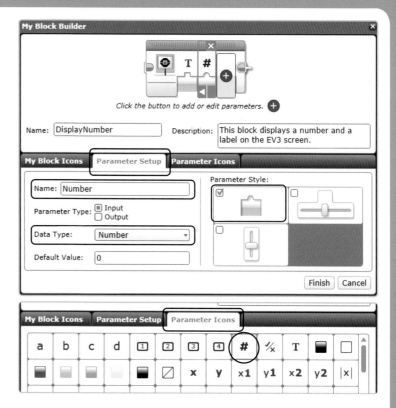

Figure 15-19: Configuring the Number input parameter

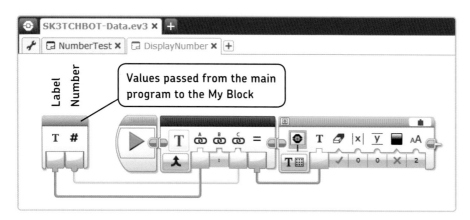

Figure 15-20: Once you click Finish in the My Block Builder, you'll see the DisplayNumber My Block on its own
tab in the SK3TCHBOT-Data project. Finish the block by connecting the wires as shown.

Congratulations, you've created a My Block with inputs! You'll find the My Block on the light blue tab of the Programming Palette. Now, finish the *NumberTest* program to test your newly created block, as shown in Figure 15-21. In this example, the main program (*NumberTest*) passes two values to the My Block (DisplayNumber). The blocks inside the My Block retrieve these values and display them on the EV3 screen.

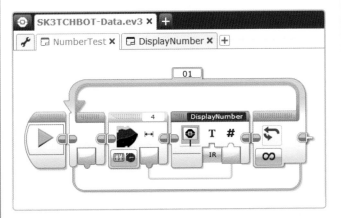

Figure 15-21: The NumberTest *program. The Infrared Sensor block sends the proximity measurement to the Number input plug of the DisplayNumber My Block, while you enter a value for the Label manually (IR). The My Block combines the Label with a colon, a space, and the sensor value, and it displays the result on the EV3 screen. For a measurement of 65, for example, the screen will show IR: 65.*

editing my blocks

Once you've created a My Block, you can alter its functionality by double-clicking the block and changing the blocks within it. For example, you can add a Sound block to the DisplayNumber block to have the robot beep each time it displays a number.

As of this writing, the LEGO MINDSTORMS EV3 software (version 1.10) doesn't allow you to edit the input and output parameters of a My Block after you create it with the My Block Builder. If you need to modify parameters or add new ones, you'll have to create a new My Block from scratch. (Once you've created the new block with its parameters, copy the contents of the old block and paste them into the newly created one to save time.)

DISCOVERY #98: MY UNIT!

Difficulty: ▢▢ **Time:** ◷◷
Create a new My Block based on the DisplayNumber My Block with an additional Text input called *Unit* (see Figure 15-22). Then append the unit of measurement to the number. For example, displaying the Rotation Sensor value by entering *MB* as the label and *Deg* as the unit should result in *MB: 375 Deg* (short for Motor B: 375 Degrees).

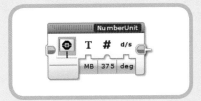

Figure 15-22: The My Block of Discovery #98

HINT You'll need an additional Text block.

DISCOVERY #99: ENHANCED DISPLAY!

Difficulty: ▢▢▢ **Time:** ◷◷
The DisplayNumber My Block you just made is useful only for displaying a single value at the top of the screen. Can you create a more useful version of this My Block with a setting to choose a line number and a setting to clear the screen, as shown in Figure 15-23?

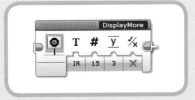

Figure 15-23: The My Block of Discovery #99

HINT There is room for up to six lines of text on the EV3 screen if you use the biggest font size. Use a Math block to convert the LineNumber input on the My Block into a value for the Row setting on the Display block: Row = LineNumber × 2.

a my block with output

Besides making My Blocks with inputs, you can also create your own blocks with output plugs. To see how this works, you'll create a My Block that detects the direction of moving objects with the Infrared Sensor. The *Direction* My Block has one Logic output plug called *Approaching*, as shown in Figure 15-24.

The output is true if an object is approaching the sensor (the distance to an object is decreasing); the output is false otherwise (either the distance to the object is increasing or unchanging). Create the Direction My Block and the *DirectionSound* test program using the steps that follow.

1. Create a new program called *DirectionSound*, and place two Infrared Sensor blocks, a Wait block, and a Compare block on the Canvas, as shown in Figure 15-25. These blocks take two proximity measurements 0.2 seconds apart and compare them to see whether the second value (*a*) is less than the first (*b*). If so, then the object is approaching the sensor, and the Compare block's output will be true.

2. Select each of the four blocks and go to **Tools ▸ My Block Builder**. Enter *Direction* as the block's name, and choose the Infrared Sensor as its icon, as shown in Figure 15-26.

3. Add one parameter, configure it as a Logic output called *Approaching*, and select an appropriate symbol, as shown in Figure 15-26. Then, click **Finish**.

4. To complete the My Block, connect the Result output of the Compare block to the Approaching plug, as shown in Figure 15-27. This value is then passed to the main program, which contains the Direction My Block.

Now, return to the *DirectionSound* program, and test your My Block by having the robot play a high-pitched tone when an object is approaching (true) and a low tone when the object is moving away or standing still (false), as shown in Figure 15-28. In this example, the blocks inside the Direction My Block calculate the direction of the moving object and pass the result to the main program (*DirectionSound*) through the Approaching plug.

Figure 15-24: The Direction My Block

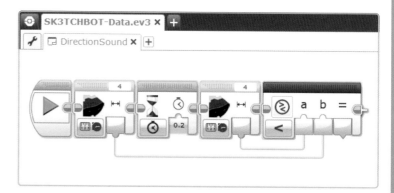

Figure 15-25: Configure the blocks for the Direction My Block as shown. When you're ready, drag a selection around the blocks and launch the My Block Builder.

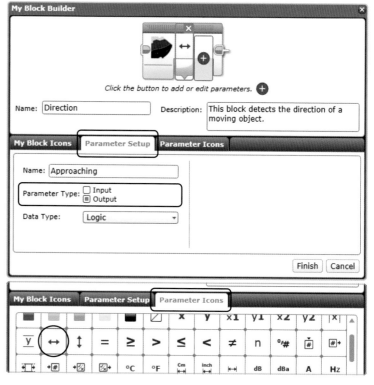

Figure 15-26: Configure the My Block and its Logic output parameter as shown.

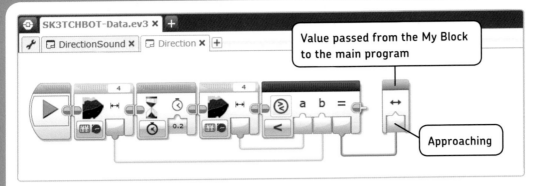

Value passed from the My Block to the main program

Approaching

Figure 15-28: Connect the Result output of the Compare block to the Approaching plug.

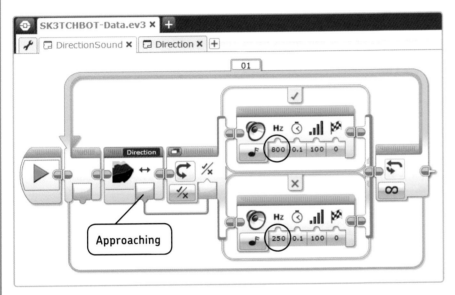

Approaching

Figure 15-27: The DirectionSound program uses the Logic value from the Approaching output to play a high-pitched sound if an object is approaching and a low tone if it's not approaching.

DISCOVERY #100: PROXIMITY AVERAGE!

Difficulty: ⬚⬚ **Time:** 🕐🕐

Create a My Block that calculates the average of two proximity measurements. Then make it output the result with a Numeric data wire. As in the Direction My Block, the measurements should be taken 0.2 seconds apart.

DISCOVERY #101: PROXIMITY RATE!

Difficulty: ⬚⬚ **Time:** 🕐🕐

Create a My Block with one Numeric output that determines the speed of a moving object by calculating the rate at which the proximity value changes. To accomplish this, have the block take two sensor measurements, as in the Direction My Block, and calculate the rate of change as follows:

$$\text{Speed} = \frac{\text{Second measurement} - \text{First measurement}}{\text{Time between measurements}}$$

To test the block, display the speed value on the screen, and change the brick status light to green if an object is approaching, orange if the object is standing still, and red if the object is moving away from the sensor. How can you determine the direction using the speed value?

a my block with input and output

In the final example, you'll create a My Block with both an input and an output plug, as shown in Figure 15-29. The *IsEven* My Block determines whether a Numeric input value called *Number* is even. If it is, a Logic output called *Even* carries out true; if it's not, it carries out false. Create the block using the following steps:

1. Create a new program called *EvenSound*, and place a Math block and a Compare block on the Programming Canvas. (For now, just leave their settings unchanged.)

2. Select the two blocks you've just placed, and open the My Block Builder. Enter *IsEven* as the block's name, and choose the icon shown in Figure 15-29.

3. Add one Numeric input plug called *Number* and one Logic output plug called *Even*, and configure the icons as shown in Figure 15-29. Click **Finish**.

4. Now, complete the My Block by configuring the two blocks and connecting the data wires as shown in Figure 15-30. The Math block uses the modulo operator (%), which calculates the remainder after dividing two numbers. In this case, it tells us the remainder of dividing the input value *a* by 2. Even numbers are divisible by 2, so the remainder will be 0 and the Compare block will carry out true. For odd numbers, the remainder isn't 0, causing the Compare block to output false. (For example, 7%2 gives a remainder of 1, and the Compare block outputs false, which tells us the input is an odd number.)

Now that you've finished creating the My Block, test it by completing the *EvenSound* program, as shown in Figure 15-31. The program displays a random number between –100 and 100 on the EV3 screen, and it determines whether the number is even or odd using the IsEven My Block. The robot says "Yes" for even numbers and "No" for odd numbers.

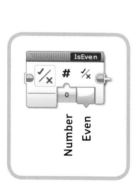

Figure 15-29: The IsEven My Block

Figure 15-30: The configuration of blocks in the IsEven My Block

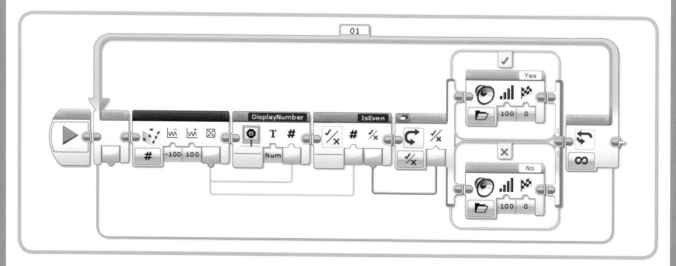

Figure 15-31: The EvenSound program

strategies for making my blocks

Creating My Blocks with inputs and outputs allows you to create a wide variety of blocks. My Blocks can be useful for the following:

Repetitive tasks: There are some things you'll do in many of your programs, such as displaying values on the screen. If you create a My Block for such a task once, you'll save time in the long run.

Program organization: You can break down large programs into My Blocks to make it easier to understand your program and to test parts of it individually. For example, if you were to create a program for an autonomous robotic arm, you could create one My Block to localize an object, another to drive toward it, and a third to lift and move the object.

Processing information: As you create programs that process values with Data Operations blocks, it can be difficult to understand how your program really works when you look at it later. To resolve this, you can break down complex calculations into smaller ones, grouped into My Blocks. For example, in the *EvenSound* program (see Figure 15-31), hiding the modulo operation in the IsEven My Block makes it easier to see how the program works.

starting points for my blocks

You can begin creating a My Block in two ways:

Convert a selection of blocks into a My Block: Use this method if you have already configured and tested the blocks in your program. This is a useful technique to break a large program into subsections, each with its own functionality. If your selection of blocks is connected to other blocks in your program with data wires, you should configure a My Block parameter for each data wire. In fact, the My Block Builder automatically configures such parameters, but you'll still need to add a descriptive name and icon for each one.

Create a My Block from scratch: Use this method if you know *what* your My Block should do but are not yet sure *how* to configure the blocks inside it to accomplish the task. For example, if you want to create a block that determines whether a number is even or odd but don't know how it will work exactly, you could start by making a My Block with one Numeric input and one Logic output. Once it's created, experiment with blocks inside the My Block until you accomplish the required task. (Note that you can't start the My Block Builder without selecting any blocks, so you'll have to start from a dummy block, like a Wait block, and delete the dummy block later.)

sharing my blocks between projects

When you create a My Block, it will be accessible for use only within your current project. To use it within another project, you'll have to copy or export it from your current project and paste or import it into the other. (See "Managing My Blocks in Projects" on page 53.)

further exploration

In this chapter, you learned about Data Operations blocks as well as how to create My Blocks with input and output parameters. These techniques will allow your robot to process and combine sensor values in order to use them as inputs for actions such as sounds and movements. You'll see many practical examples of this in the remaining chapters of this book. For now, practice your newly acquired skills with the following Discoveries.

DISCOVERY #103:
IS IT AN INTEGER?

Difficulty: ⬜⬜⬜ **Time:** 🕐🕐

Can you create a My Block that determines whether a Numeric input value is an integer (a whole number) rather than a decimal? Create a My Block called *IsInteger* with parameters similar to those of the IsEven My Block, but use different blocks on the inside.

HINT Round the input value and compare it to the original input. What does it mean if the input is equal to its rounded value?

DISCOVERY #104:
DOUBLE STALLED!

Difficulty: ⬜⬜⬜ **Time:** 🕐🕐

Can you create a program that has both motor B and motor C turn until either one of them is stalled? When you're ready, turn the blocks that make the program wait until a motor is stalled into a My Block called *WaitForStall* so you can use it in any of your programs. (It would be especially useful for vehicles like EXPLOR3R and the Formula EV3 Racer).

HINT Part of your program is similar to the *LogicOr* program in Figure 15-10. Use two Motor Rotation blocks in Compare – Current Power mode.

NOTE Remember that you can find the solutions to many of the Discoveries on the book's companion website (*http://ev3.robotsquare.com/*).

DISCOVERY #105:
REFLEX TEST!

Difficulty: ⬜⬜⬜ **Time:** 🕐🕐🕐

Can you create a program to test your reaction time? Make the brick status light turn green for a random amount of time and then turn red. As soon as you see the red light, you should press the Touch Sensor. The program should then display the time it took for you to see the red light and press the sensor. Place the whole program in a loop to see whether you can improve your reaction time! When you're ready, expand the program to prevent people from cheating by pressing the Touch Sensor early.

HINT The actions of the program are as follows: Set the light to green. Wait for a random amount of time. Set the light to red. Reset the timer. Wait for a Touch Sensor press. Display the timer value on the screen. Wait 3 seconds so you can see the value.

DESIGN DISCOVERY #26:
ROBOT CLOCK!

Building: ✸✸✸ **Programming:** ⬜⬜⬜

Can you build your own working EV3 clock? Use the three motors in the EV3 set to control the hour, minute, and second hands on your clock. Use the timer block and calculate the position of each hand based on the number of seconds elapsed since the start of the program.

HINT Multiply the timer value by a factor of 10 or more while testing the robot to make it easier to test the behavior of the hour hand.

16

using constants and variables

If you've made it this far, you're just a few steps away from mastering all of the programming skills in this book. You've learned how to use many different programming blocks as well as how to work with essential tools, such as data wires. This chapter completes the programming section of this book by teaching you how to use constants as well as how to use the EV3's memory with variables.

using constants

The *Constant block* provides a starting point for a data wire whose value you can choose manually. You choose the data wire type by selecting a mode (Text, Numeric, or Logic), and you enter its value in the Value field, as shown in Figure 16-1.

The Constant block is useful if you want to configure the settings of multiple blocks with the same value. For example, the *ConstantDemo* program (see Figure 16-1) makes the motors on SK3TCHBOT turn at the same speed, but in opposite directions. To accomplish this, the Constant block sends

the number 50 to both the first Large Motor block and the Math Block, which multiplies the value by –1 before passing it on to the second Large Motor block. To change the speed of both motors at once, simply change the value in the Constant block. (Without this block, you'd have to change two values).

using variables

Think of a *variable* as a kind of suitcase that can carry information. When a program needs to remember a value (such as a sensor reading) for later use, it puts that value in the suitcase and stows it away. When the program needs to use the value, it opens the suitcase and uses the stored value. The variable is stored in the EV3's memory until it's needed.

Once information is stored in a variable, you can access it from other parts of your program. For example, you could store an Infrared Sensor's proximity value in the suitcase and use it to control the speed of a motor 5 seconds later.

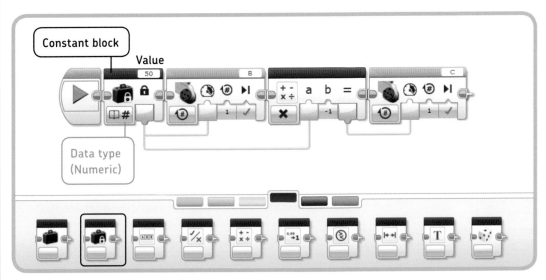

Figure 16-1: The ConstantDemo program. (Create a new project called SK3TCHBOT-Variable for all of the programs in this chapter.)

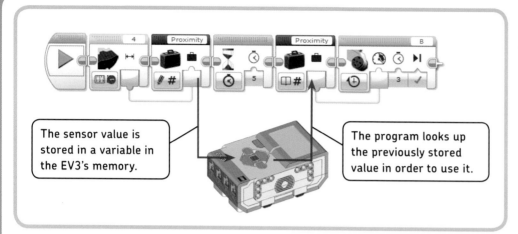

The sensor value is stored in a variable in the EV3's memory.

The program looks up the previously stored value in order to use it.

Figure 16-2: Values, like the proximity reading shown here, can be stored in variables in the EV3's memory. (You'll learn what this program does and how to create it later in this chapter.)

The program can access this stored information at any time while it's running, but the data is lost once the program stops. To store and access variable information, use a Variable block, which you'll recognize by the suitcase icon on the block. Figure 16-2 shows an overview of what happens when you use variables.

defining a variable

Each variable has a name, has a data type, and contains a value. For example, a Numeric-type variable might be called *Proximity* and have a value of 56. Besides Numeric variables, there are Logic variables (containing true or false) and Text variables (containing a text line, such as *Hello*).

But before you can use a variable in your program, you'll need to *define* it by choosing a name and a type for the variable. You can do this on the Project Properties page, as shown in Figure 16-3, or you can use the Variable block to define a variable, as you'll see in a moment. When you define a variable in a project, you can use that variable in each program inside the project.

To delete a variable, open the Variables tab on the Project Properties page, select the variable you want to delete, and click **Delete**.

using the variable block

Once you've defined a variable, you can use it in a program with the *Variable block*. The Variable block can either read values from or write (store) values to a variable in the EV3 memory, as shown in Figure 16-4.

To configure a Variable block, first use the Mode Selector to choose whether to read from a variable (Read; book icon) or write a value to it (Write; pencil icon). Next, select the type of variable you want to read or write (Numeric, Logic, or Text). Finally, select the variable you want to use from the Name list.

Figure 16-3: Defining a variable on the Project Properties page. Step 1: Open the Project Properties page. Step 2: Open the Variables tab. Step 3: Click **Add** to make a new variable. Step 4: Enter a name for the variable (Proximity) and select the data type (Numeric). Then, click **Ok**.

NOTE The Name list contains only the variables with the type that matches the block's mode. For example, to see the Numeric variable you've just defined (*Proximity*), the block must be in Read – Numeric or Write – Numeric mode.

The Variable block has one parameter called *Value*. In Write mode, the Value input allows you to enter the value you want to store. If you connect a data wire rather than entering a value, the value carried by this wire is stored. If a value was stored in this variable previously, the old value is erased, and the new one is stored instead.

In Read mode, the Variable block retrieves the information from the EV3's memory and outputs it with the Value plug so that you can send it to other blocks with a data wire. When a variable's value is read, the value doesn't change, so if you read it again with another Variable block, you'll get the same value.

defining variables with the variable block

A second way to define a variable is to click **Add Variable** in the Name list in the Variable block, as shown in Figure 16-5. Doing so creates a new variable of the same type as the block's mode. For example, because this block is in Read – Numeric mode, the new variable will be Numeric. (To create a new Logic variable using this method, first change the mode to Read – Logic or Write – Logic.)

creating a program with a variable

Now that you know the essentials of defining and using variables, you're ready to create the *VariableDemo* program in Figure 16-6. The program stores the Infrared Sensor's proximity value in a variable called *Proximity*. After 5 seconds, it retrieves the value from the variable and uses it to control the speed of motor B, which means that the motor speed is based on what the sensor measured 5 seconds ago. (Before configuring this program, define a variable called *Proximity* if you haven't already done so, as shown in Figure 16-3.)

VariableDemo demonstrates the concept of using variables, but it's a very basic program. Once you've created it, continue practicing with variables in Discoveries #106 and #107 on the next page.

NOTE Don't confuse the Variable block with the Constant block. Both blocks have a suitcase icon on them, but the Constant block also shows a padlock, which serves as a reminder that constants can't change while a program is running.

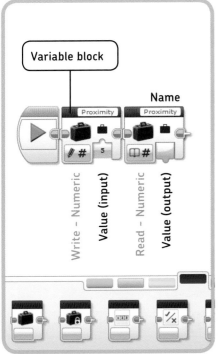

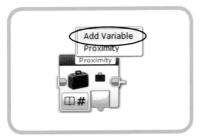

Figure 16-5: To define a variable directly from a Variable block, click **Add Variable**. Enter a name for your variable in the dialog that appears and then press **Ok**.

Figure 16-4: Storing and reading values with the Variable block

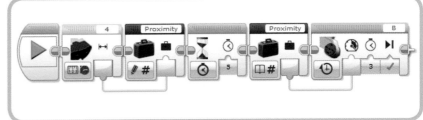

Figure 16-6: The VariableDemo program

DISCOVERY #106:
OLD VS. NEW!

Difficulty: ⬜ **Time:** ⏱

Can you create a program that repeatedly compares new sensor readings to the sensor value stored in a variable at the beginning of the program? If the new reading is lower, the robot should say "Yes"; otherwise, it should say "No." Figure 16-7 shows part of the program.

HINT What is your first step when creating a program with variables? Which Variable blocks must read or write a value? How do you connect the data wires?

DISCOVERY #107:
PREVIOUS VS. NEW!

Difficulty: ⬜⬜ **Time:** ⏱⏱

The program in Discovery #106 compares the new sensor value to the sensor value measured at the start of the program. Can you create a new program that compares each new measurement to the previous one? If the new one is lower, the robot should say "Yes"; otherwise, it should say "No." As a consequence, the robot will say "Yes" whenever an object is approaching the Infrared Sensor from the front.

HINT Create a program similar to the one shown in Figure 16-7, with one variable called *Previous*. Each time through the loop, the robot should compare the *Previous* value to a new sensor reading. The last block in the loop should be a Variable block configured to store the newer sensor value in *Previous*; this way, the next time through the loop, *Previous* contains the measurement found during the previous repetition of the loop.

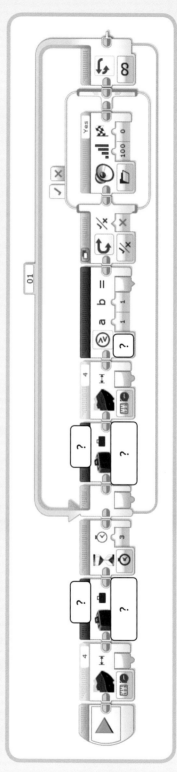

Figure 16-7: A starting point for Discovery #106

changing and incrementing variable values

Sometimes you'll want to change a variable's value—for example, when you want to use a variable to track a high score or a total number of Touch Sensor presses. Often, you'll want to increase a variable's value by 1, which is called *incrementing*. The *TouchCount* program you'll now create demonstrates how to use a variable to track the number of times a Touch Sensor is bumped (pressed and then released).

You begin by defining a new Numeric variable called *Count* to store the number of Touch Sensor bumps. The program will wait until the Touch Sensor is bumped, at which point the *Count* value will increase by 1. To make the program keep counting, you'll also use a Loop block.

But how do you increase the variable's value by 1? As shown in Figure 16-8, you use a Variable block to read the *Count* value. Then, you transfer this value to a Math block, which adds one to the value. The result of this addition is wired into another Variable block configured to write (store) the new value in the *Count* variable, which is therefore now increased by 1. The Loop block repeats this process for 5 seconds, after which the total *Count* value is displayed on the EV3 screen. (The loop has to run at least once, which means that the final value will always be 1 or more.)

Using this method, you can change a variable's value however you want. This example has shown you how to add 1 to a value, but you could use the same method to subtract from a value, too.

initializing variables

When programming with variables, it's important to *initialize* them by giving them a starting value. You did this in the *TouchCount* program by setting the *Count* value to 0 at the beginning of the program. Initializing variables makes a program more reliable by making sure that each time you run the program, it will function the same way because it starts at the same place.

The starting value doesn't have to be 0, though. For example, initializing the *Count* variable to 5 makes the program begin counting at 5.

If you don't choose an initial value for a Numeric variable in your program, the EV3 software will initialize it to 0. However, it's good practice to always initialize variables in your programs, even if you want them to start at 0.

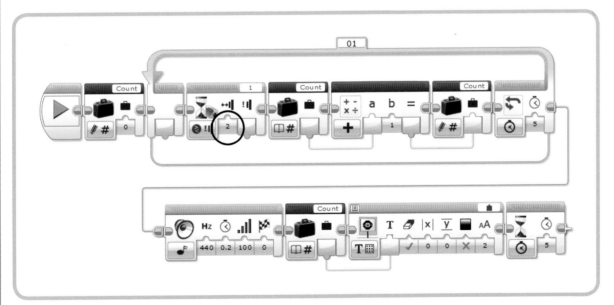

Figure 16-8: The TouchCount program counts the number of Touch Sensor presses during 5 seconds and displays the total number on the screen. Note that I connected the Sound block to the Loop block using a Sequence Wire for better visibility, but you should simply place the Sound block directly after the Loop block.

calculating an average

In the next example, you'll use a variable to calculate the *average value* of 50 sensor measurements. To calculate an average measurement (the mean value), divide the sum of all measurements by the number of measurements (50, in this case). The *Average* program in Figure 16-9 calculates this sum by repeatedly adding the sensor value to a variable called *Sum*. The variable is initialized at 0, and the blocks in the Loop block add the latest sensor measurement to *Sum* each time. After 50 repetitions, the sum is divided by 50 to calculate the average sensor measurement, which is in turn displayed on the screen.

Calculating an average value can be useful to get a more accurate sensor value. For example, you might have an obstacle-avoiding robot that turns around if the proximity value is less than 50. Sometimes, the sensor appears to detect an obstacle up close (at a proximity of 40, say) when there is actually nothing in its path. Normally, this measurement would trigger the robot to turn, but if you take the average of three measurements, giving (100 + 100 + 40) / 3 = 80, this false positive won't trigger the robot.

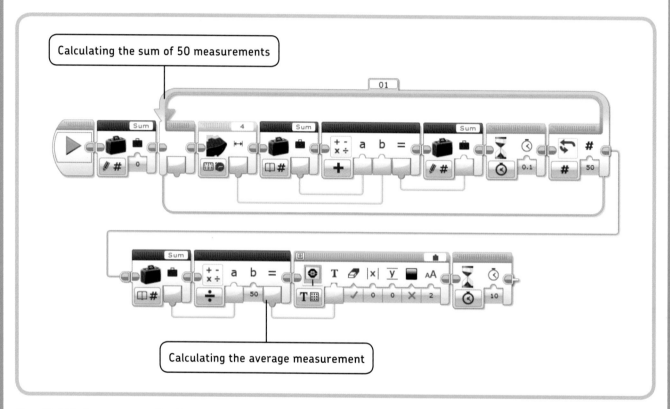

Figure 16-9: The Average *program. The Wait block in the Loop causes the program to pause for 0.1 seconds between each measurement so that the program calculates the average sensor value across a period of 50 × 0.1 = 5 seconds.*

further exploration

Congratulations! You've completed all of the programming theory in this book. Now you're ready to go on to the next chapter, where you'll create a larger program that turns SK3TCHBOT into an Etch-A-Sketch-like device. But before you do, take a look at the following Discoveries. These may be more difficult than the other ones you've done so far, but remember that there are multiple solutions to each. Give them a try, and check out solutions to many of the Discoveries at *http://ev3.robotsquare.com/* for inspiration.

DISCOVERY #108:
COUNTING UP
AND DOWN!

Difficulty: ▭▭ **Time:** ◷◷

Can you create a program based on the *TouchCount* program that uses a variable to track the number of times the Left and Right EV3 buttons are pressed? If you press the Right button, the variable's value should increase by 1; if you press the Left button, it should decrease by 1. Display the variable's value on the EV3 screen.

HINT Use a Wait block to wait until either the Left or Right button has been pressed. Then use a Switch block to determine which button was pressed. Inside the Switch block, add the blocks required to modify the variable's value and blocks that wait until the EV3 button is no longer pressed.

DISCOVERY #109:
LIMITED AVERAGE!

Difficulty: ▭▭ **Time:** ◷◷

Can you expand the *Average* program in Figure 16-9 to calculate the average of all proximity measurements taken between the start of the program and the moment that you press the Touch Sensor?

HINT Configure the loop to repeat until the Touch Sensor is pressed. Define a variable called *Measurements* to keep track of the number of sensor measurements that have been added to the *Sum* variable. When the loop ends, calculate the average by dividing *Sum* by *Measurements*.

DISCOVERY #110:
RANDOM VERIFICATION!

Difficulty: ▭▭▭ **Time:** ◷◷

In this Discovery, you are challenged to verify how well the Random block in Logic mode works. To do this, you'll have the robot generate 10,000 random logic values and determine how many of those are true and how many of them are false. Experiment with the Probability of True setting and see if you get the expected result.

HINT Create two Numeric variables, called *TrueCount* and *FalseCount*. Generate a random Logic value with the Random block. If it's true, increment *TrueCount*; if it's false, increment *FalseCount*. Display the values of each variable after 10,000 repetitions. What values do you expect?

DISCOVERY #111: CLOSEST APPROACH!

Difficulty: ▫▫▫ **Time:** ◷◷◷

Can you create a program that determines the lowest recorded proximity measurement out of 50 measurements? Have the program pause for 0.1 seconds between each measurement and display the lowest value on the screen.

HINT Create a variable called *Lowest* to hold the lowest recorded measurement. Repeatedly compare the stored measurement to a new sensor measurement. If the new measurement is lower, store the new value in *Lowest*. What value should you initialize the *Lowest* variable to at the start of the program?

DESIGN DISCOVERY #27: CUSTOMER COUNTER!

Building: ✸✸✸ **Programming:** ▫▫

Can you create a robot that counts the number of people in a room? Design a contraption that opens the door when guests press the Touch Sensor to enter the room, and have them place their hand up close to the Infrared Sensor to open the door when they want to leave. Then, modify the program you made in Discovery #108 on page 251 to count the number of people in the room: A Touch Sensor press should increase the counter; a wave at the Infrared Sensor should decrease the counter.

HINT Rather than designing a robot that actually turns a door handle, it's much easier to design a vehicle robot that opens the door by pushing it forward and closes the door by pulling it backward. Be sure not to close the door all the way. Alternatively, use some tape or string to prevent the door handle from clicking into the closed position.

17

playing games on the EV3

You'll now create a program that combines many of the programming techniques you've learned throughout this book. I've introduced each programming technique and block previously with short example programs, but now you'll see how to combine them in a larger, more sophisticated program.

In this chapter, you'll create a program that lets you play an Etch-A-Sketch-like game on the EV3. You'll create drawings on the EV3 screen using the Large Motors as input knobs, and you'll use the Touch Sensor and Color Sensor as additional input controls (see Figure 17-1).

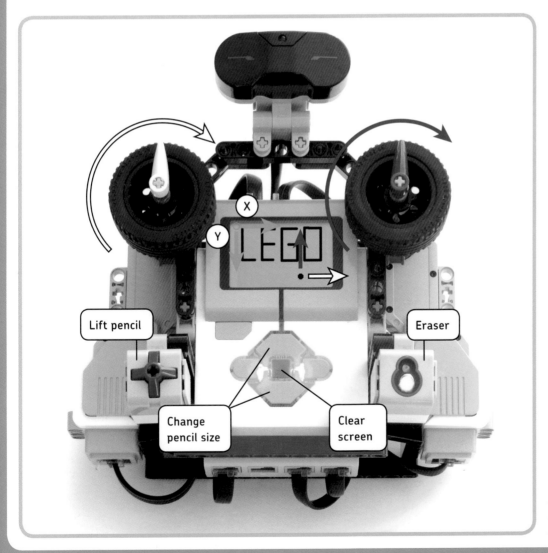

Lift pencil

Eraser

Change pencil size

Clear screen

Figure 17-1: Playing an Etch-A-Sketch-like game on SK3TCHBOT

step 1: creating basic drawings

An Etch-A-Sketch device draws lines on its screen as you turn the knobs, as if moving a pencil across a piece of paper. The left knob controls the *horizontal position* of the pencil (*X*), while the right knob controls the *vertical position* (*Y*). This behavior is surprisingly easy to re-create in an EV3 program, as shown in Figure 17-2.

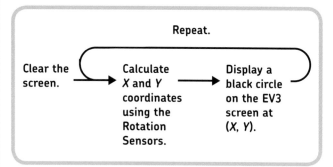

Repeat.

Clear the screen. → Calculate *X* and *Y* coordinates using the Rotation Sensors. → Display a black circle on the EV3 screen at (*X, Y*).

Figure 17-2: Program flow for the basic Etch-A-Sketch *program*

The program repeatedly displays a new dot (small circle) on the EV3 screen at a location specified by an *X*-and-*Y* coordinate. The program determines these coordinates using the Rotation Sensors in the Large Motors. The white dial (motor B) controls the *X* position, while the red dial (motor C) controls the *Y* position. As the program adds dots to the screen, you can create drawings by "moving the pencil" using SK3TCHBOT's dials.

To begin, create a new project called *SK3TCHBOT-Games* with a program called *Etch-A-Sketch*. To display the dot, you'll simply use a Display block in Shapes – Circle mode, but you'll need to create two My Blocks to clear the screen and to calculate the coordinates where the dot must be displayed.

my block #1: clear

The Display block doesn't have an option to clear the screen without displaying something new, but you can mimic this action by displaying a single white pixel while clearing the rest of the screen. Configure a Display block as shown in Figure 17-3, and turn it into a My Block called *Clear*.

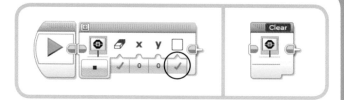

Figure 17-3: The Clear My Block clears the EV3 screen. The completed My Block is shown on the right.

my block #2: coordinates

You'll now create a My Block called *Coordinates*, as shown in Figure 17-4. It has two Numeric output values (*X* and *Y*) that you'll use to control the position of each new dot. Let's have a look at how this block calculates the *X* coordinate.

Recall that a Rotation Sensor gives you the number of degrees a motor has turned since the start of the program. In principle, you could use the sensor value from motor B to control the *X* coordinate directly. However, this would make the pencil move a bit too fast: Turning the white dial for only 180 degrees would move the pencil across the entire screen. (The screen has 178 pixels in the *X* direction.)

To make more precise movements, divide the number of degrees by 3 so that turning the white dial for 180 degrees draws a line across only a third of the screen. Add 89 (half of 178) to the result of this division so that the dot is displayed in the middle of the screen when the degrees value is 0 at the start of the program.

The procedure for the *Y* coordinate is similar, except that you'll use motor C and you'll add 64 to the division (there are 128 pixels in the *Y* direction). Because motor C is mounted upside down, you get a negative degrees value if you turn it in the direction of the red arrow in Figure 17-1. Consequently, the pencil moves upward (vertical red arrow), which is opposite to the positive *Y* direction (vertical blue arrow).

To create the My Block, place two Motor Rotation blocks and two Math blocks on the canvas, and turn them into a My Block with two Numeric outputs called *Coordinates*, as shown in Figure 17-4.

completing the basic program

Now that you've created the essential components of the flow diagram in Figure 17-2, finish the program as shown in Figure 17-5. Besides the two My Blocks, the program contains a Display block configured to display the dot (small circle) at the required coordinates and a Loop block to repeat the program. Download the program to your robot and test it before expanding the program in the next section.

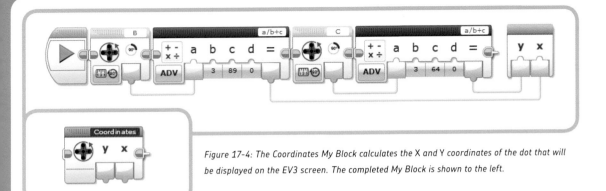

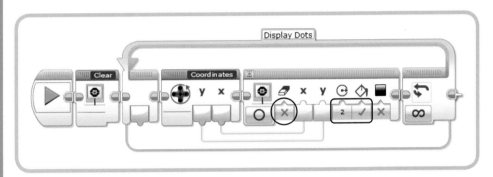

Figure 17-4: The Coordinates My Block calculates the X and Y coordinates of the dot that will be displayed on the EV3 screen. The completed My Block is shown to the left.

Figure 17-5: The basic Etch-A-Sketch program. Note that the Display block is configured not to clear the screen; this ensures that previously printed dots remain visible when another dot is added to the screen.

NOTE If you're experiencing problems while creating this program, download the finished program from *http://ev3.robotsquare.com/* and compare it to your own program.

step 2: adding pencil controls

The basic *Etch-A-Sketch* program is fun, but you can give it more functionality by expanding it, as shown in Figure 17-6. You'll use the Touch Sensor to temporarily stop drawing, the Color Sensor to turn the pencil into an eraser, the Center EV3 button to clear the screen, and the Up and Down buttons to change the pencil size.

To make it easier to spot potential problems, be sure to test your program each time you add a new feature. When you're ready, you'll be challenged to further expand the program with new Discoveries.

moving the pencil without drawing

While drawing, you'll sometimes want to lift the pencil in order to move to a new position without creating any lines. To accomplish this, make the program draw new dots only if the Touch Sensor is *released* (see Figure 17-7). If the Touch Sensor is pressed, nothing happens. (The false tab of the Switch block doesn't contain any blocks.)

turning the pencil into an eraser

Now you'll add an eraser tool to your program by making it draw white dots when you trigger the Color Sensor, as shown in Figure 17-8. By drawing white dots, you can effectively erase parts of your drawing, which is helpful in case you make mistakes.

When you cover the Color Sensor with your finger, the reflected light intensity measurement increases beyond 10%, and the Compare Result of the Color Sensor block becomes *true*, causing the Display block's Color setting to become white. When you remove your finger from the sensor, the output becomes *false*, and the Display block continues to add black circles to the screen.

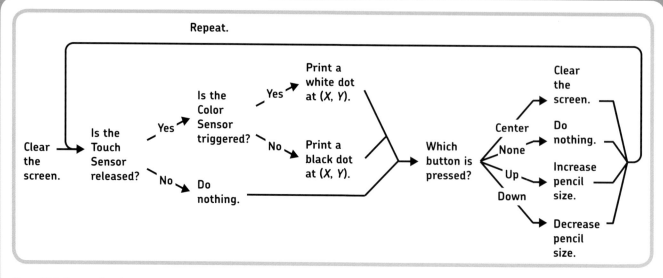

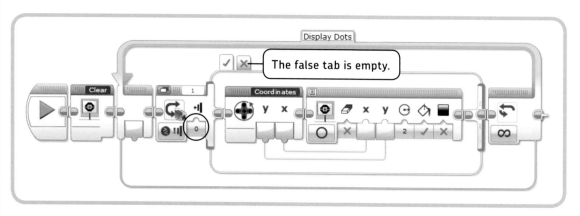

Figure 17-6: Program flow for the extended Etch-A-Sketch program

Display Dots

The false tab is empty.

Figure 17-7: The program draws new dots only if the Touch Sensor is released. To move the pencil to a new position without drawing lines, press the Touch Sensor, turn the input knobs, and release the Touch Sensor to continue drawing.

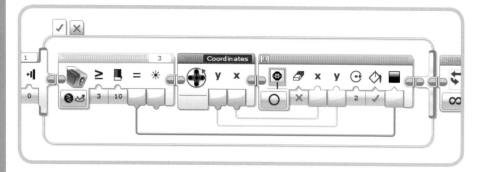

Figure 17-8: Add a Color Sensor block configured as shown, and connect the Logic data wire to the Display block. Your pencil now acts as an eraser whenever you cover the Color Sensor with your finger.

clearing the screen

To clear the screen when you press the Center button, add a Switch block and a Clear block (see Figure 17-9). If none of the EV3 buttons are pressed, nothing happens. (The No Buttons case, which is also the default case, does not contain any blocks.)

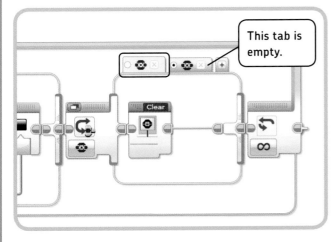

Figure 17-9: These blocks erase the screen when you press the Center button.

setting the pencil size

Finally, you'll expand the program to set the pencil size with the Up and Down buttons on the EV3 brick. To do so, you'll use a variable called *Size* to control the value of the Radius setting on the Display block. Then, you'll add blocks that allow you to modify the *Size* variable using the EV3 buttons.

To begin, define a Numeric variable called *Size* and add two Variable blocks to your current program, as shown in Figure 17-10. The first Variable block initializes the value to 1. The second block sends the *Size* value to the Radius setting of the Display block. Consequently, the Radius is 1 as long as you don't change the variable's value.

Next, add two cases (tabs) to the Switch block that you've configured previously. Add one case for the Up button and another for the Down button.

If you press the Up button, the *Size* variable should be incremented, as shown in Figure 17-11. Because the screen is quite small, you'll limit the *Size* value to 25. In other words, you'll increment the variable only if its current value is less than 25, as determined by a Compare block. If the *Size* value is 25 and you try to increase it further, the Compare block outputs false, you'll hear a beep, and the value isn't changed. Finally, a Wait block pauses the program until you release the Up button. (If you do not wait for a release, the program continues to repeat, incrementing the variable each time it goes through the loop.)

The variable should be decremented (lowered by 1) if you press the Down button, but only if the current *Size* value is greater than 1 so you don't end up displaying a circle with a negative or 0 radius (which wouldn't work). If the *Size* value is 1 and you try to decrease it further, the Compare block outputs false, you'll hear a beep, and the variable's value isn't changed (see Figure 17-12).

Congratulations! You've finished the *Etch-A-Sketch* program. Once you've verified that your program works, see whether you can further expand the program using Discoveries #112 through #114.

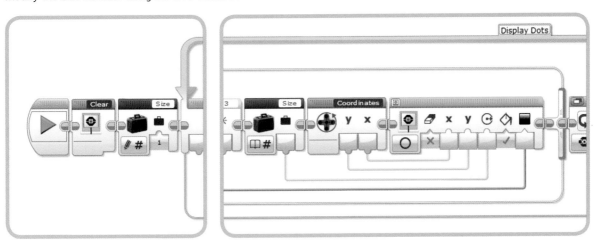

Figure 17-10: Add two Variable blocks to the program as shown.

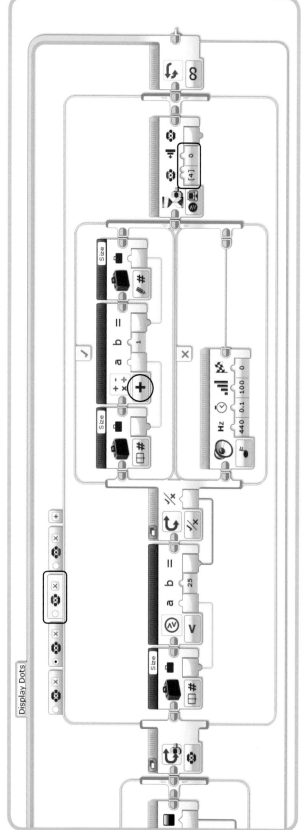

Figure 17-11: When you press the Up button and the current value in the Size variable is less than 25, the value is increased by 1. Note that the second tab (No Buttons) remains the default case.

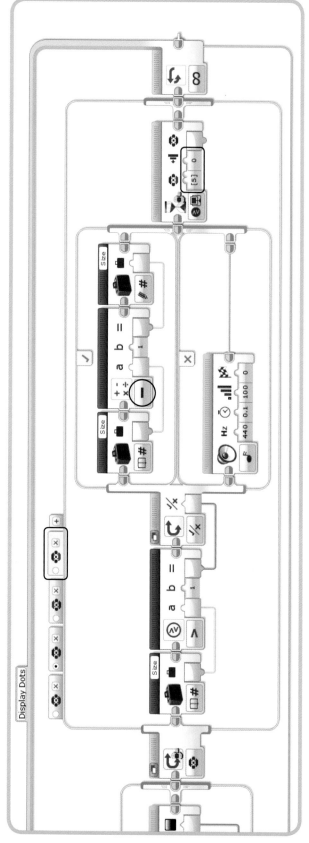

Figure 17-12: When you press the Down button and the current value in the Size variable is greater than 1, the value is decreased by 1.

DISCOVERY #112:
ROBOT ARTIST!

Difficulty: ▢ **Time:** 🕐🕐

Creating a nice drawing with the *Etch-A-sketch* program isn't easy, but there's an easy way to cheat. You can make the program draw perfectly straight lines by making motors B and C move with programming blocks. Can you preprogram the robot to draw the image of a house on the screen? How can you draw diagonal lines?

HINT Control the movement of the motors in parallel to the *Etch-A-Sketch* program. (See "Multitasking" on page 56.)

DISCOVERY #113:
FORCE FEEDBACK!

Difficulty: ▢▢ **Time:** 🕐🕐

Can you expand the Coordinates My Block to actively prevent the user from moving the pencil off the screen? The pencil should automatically move back to the screen if the user turns the knobs too far.

HINT Make motor B move in On mode at −5% speed if *X* becomes larger than 178, and make it move at 5% speed if *X* becomes less than 0. If *X* is between 0 and 178, make the motor stop and set Brake at End to false. Implement similar behavior for the *Y* direction.

DISCOVERY #114:
PENCIL POINTER!

Difficulty: ▢▢▢ **Time:** 🕐🕐

The smallest dots that the *Etch-A-Sketch* program can currently show on the screen are 1 pixel in radius, or 2 pixels in diameter. Can you expand the program to print single pixels as the smallest dots? This lets you add the finest details to your drawings!

HINT You can show a single dot on the screen by displaying a pixel at (*X*, *Y*) using a Display block in Shapes – Point mode.

further exploration

In this chapter, you created a program that combines many of the programming techniques you've learned throughout this book. If you enjoyed creating this program, I recommend you explore Discoveries #115 and #116. In the next part of this book, you'll learn how to build and program two more cool robots.

DISCOVERY #115:
ARCADE GAME!

Difficulty: ▢▢▢ **Time:** 🕐🕐🕐

In this Discovery, you'll explore an arcade game program for SK3TCHBOT. Begin by downloading the program from *http://ev3.robotsquare.com/* and running it to see what happens. You should see a randomly positioned *target* and a user-controlled *player* on the EV3 screen (see Figure 17-13).

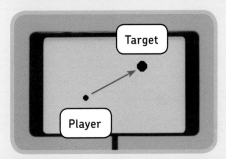

Figure 17-13: The arcade game

If you hit the target within 4 seconds by turning SK3TCHBOT's dials, you score 1 point; if you need more than 4 seconds, you lose 1 point. Each time you hit the target, a new target appears somewhere on the screen. After 10 turns, the program tells you how well you did.

Explore the program to examine how it works, using the flow diagram of Figure 17-14 as a guide. (If you're up for an extra challenge, try creating the program yourself!)

(continued)

(Discovery #115, continued)

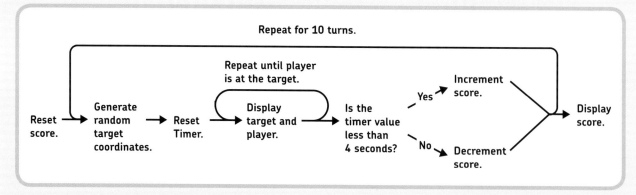

Figure 17-14: The program flow of the Discovery #115 arcade game

DISCOVERY #116: BRAIN TRAINER!

Difficulty: ▢▢▢▢ Time: ◷◷◷

In this Discovery, you'll explore a brain-training program you can play on the EV3 brick, available for download at *http://ev3.robotsquare.com/*. The program displays a random addition, subtraction, multiplication, or division problem on the EV3 screen (for example, 7 × 3). The user decides whether the displayed answer is correct by pressing the Left button (incorrect) or the Right button (correct), as shown in Figure 17-15. The answer is correct roughly half of the time.

The program is made using only the programming blocks discussed in this book, but it may still be challenging to understand at first. Use the comments placed throughout to explore the program.

Figure 17-15: The brain-trainer game. Pressing X scores 1 point in this case because the given answer is incorrect.

Once you understand how it works, expand the program by having it keep track of how many times you answered correctly and incorrectly within 30 seconds. Can you improve your score by practicing?

DESIGN DISCOVERY #28: PLOTTER!

Building: ✳✳✳ Programming: ▢▢▢

Can you design a robot that makes drawings on a sheet of A4 or US letter paper? Use one Large Motor to move the pencil across the width of the paper to draw horizontal lines (*X*), and use another Large Motor to move the paper back and forth to draw vertical lines (*Y*). Finally, use the Medium Motor to lift and lower the pencil so that you can move the pencil and continue drawing elsewhere. To make it easier to control the machine, create My Blocks for moving the pencil, with input parameters to specify *X* and *Y* coordinates.

HINT For inspiration, you can study the mechanics of an actual inkjet printer. The printer moves the cartridge left and right across a fixed beam (this is how you can create horizontal lines), and it moves the paper back and forth using wheels (this is how you can draw vertical lines).

machine and humanoid robots

18

the SNATCH3R: the autonomous robotic arm

Throughout the previous chapters, you've learned a great deal about programming EV3 robots. Now you're ready to build and program the more sophisticated robots in this part of the book. This chapter will teach you to build and program the *SNATCH3R*, a robotic arm that can find and pick up objects, as shown in Figure 18-1.

You'll first create a program that lets you control the robot remotely so you can test its mechanical functions, and then you'll program it to find and grab the infrared beacon autonomously. You'll use data wires and variables to make the robot scan its surroundings so that it can find the beacon from up to 2 meters (6 feet) away, even if it's behind the robot.

The Discoveries throughout this chapter will help you see how the programs really work so you can expand on them with more features. For example, you'll be challenged to make the robot follow lines with the Color Sensor and pick up objects in its path.

understanding the grabber

The SNATCH3R uses two Large Motors to control a set of treads for driving, but the really cool part of this robot is its multifunctional grabber. Normally, grabbing and lifting objects requires two motors: one to grab the object and another to lift it. The SNATCH3R requires just one Medium Motor to accomplish both tasks because of a specialized construction of LEGO beams, axles, and gears. To better understand how the robot works, you'll build a simplified mechanism using the instructions on the next page.

Figure 18-1: The SNATCH3R can grab and lift lightweight objects, such as empty water bottles.

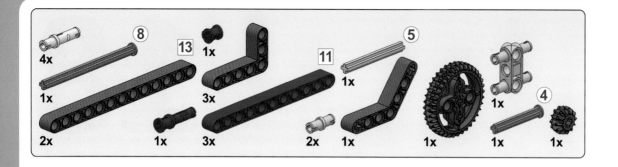

1

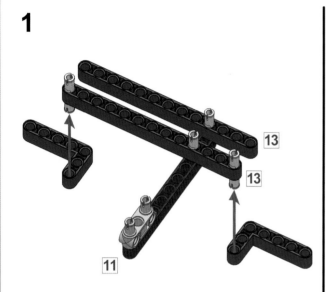

2

3

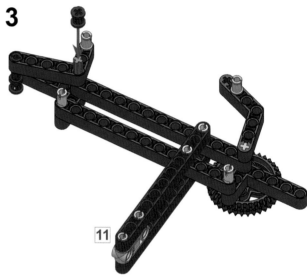

4

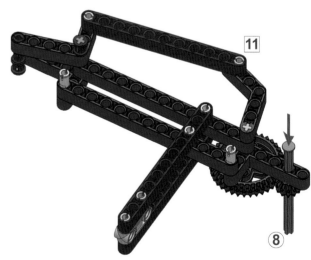

the grabber mechanism

To see how the grabber mechanism works, hold the red beams of the sample mechanism with one hand and use your other hand to turn the 12T gear by rotating its axle (see Figure 18-2). Try not to push the axle up or down while rotating it; the mechanism should do this on its own.

When you turn the small gear, the 36T gear will begin to pull the red beam upward so that the "claw" closes. Something similar happens in the SNATCH3R mechanism. The Medium Motor drives a 24T gear with a worm gear, which starts a chain reaction that ultimately causes the grabber to grasp objects positioned between its claws. When the motor spins backward, the reverse happens and the grabber opens.

the lifting mechanism

Once the grabber is closed, the beams and gears that caused it to close can no longer move. In the simplified mechanism (see Figure 18-3, top), this means that the 36T gear (a) hardly moves relative to the beam shown in blue (b). As a consequence, you can rotate the blue beam relative to the green beams (c and d) by rotating the 36T gear, and doing so raises the grabber into the air.

Similarly, as shown in Figure 18-3 (bottom), the 24T gear in the SNATCH3R (a) becomes locked to the beams in the grabber mechanism (b), and rotating it relative to the motor compartment (c) by turning the Medium Motor forward causes

the grabber (d) to rise. When the motor rotates backward, the grabber lowers.

Note that the two green beams in the sample mechanism (c and d) are always positioned horizontally, regardless of the arm's position. Similarly, the motor compartment (c) and the grabber (d) in the SNATCH3R remain positioned horizontally.

A Touch Sensor in the base of the SNATCH3R detects whether the grabber is lifted all the way up. Therefore, to grab and lift an object, you rotate the Medium Motor forward until the Touch Sensor is pressed. If the arm starts out in the lowered position with its claws open, the motor will have turned for 14.2 rotations when it reaches the Touch Sensor. Therefore, to lower and release the object, you rotate the motor backward for 14.2 rotations.

But how does the mechanism know to close the grabber before raising it and, when reversing, to lower it before opening the grabber? The mechanism doesn't "know" to do this, of course, but it's designed so that this behavior results naturally because of gravity. Closing the claws requires less energy than raising the grabber, so the claws always close first when the motor turns forward. Similarly, lowering the grabber requires less energy than opening the claws, so the grabber always lowers first when the motor turns backward. The details are beyond the scope of this book, but you can see that gravity plays a role by holding the sample mechanism on its side. You should find that without gravity to pull the claw downward, the mechanism doesn't work properly.

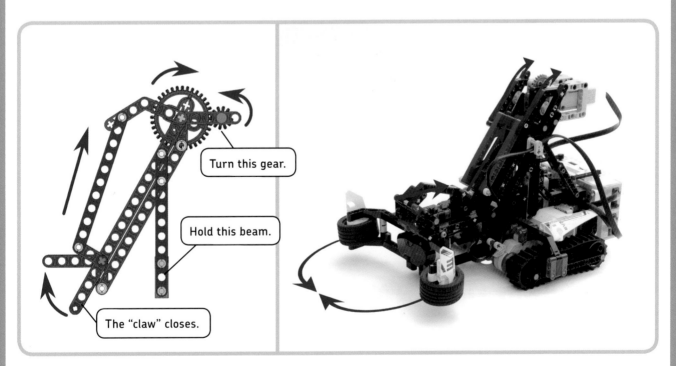

Turn this gear.

Hold this beam.

The "claw" closes.

Figure 18-2: Rotate the axle with the small gear to make the "claw" of the example mechanism close (left). Similarly, rotating the Medium Motor causes the SNATCH3R to grasp objects between its claws (right).

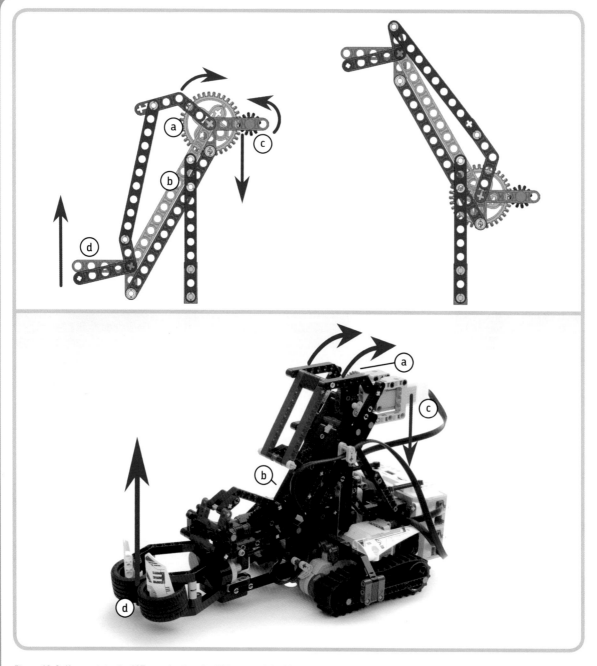

Figure 18-3: If you rotate the 12T gear further, the 36T gear and the blue beam turn relative to the green beam so that the grabber is lifted into the air (top). Similarly, rotating the Medium Motor further forward causes the SNATCH3R's grabber to rise (bottom).

building the SNATCH3R

Now that you've gotten a sense of how the SNATCH3R's grabber mechanism works, it's time to build the robot to see the full mechanism in action. Follow the directions on the next pages, but first take the sample mechanism apart and select the pieces you'll need for the complete model, as shown in Figure 18-4.

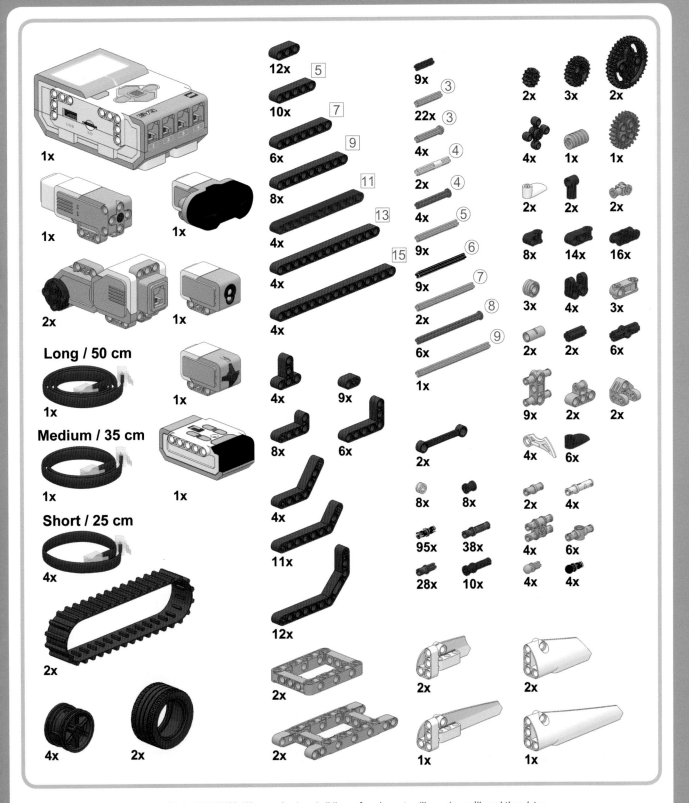

Figure 18-4: The pieces needed to build the SNATCH3R. When you're done building, a few elements will remain; you'll need them later.

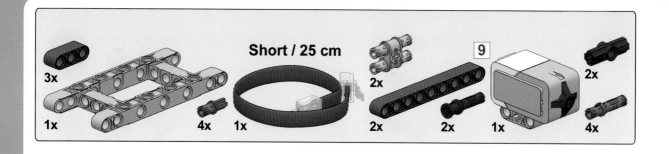

Short / 25 cm

3x · 1x · 4x · 1x · 2x · 9 · 2x · 2x · 1x · 2x · 4x

1

2

9 · 9

3

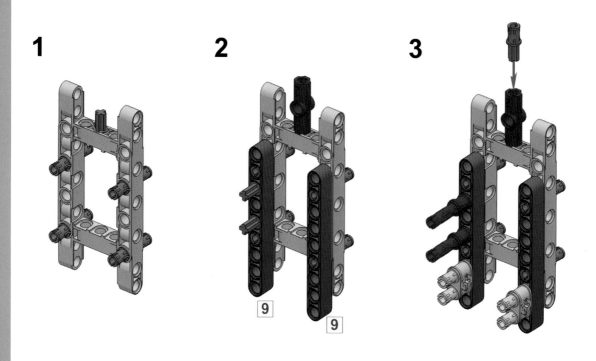

4

5

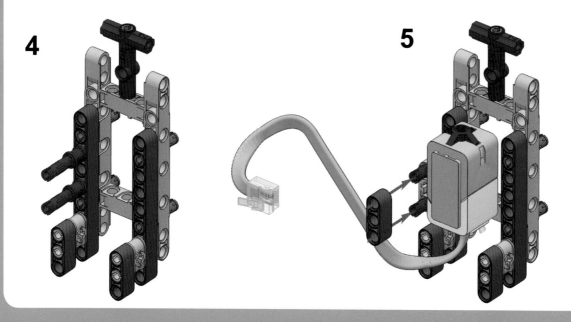

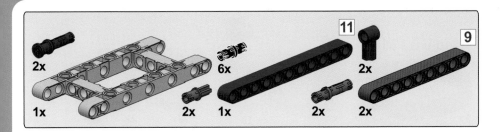

1

2

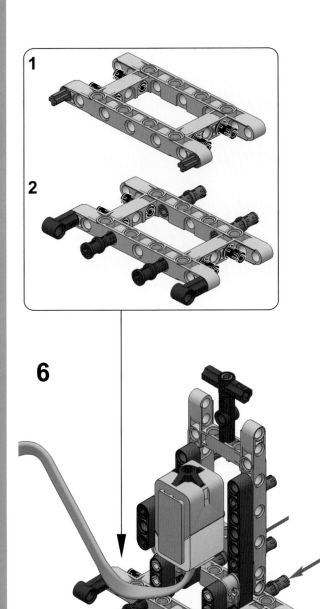

6

7

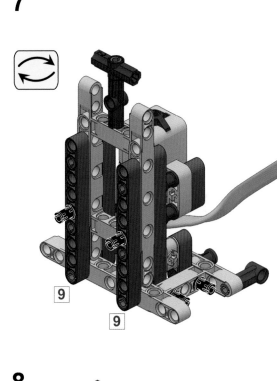

8

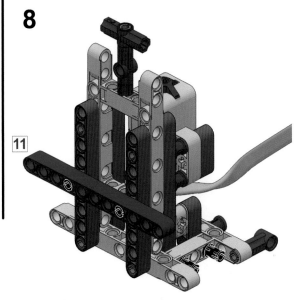

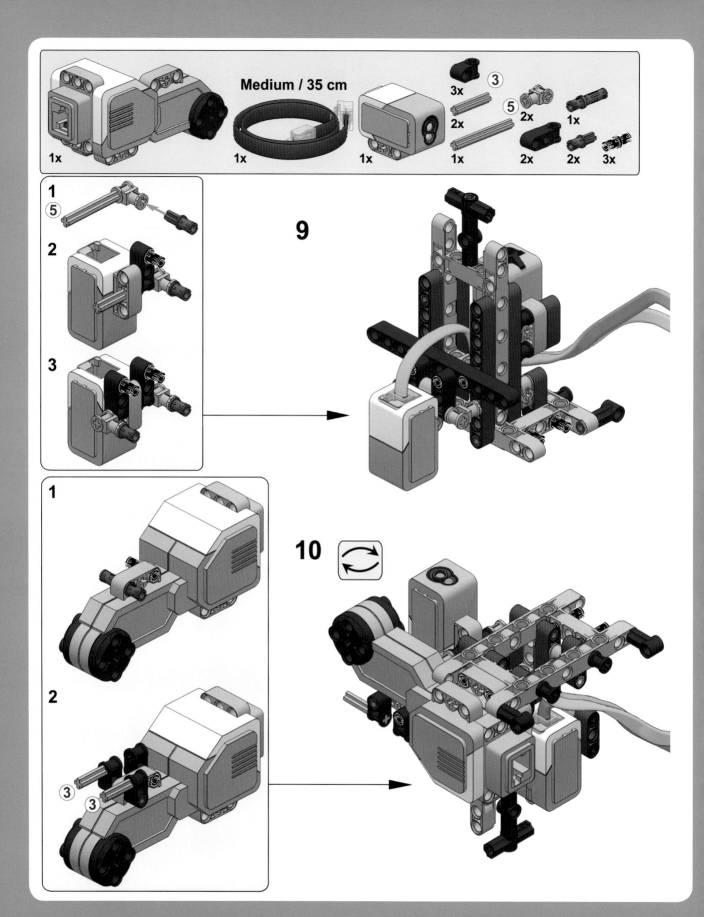

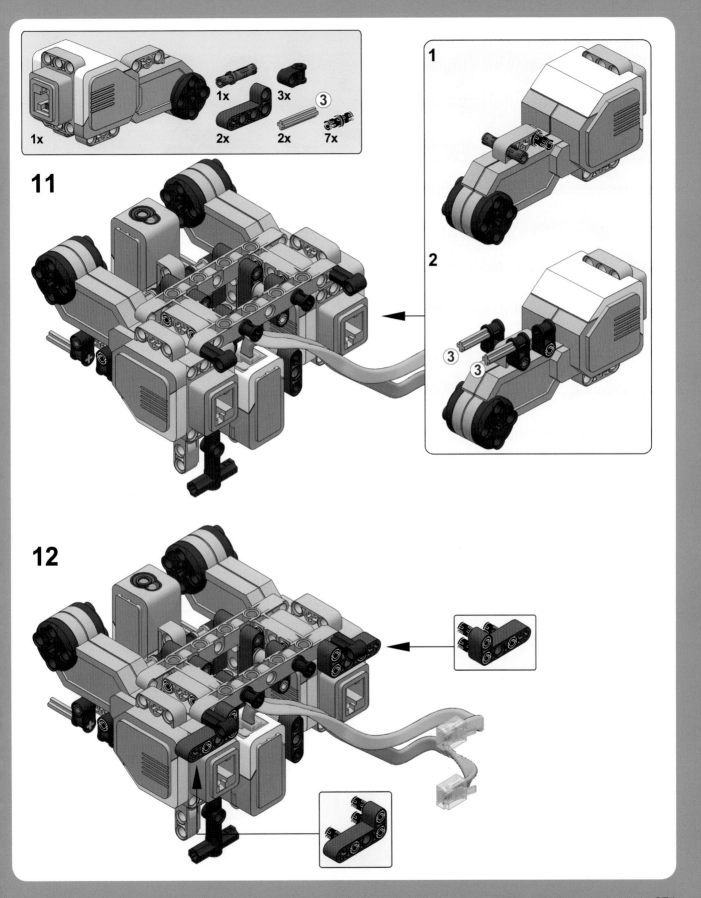

11

12

1x · 1x · 2x · 3x · 2x · 7x · 3

1 · 2 · 3 · 3

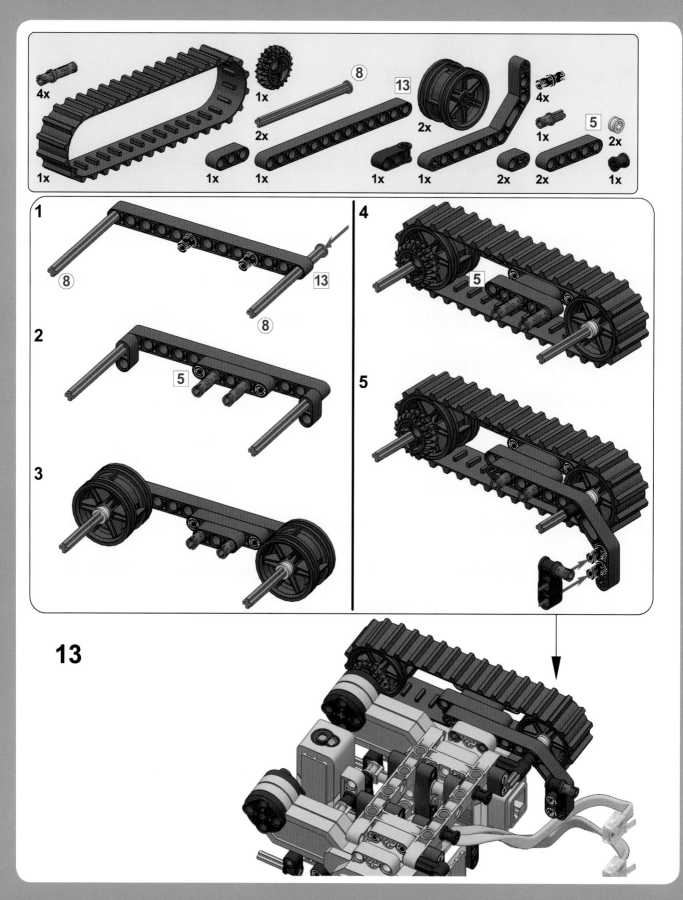

13

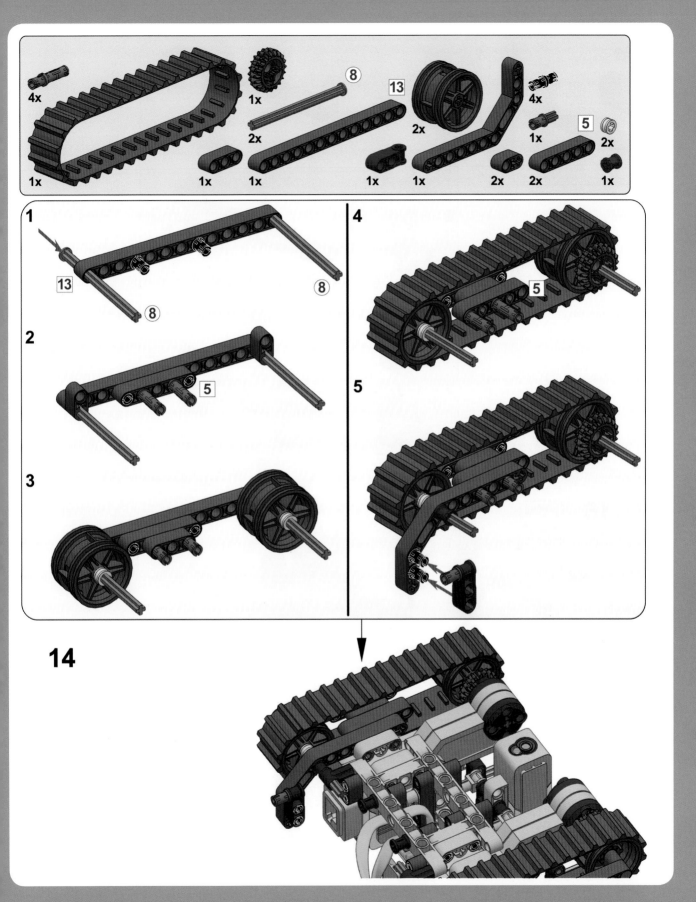

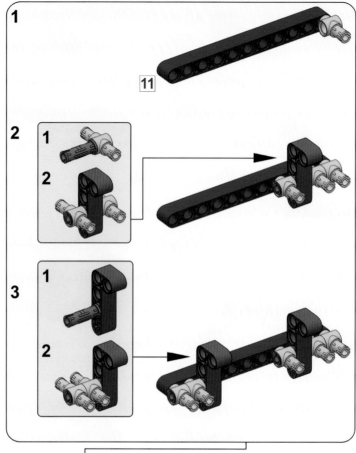

15

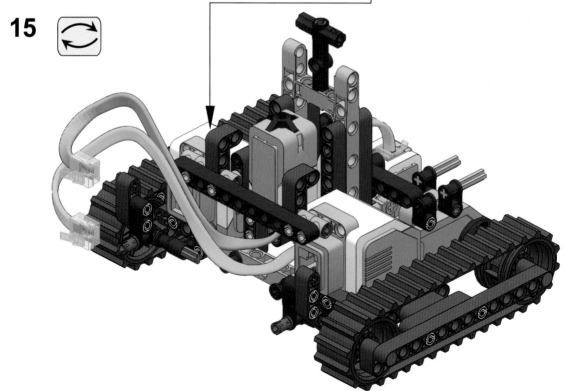

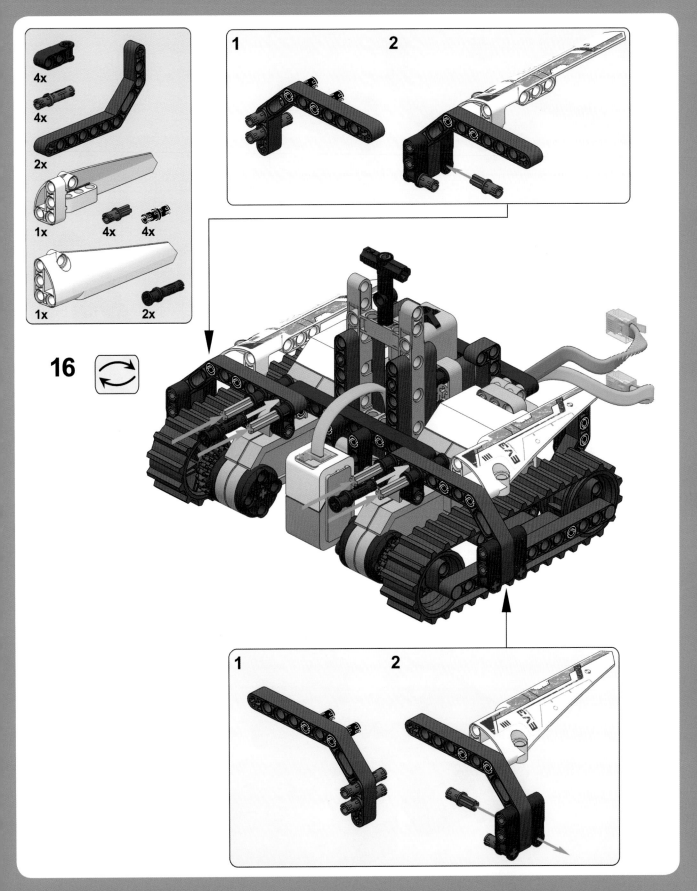

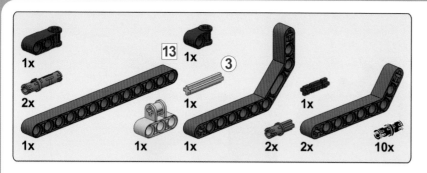

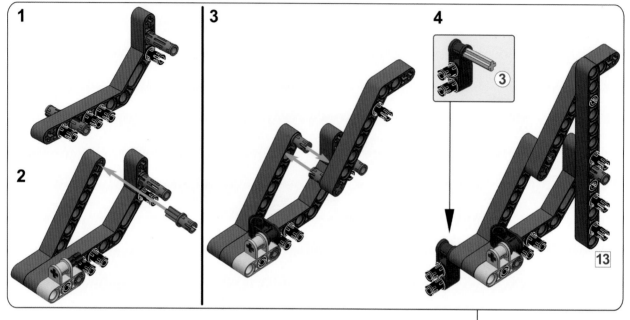

17

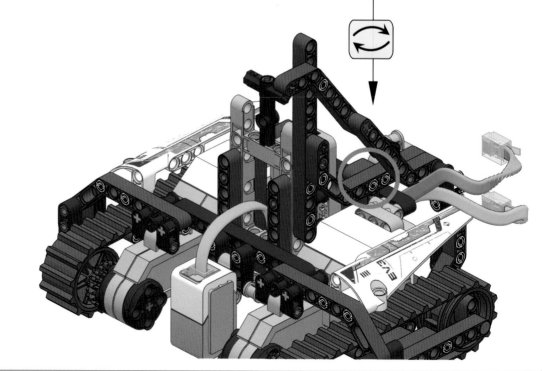

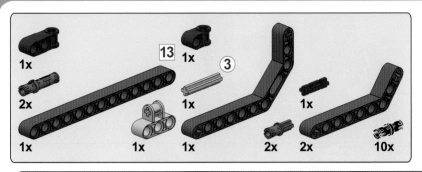

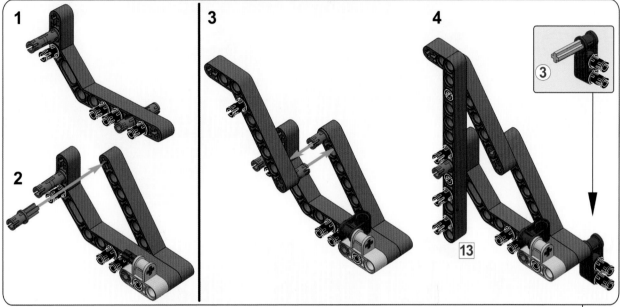

18

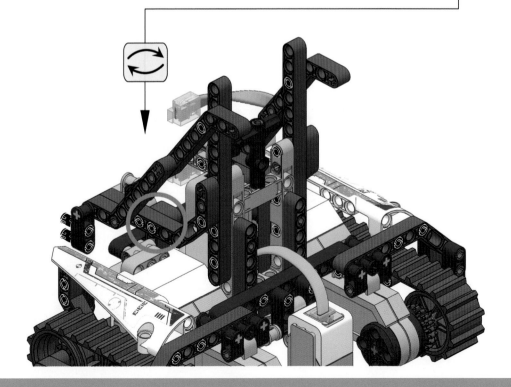

4x   2x

Short / 25 cm

2x

19

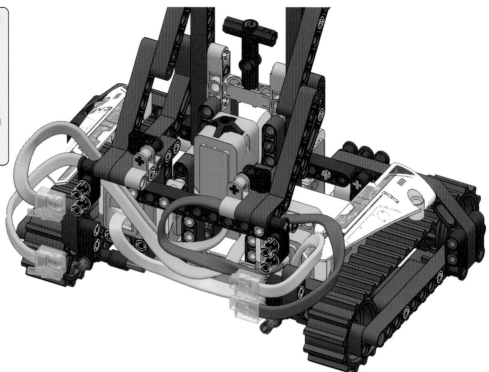

20

x2

B

1

3

C

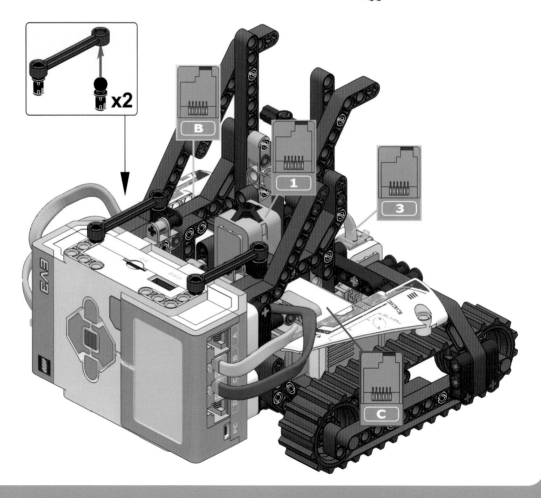

21

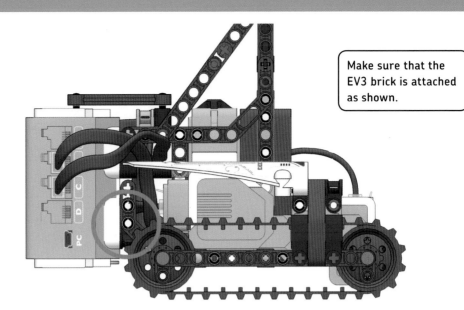

Make sure that the EV3 brick is attached as shown.

22

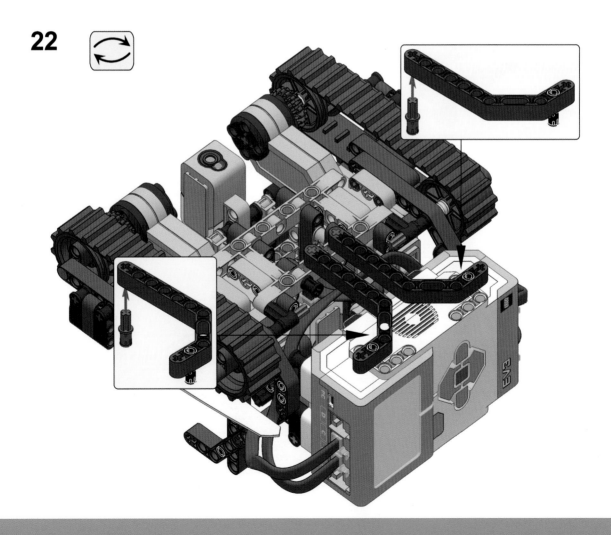

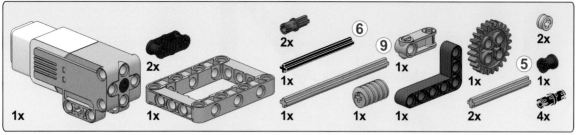

1

2

3

4

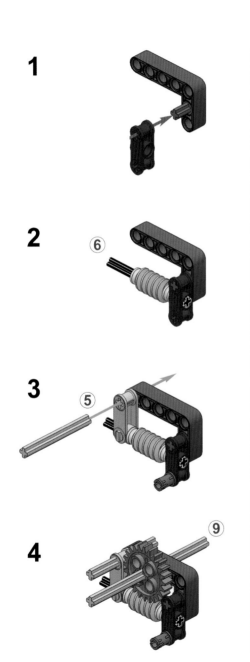

5

6

7

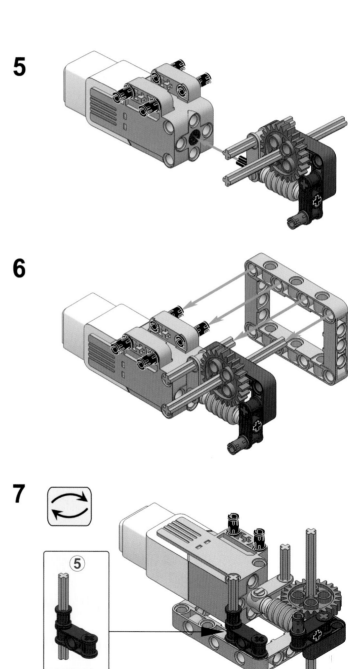

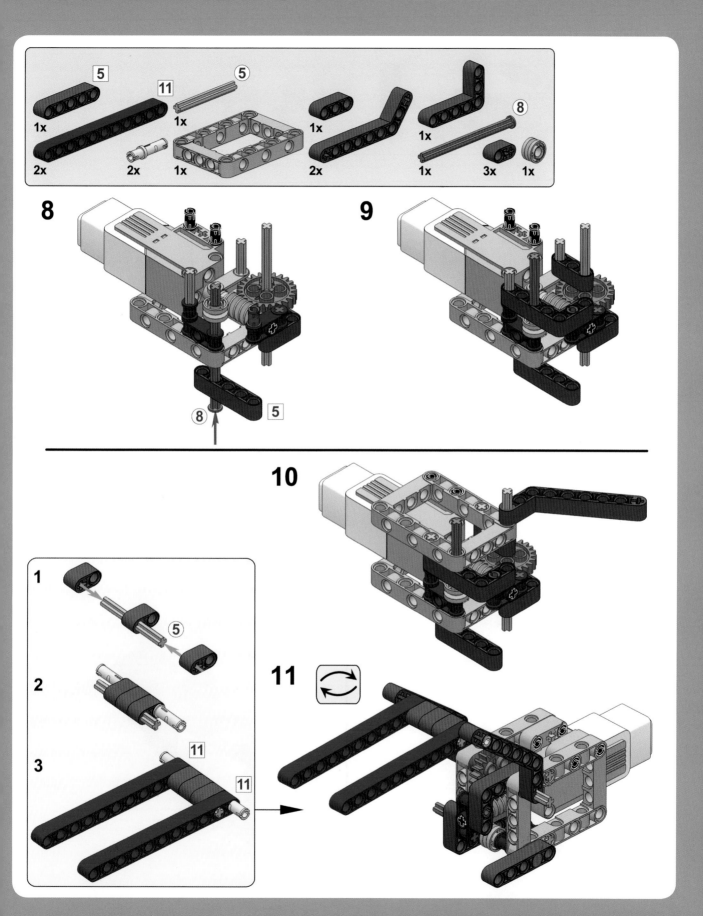

9 3x 1x

7 1x 2x 1x 9x 2x

4 1x 2x

3 1x 1x 2x

1

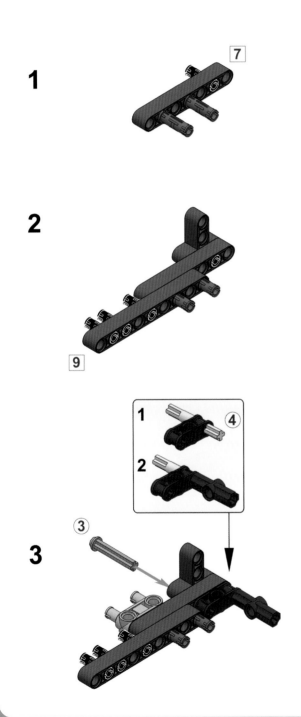

2

3

4

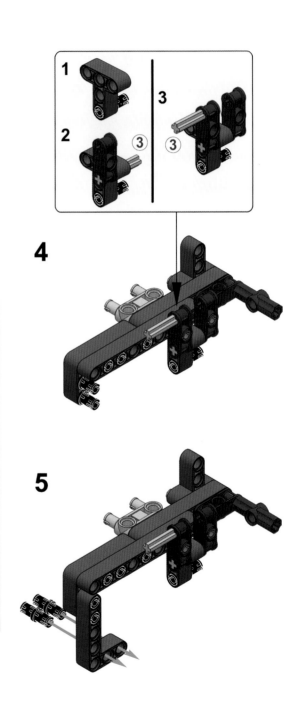

5

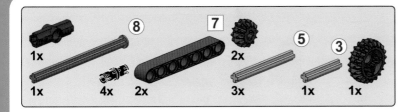

6

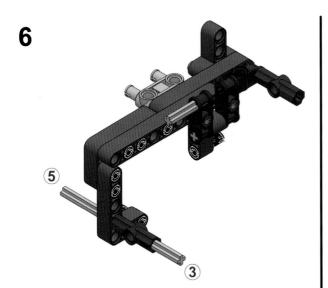

7

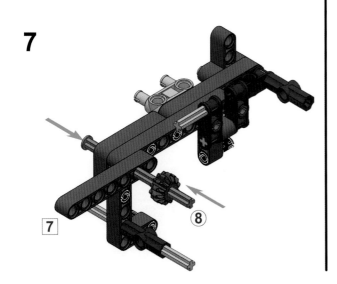

8

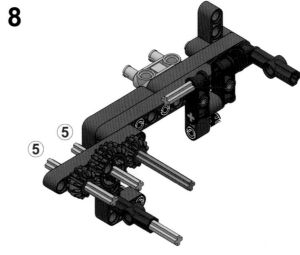

9

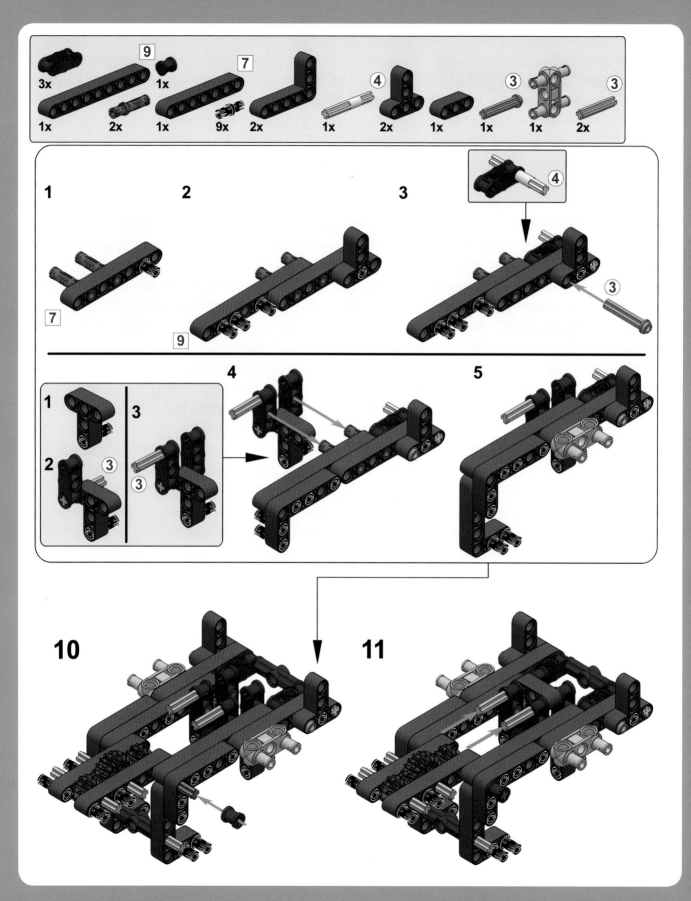

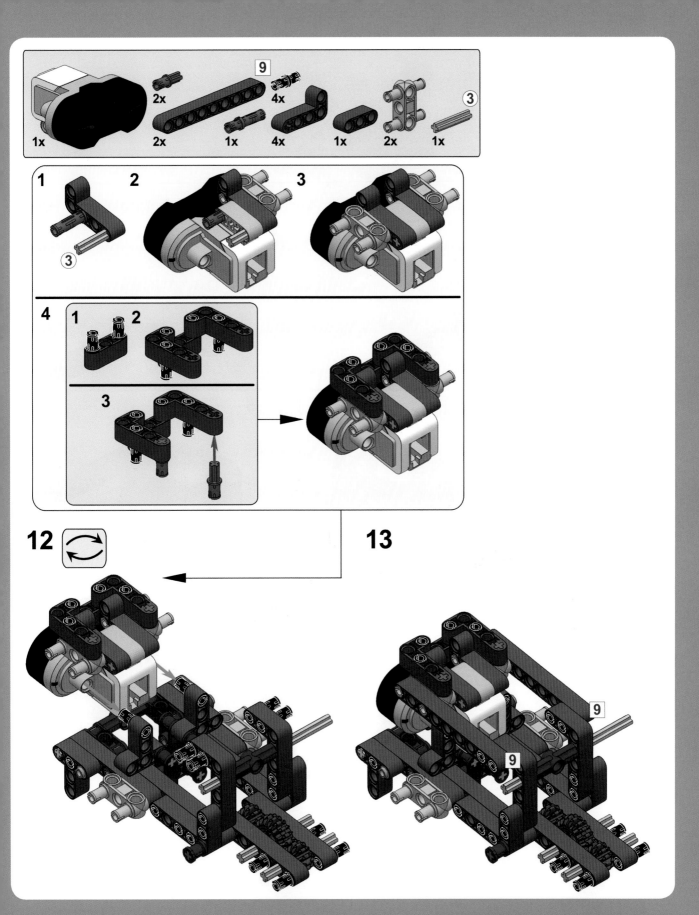

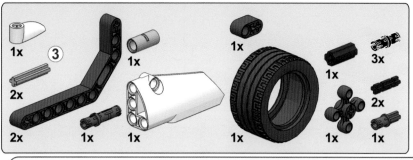

1

2

3

4

5

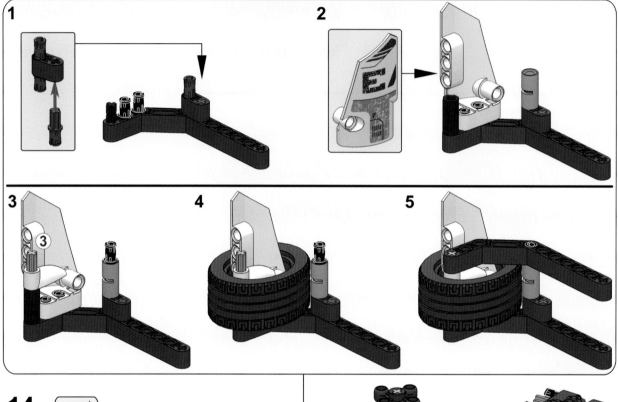

14

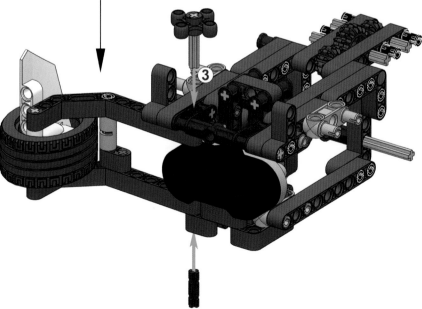

1

2

3

4

5

15

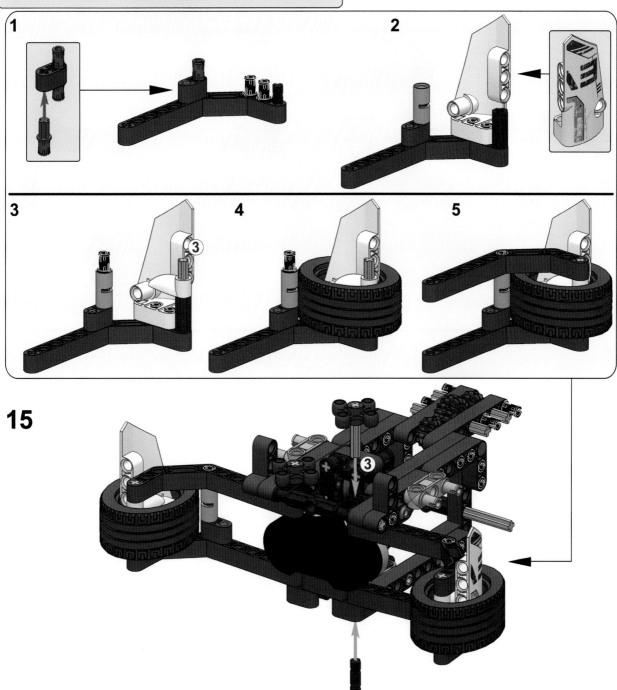

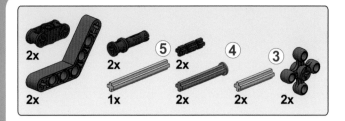

16

These helper pieces keep the claws in place while you build the robot arm. You'll remove them later.

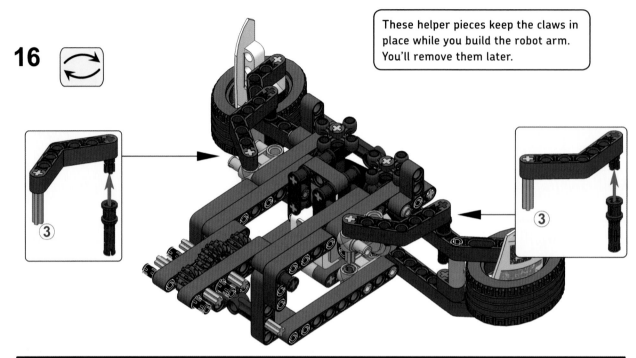

17

18

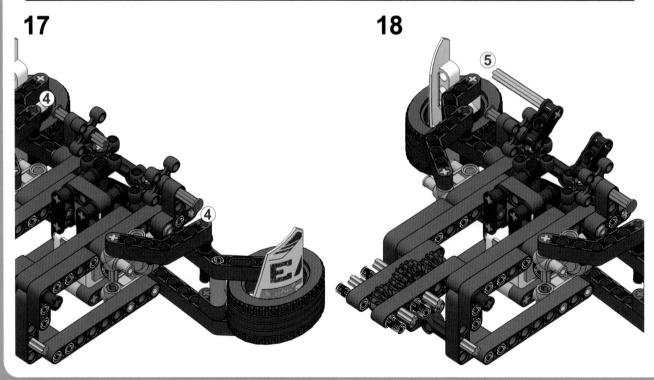

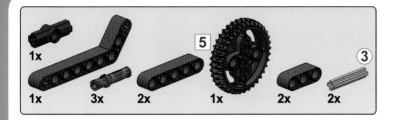

1x 1x 3x 2x 5 1x 2x 3 2x

1 2 5 3 4

19

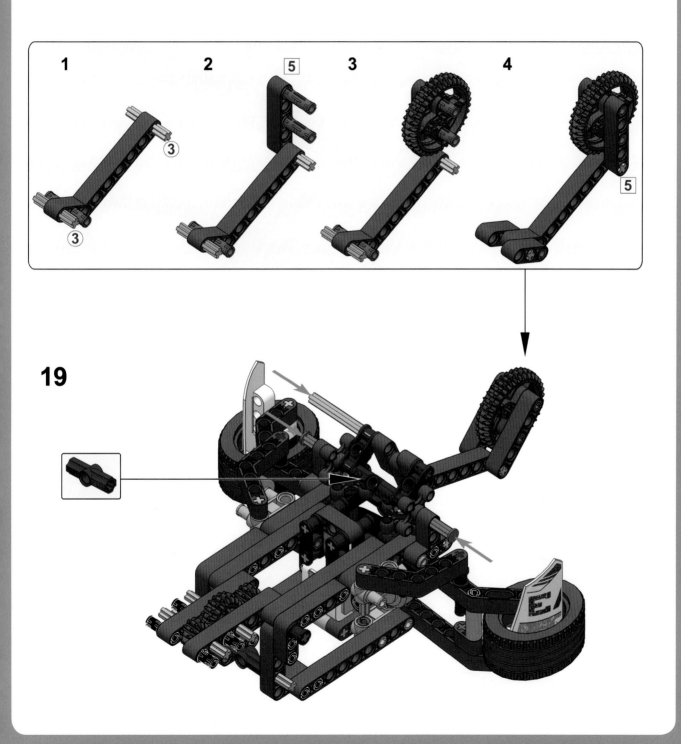

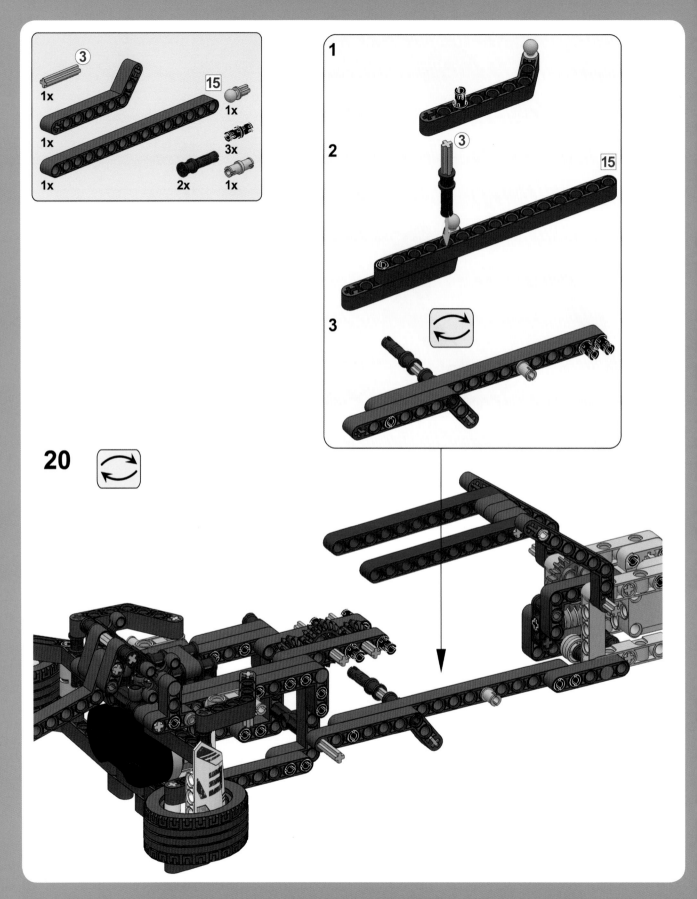

20

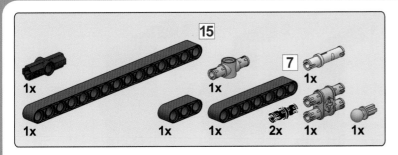

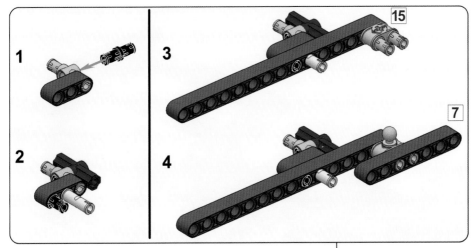

21

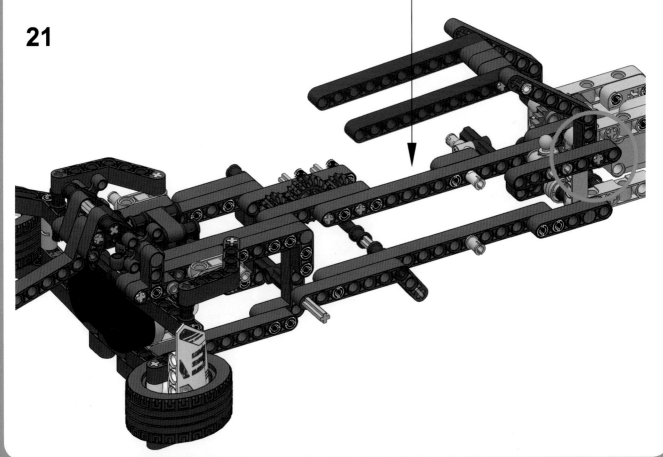

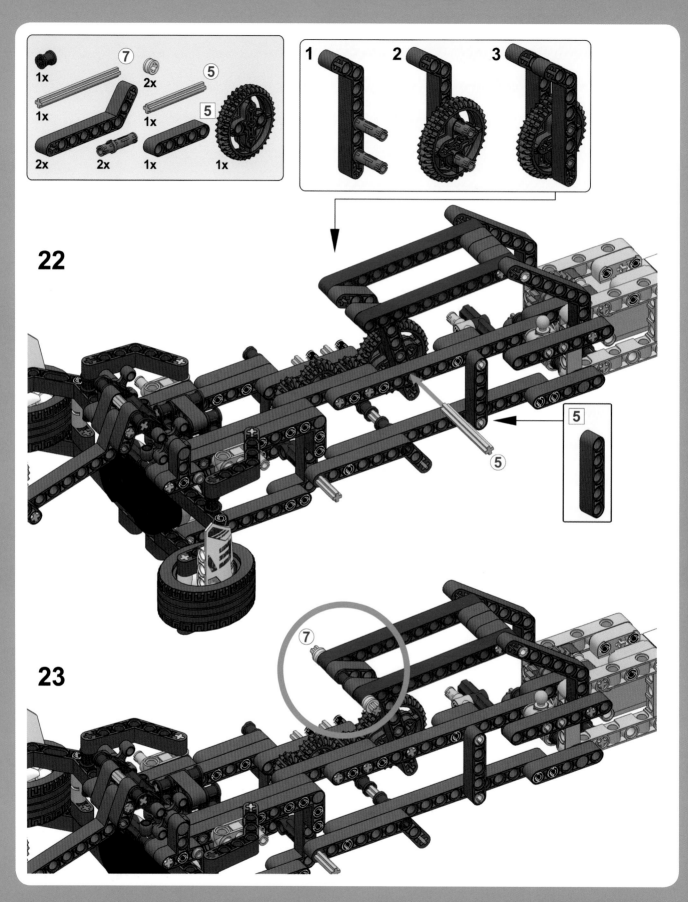

22

23

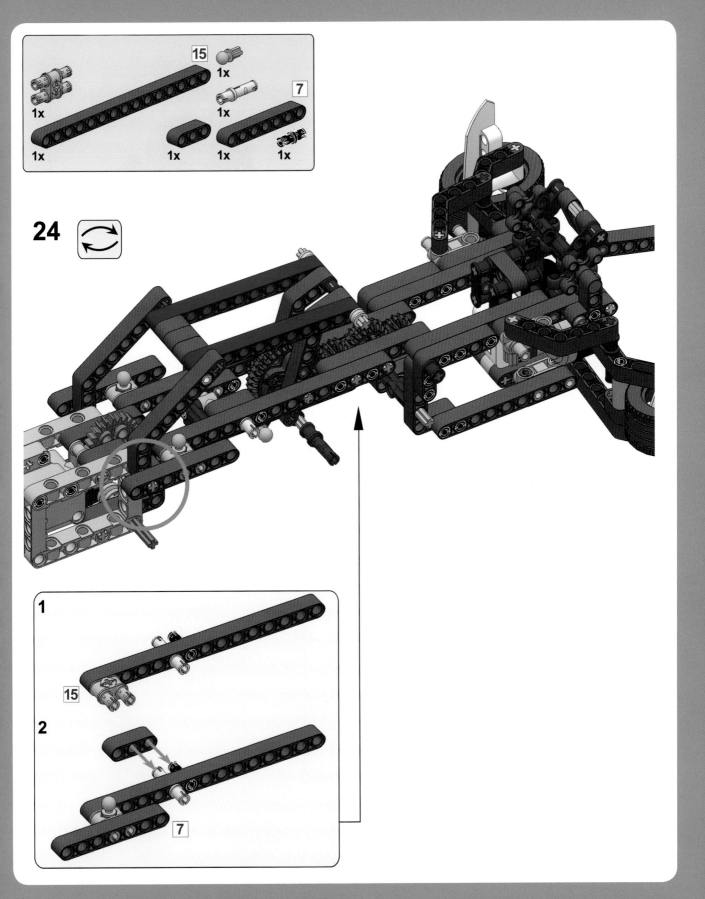

24

1x
15 1x
1x
7 1x
1x 1x 1x

1

15

2

7

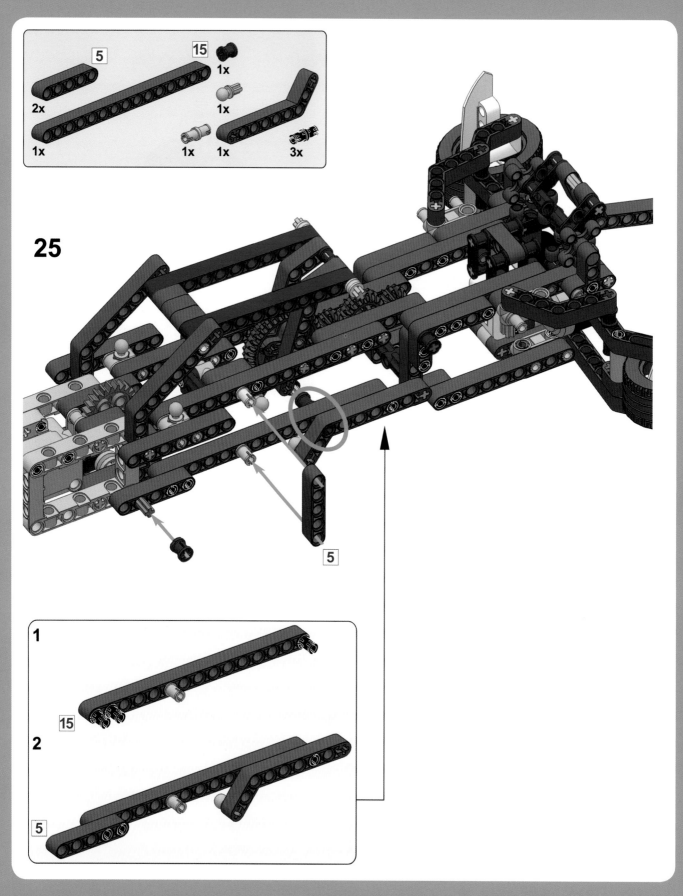

25

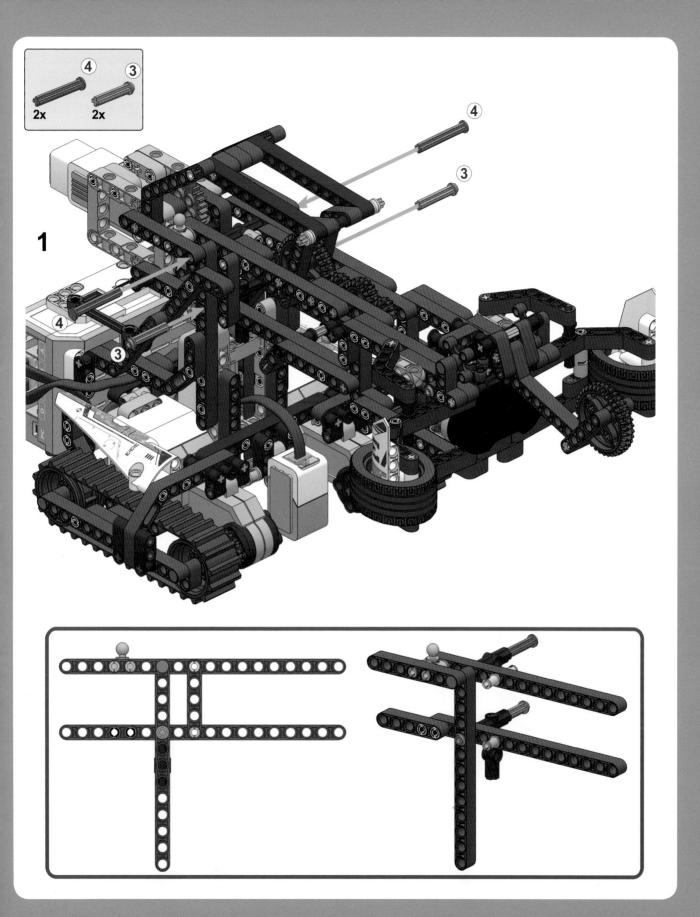

1

2x 4
2x 3

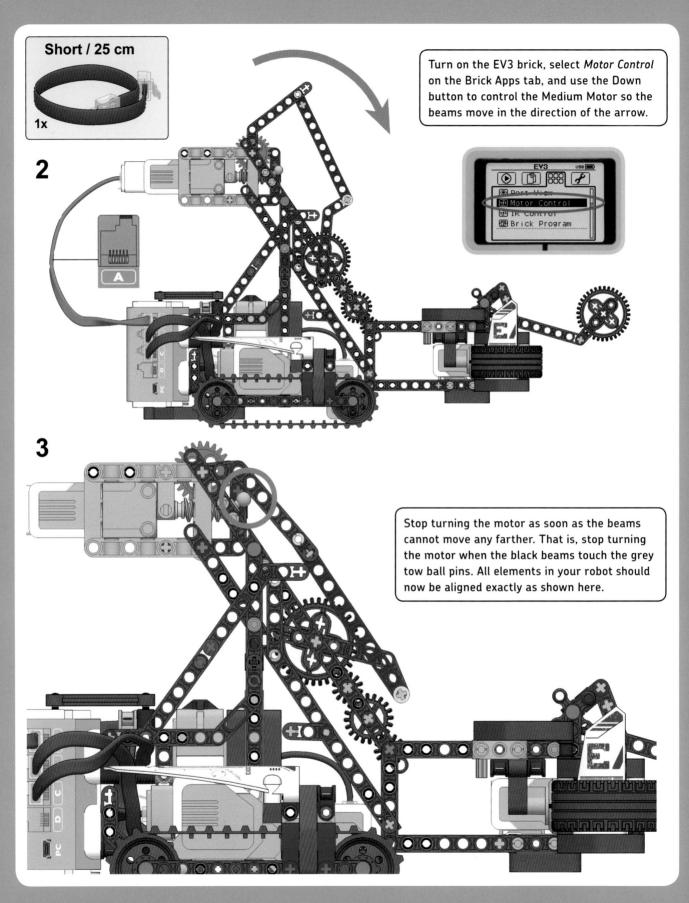

Short / 25 cm

1x

Turn on the EV3 brick, select *Motor Control* on the Brick Apps tab, and use the Down button to control the Medium Motor so the beams move in the direction of the arrow.

2

EV3   USB

▶ Port View
□□□ Motor Control
IR Control
Brick Program

A

3

Stop turning the motor as soon as the beams cannot move any farther. That is, stop turning the motor when the black beams touch the grey tow ball pins. All elements in your robot should now be aligned exactly as shown here.

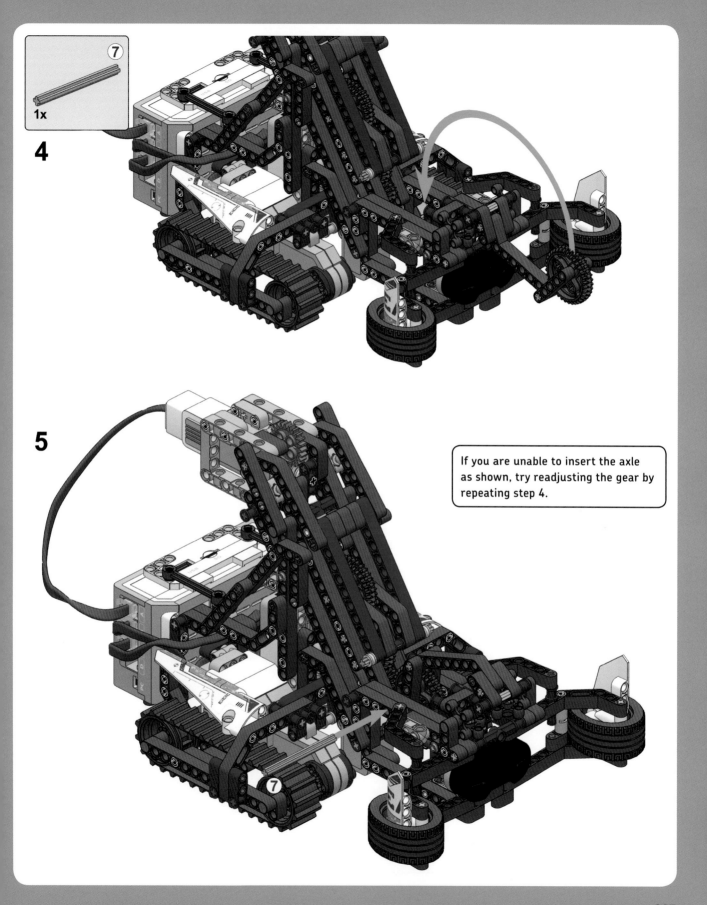

4

1x

7

5

If you are unable to insert the axle as shown, try readjusting the gear by repeating step 4.

7

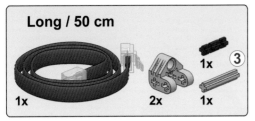

Long / 50 cm

1x   2x   1x   ③ 1x

Remove each of these helper pieces
from the claws.

6

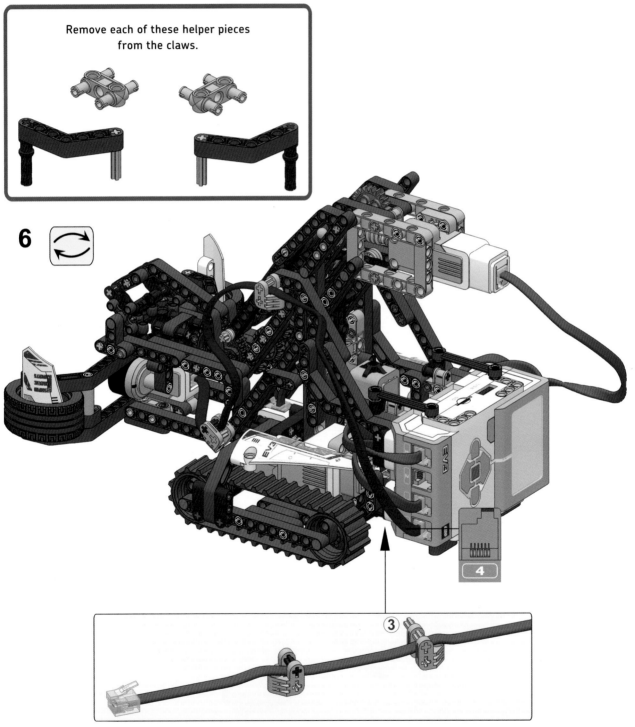

4

③

controlling the grabber

Now that you've built the robot, you're ready to test its mechanical functions by creating a remote control program. You'll drive the robot using Move Steering blocks, and you'll create three My Blocks to control the grabber.

my block #1: grab

To grab an object and raise the grabber, the Medium Motor must rotate forward until the Touch Sensor inside the robot's base is pressed. You'll limit the motor speed to 40% to reduce the amount of power required to drive the motor.

Create a new EV3 project called *SNATCH3R*, place three blocks on the Canvas as shown in Figure 18-5, and turn them into a My Block called *Grab*. Because the worm gear in the mechanism prevents the motor from turning when the motor isn't powered, setting the Brake at End to false is sufficient to keep the motor in place, and doing so saves some battery power.

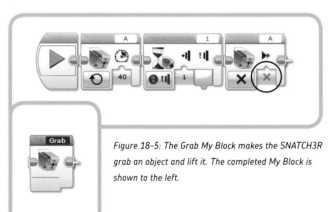

Figure 18-5: The Grab My Block makes the SNATCH3R grab an object and lift it. The completed My Block is shown to the left.

my block #2: reset

When the program starts to run, the grabber can either be lowered with its claws open, lifted all the way up with its claws closed (see Figure 18-1), or in any intermediate position. To prevent damage to the mechanism and the motor, it's important keep the grabber mechanism between these two boundaries by limiting how far the motor can turn.

You'll control the upper boundary with the Touch Sensor: If the sensor is pressed, the motor shouldn't move any farther forward. You'll control the lower boundary using the Rotation Sensor of the Medium Motor: If the sensor value is less than 0 degrees, the motor shouldn't move any farther backward.

To make this work, you'll need to make sure the Rotation Sensor value is 0 degrees when the grabber is in the lowered position before the program begins. To accomplish this, raise the grabber until the Touch Sensor is pressed with the Grab My Block, and then lower it with a Medium Motor block configured to rotate the motor backward for 14.2 rotations. Finally, reset the Rotation Sensor value to 0. The distance between the upper and lower boundary is 14.2 rotations only when the claws are fully closed while the grabber is raised, so you need to make sure there aren't any objects between the robot's claws during this reset procedure.

Create the *Reset* My Block using the instructions in Figure 18-6. You'll place this block at the start of every program for the SNATCH3R.

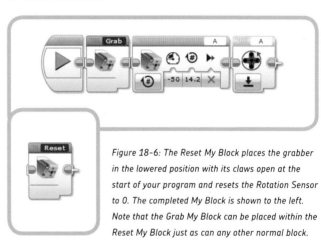

Figure 18-6: The Reset My Block places the grabber in the lowered position with its claws open at the start of your program and resets the Rotation Sensor to 0. The completed My Block is shown to the left. Note that the Grab My Block can be placed within the Reset My Block just as can any other normal block.

my block #3: release

To lower the grabber and release the object, the Medium Motor should turn backward until the grabber is in the lowered position or, in other words, until the Rotation Sensor is at 0 degrees. Create the *Release* My Block, as shown in Figure 18-7.

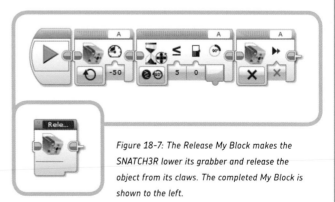

Figure 18-7: The Release My Block makes the SNATCH3R lower its grabber and release the object from its claws. The completed My Block is shown to the left.

creating the remote control program

Now create and run the *RemoteControl* program, as shown in Figure 18-8. Drive the robot around, and make it grab and move objects using the infrared remote control, as shown in Figure 18-9. The SNATCH3R should be able to grab, lift, and move lightweight objects, such as empty soda cans and water bottles.

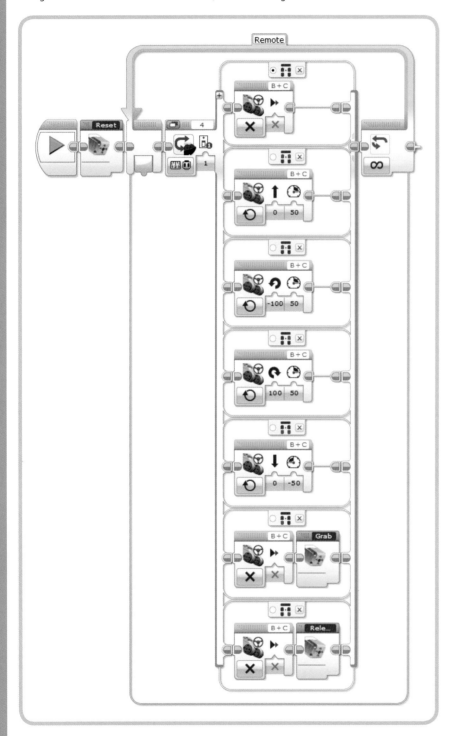

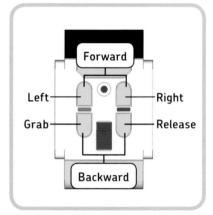

Figure 18-9: The commands for the RemoteControl *program*

Figure 18-8: The RemoteControl *program. The default case makes the motor stop moving so that the robot stops when you release the buttons.*

DISCOVERY #117: EXTENDED REMOTE!

Programming: 🖳🖳 **Time:** ⏲⏲

The *RemoteControl* program is great for testing the SNATCH3R's functionality, but it's a very basic program. Can you expand the program so you can make the robot drive in any direction, even while it's busy grabbing or releasing an object?

HINT Create two parallel sequences of blocks, with one to control the driving on channel 1 and another to control the grabber on channel 2. This allows you to control driving and grabbing simultaneously by switching between the two channels of the remote.

DISCOVERY #118: REMOTE SPEED CONTROL!

Programming: 🖳🖳🖳 **Time:** ⏲⏲

The Move Steering blocks in the *RemoteControl* program make the motors turn at 50% speed. However, sometimes you need to go faster (75%) to drive a large distance, and sometimes you need to go slower (25%) to accurately position the grabber in front of an object. Can you add remote control commands to change the speed of the motors?

HINT Define a numeric variable called *Speed*, and use it to control the speed of each Move Steering block. Then, add two extra cases to the Switch block, and add blocks that increment or decrement the value of the *Speed* variable by 10 each time you press a button.

troubleshooting the grabber

If you experience any problems with the SNATCH3R's grabber when running the *RemoteControl* program, you should solve them before proceeding to the next section. If you're not sure how the SNATCH3R should work exactly, watch a video of the robot in action at *http://ev3.robotsquare.com/*. The following problems and solutions may help you troubleshoot your robot.

* *The grabber doesn't close its claws before lifting.* This can happen when you have accidentally misaligned the gears in the SNATCH3R's arm. You can solve this problem by repeating the final building steps. To begin, remove the 7M axle you added on page 297 by pushing it out of the robot with another axle. Then, reattach the helper elements to the claws (step 16 on page 288). You can then continue normally from page 296 and test your robot again. Be sure to carefully observe the side views of the mechanism on page 296; each element of your robot should be aligned exactly as shown.
* *The cable connected to the Infrared Sensor prevents the claws from closing.* This can occur if the cable is in the way of the 36T gear in the grabber. To check whether this is the problem you're experiencing, try removing the cable completely and run the Reset My Block. If the grabber works fine now, you know the cable is the problem. Reattach the cable to your robot in such a way that it does not interfere with any of the gears.
* *The Medium Motor isn't aligned parallel to the ground.* This happens if the grabber isn't correctly mounted to the robot's driving base. To resolve this, remove the axles you added on page 295 and carefully reattach them according to the instructions. The side views on that page show exactly in which holes you should mount these axles (some elements have been removed for better visibility).

searching for the IR beacon

You'll now create a program that makes the SNATCH3R find, grab, lift, and move the infrared beacon. Each task should run *autonomously*, which means that all tasks are performed without human interaction.

building the IR bug

Before you create the program, you'll need to add some elements to the infrared beacon to make it easier for the robot to grab and lift it. Follow the instructions on the next page to build the IR "bug" using the remaining elements of the EV3 set. (If you used the Bill of Materials from Figure 18-4 to select the elements for your robot, you should have the required pieces at hand.)

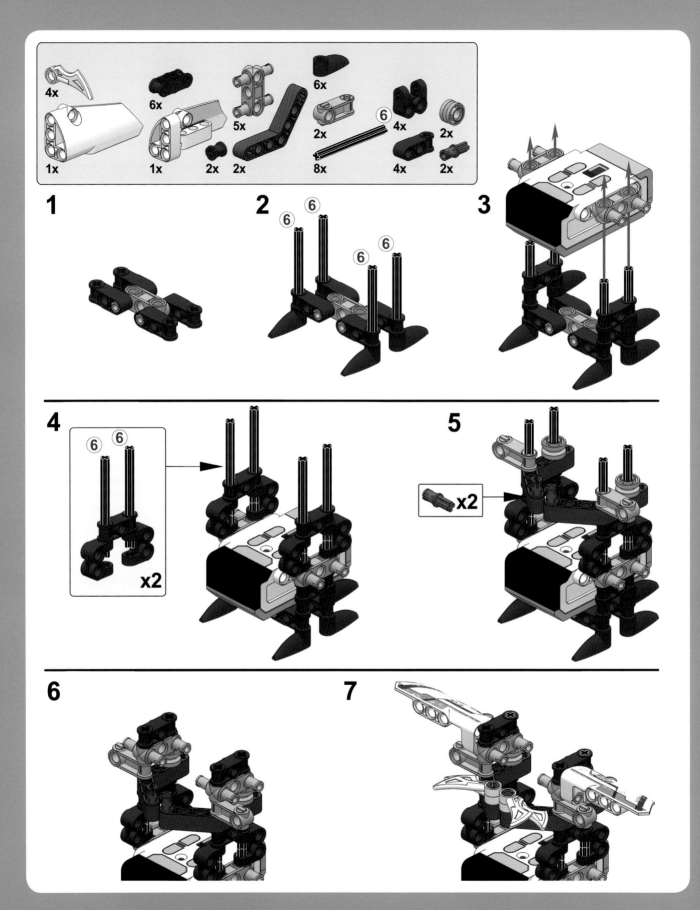

1

2

3

4

5

6

7

my block #4: search

You learned earlier that the robot can find the infrared beacon by driving to the left if the Infrared Sensor's Beacon Heading is negative or to the right if it's positive. Now that you've learned to use data wires and variables, you can create a more sophisticated program that makes the robot actually search for the beacon so that it will find the beacon up to 2 meters (6 feet) away, even if the beacon is behind the robot.

You'll create a My Block called *Search* that has the robot scan its surroundings while making one complete turn to the left. Then, it turns right until it's back at the position where it saw the beacon and its claws point toward the beacon. In principle, the robot could now find the beacon simply by driving straight ahead, but because the sensor measurement is not that accurate, you'll create a My Block to make it easy to search for the beacon again if necessary.

understanding the sensor measurements

To understand the search algorithm, you need to understand the Beacon Heading measurements that the robot takes as it makes one turn, as shown in Figure 18-10. When the sensor points in the direction of the beacon, the heading value (H) is 0 or close to 0. When the sensor points about 90 degrees away from the beacon, the heading is 25 or –25. Additionally, the sensor value is 0 when the robot faces away from the beacon because the SNATCH3R's body blocks the view of the beacon and the sensor can't determine the signal's direction.

To find the beacon, we should therefore look for a value *near* 0 but not exactly 0. We should ignore 0 measurements because they might indicate that the beacon is behind the robot. Doing so will reduce the accuracy slightly, but at least you'll be sure that the robot doesn't drive away from the beacon. (In the final search phase, you'll no longer ignore 0 values.)

Finally, it doesn't matter whether we detect positive or negative values, so we can take the absolute value of the measurement. For example, –3 and 3 are equally close to 0.

To sum up, we have to look for the *lowest measured absolute value that is not 0* and store it in the robot's memory.

Besides knowing the lowest heading value, the robot should have a sense of where this detection was made. The robot uses the Rotation Sensor in motor C to keep track of the robot's position, as shown in Figure 18-10. The Rotation Sensor value (R) is 0 at first, and it increases as the robot turns to the left. Each motor has to turn about 1800 degrees in order to have the robot make one complete turn.

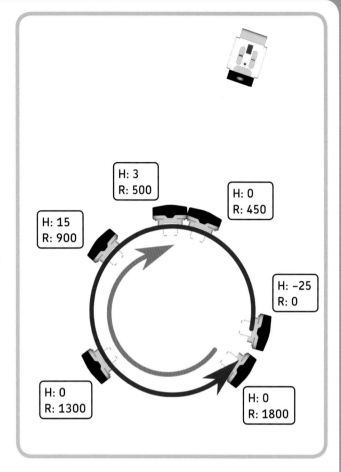

Figure 18-10: As the robot turns left (blue arrow), the Infrared Sensor continuously measures the Beacon Heading value, but this diagram shows only six measurements. If we ignore 0 values, the lowest absolute heading value (H) is 3 in this case. At this position, the Rotation Sensor value of motor C (R) is 500 degrees. When the robot completes the circle (when the Rotation Sensor measures 1800 degrees), the robot turns right until it measures 500 degrees again (green arrow) and the robot points (roughly) in the direction of the beacon.

By storing the Rotation Sensor value measured at the time the lowest Beacon Heading value was detected, the robot is able to return to this position later; it just has to turn to the right until motor C is back at the stored position (500 degrees in the example).

understanding the search algorithm

The flow diagram in Figure 18-11 shows how the robot can determine the lowest valid heading value and the corresponding Rotation Sensor value. In the diagram, the variable called *Reading* represents the absolute value of the Beacon Heading measurement, updated with a new measurement each time the loop repeats.

Rather than storing all measurements and picking the lowest one, the program stores only one value in a variable called *Lowest*. Each time a new valid reading is less than the value in *Lowest*, the new reading is stored in *Lowest* and the

Rotation Sensor value is stored in a variable called *Position*. Ultimately, *Lowest* contains the lowest recorded valid sensor measurement, and *Position* contains the Rotation Sensor value recorded when this lowest detection was made.

creating and testing the search my block

You can now implement the flow diagram using the programming instructions shown in Figures 18-12 through 18-16. You'll implement the two questions in the flow diagram using Switch blocks, marked *a* and *b* respectively.

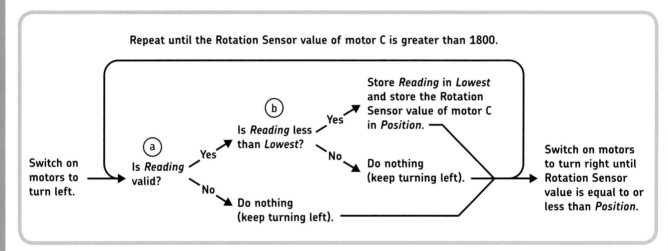

Figure 18-11: The search algorithm. The Reading *variable contains the absolute value of the Beacon Heading and is considered valid if it isn't 0.*

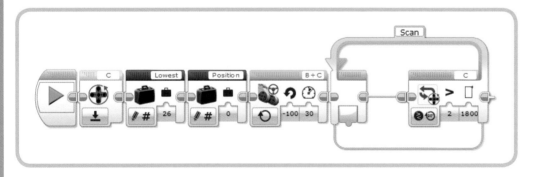

Figure 18-12: Step 1: Define two Numeric variables called Lowest and Position, and initialize them as shown. Initialize Lowest to 26 so that any successful measurement (25 or less) will be the next lowest value. Also, reset the Rotation Sensor of motor C, switch on the motors to turn left, and add the Loop block you'll use to scan for the beacon. The loop runs until motor C has turned 1800 degrees. (This should result in the robot turning roughly 360 degrees.)

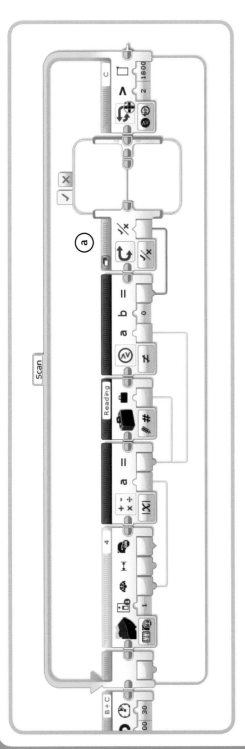

Figure 18-13: Step 2: Now add the blocks to the loop that make the robot store one measurement, take its absolute value, and determine whether it's nonzero. Define a Numeric variable called Reading, and configure the blocks as shown. You use the Compare block to check whether the value is nonzero, and the blocks on the true tab of the Switch block (a) will run if this is the case. Nothing should happen if the value is zero, so the false tab should remain empty.

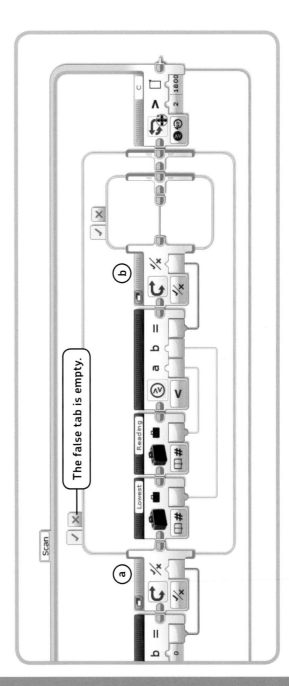

Figure 18-14: Step 3: Now that you have a valid measurement, you can compare it to the lowest value recorded so far (a). If Reading is less than Lowest, the output of the Compare block is true, and the blocks on the true tab of the Switch block (b) will run.

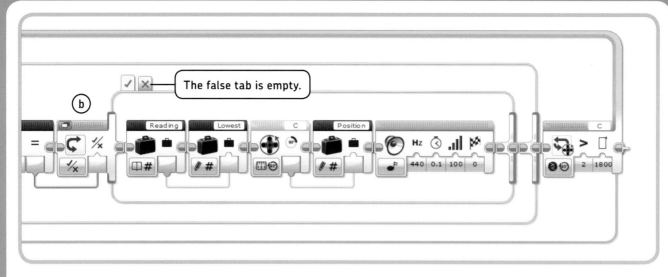

Figure 18-15: Step 4: Now that you know that the new value in Reading is less than Lowest, you store Reading in Lowest, and you store the current Rotation Sensor value in Position. You add the Sound block so that you hear a beep each time the robot updates Lowest and Position with new values. Nothing should happen if the new Reading is not less than Lowest, so the false tab should remain empty.

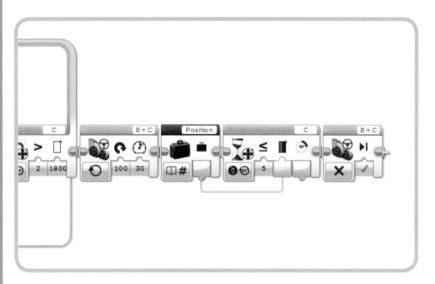

Figure 18-16: Step 5: Having completed its full turn to the left, the robot should now turn right until the Rotation Sensor is back at the point where the lowest measurement was detected (the value stored in Position). The motors are switched off, and the robot should now face in the direction of the beacon.

Figure 18-17: Step 6: Turn all of the blocks into a My Block called Search.

When you're ready, turn all the blocks into a My Block called *Search*, as shown in Figure 18-17.

Place the IR bug about 1 meter (3 feet) away from the robot, and make the beacon continuously send a signal by pressing the button at the top of the remote (Button ID 9). The robot can face in any direction when the program begins, but the infrared beacon should point toward the robot, as shown in Figure 18-18. If the grabber is not in the reset position with its claws open, run the Reset My Block you made earlier.

Now run the Search My Block to test it. The robot should turn around and beep each time it sees a lower measurement than the previously stored lowest value. After making one complete turn to the left, it should turn to the right until it faces the beacon. This should be the point where you last heard the robot beep. If the robot isn't successful on its first try, place the IR bug closer to the robot and run the Search My Block again.

Figure 18-18: Place the IR bug about 1 meter (3 feet) away from the robot. The robot can face in any direction, but the beacon should point toward the robot.

DISCOVERY #119: SIGNAL VERIFICATION!

Programming: 🖧🖧 **Time:** 🕐

If you forget to switch on the infrared beacon, the robot won't detect a signal at all. Can you add blocks at the end of the Search My Block that make the robot say "Error" if the robot wasn't able to get the beacon's direction any time during the loop?

HINT If the robot didn't detect anything during the loop, what value will be stored in *Lowest* after the loop completes? Look for the answer in Figures 18-11 and 18-12.

creating the final program

The final program will make the robot search for the IR bug, drive toward it, grab and lift it, move it to a new position, and then lower and release it. To begin, create a new program called *Autonomous* and implement each of these actions using the instructions that follow.

finding the beacon

The first part of the program will make the SNATCH3R search for the beacon with the Search My Block and then drive forward to approach it. The robot will continue to search and drive forward until the beacon proximity is less than 50%, indicating that the beacon is nearby, as shown in Figure 18-19.

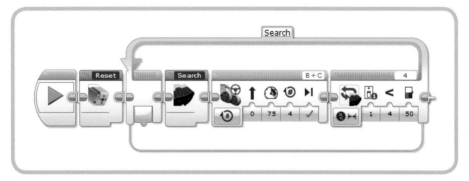

Figure 18-19: Step 1: The robot searches for the beacon until it's nearby.

driving toward the beacon

Once the beacon is in sight, the robot can drive toward it by going forward while adjusting its steering to the Beacon Heading value. Rather than using a fixed steering setting, as you did in Chapter 8, you'll make the amount of steering proportional to the Beacon Heading. The farther the beacon is to the left, the more the robot steers left; the farther the beacon is to the right, the more the robot steers right.

Because the previous search loop ended only once the beacon proximity was less than 50%, you know that the robot faces at least roughly in the direction of the beacon. Therefore, 0 values should indicate that the beacon is right in front of the robot, and you no longer have to ignore them. In fact, because steering is proportional to the heading, steering will be 0 if the heading is 0, and the robot will drive straight ahead.

The robot will continue to adapt its steering setting until the beacon proximity is 1%, when the IR bug is nearly positioned between the robot's claws, as shown in Figure 18-20. Test this section of the program by selecting only the Loop block and clicking the **Run Selected** button.

lifting and moving the IR bug

The robot is almost ready to grab the beacon, but first it drives forward for one more rotation to ensure the IR bug is properly positioned between the robot's claws. Once the grabber is raised with the Grab My Block, the robot turns around, drives forward for a short while, and then lowers and releases the object in a different position with the Release My Block, as shown in Figure 18-21. When you're ready, run your program to see how well the SNATCH3R can autonomously find the IR bug.

further exploration

You've just completed one of the most complex robots in this book. Congratulations! In this chapter, you've seen how you can combine many advanced building and programming techniques to create a truly autonomous robot. Now that you've built and programmed the SNATCH3R, see what else you can do with this robot. To get started, explore some of the Discoveries that follow to put your robotics skills to the test.

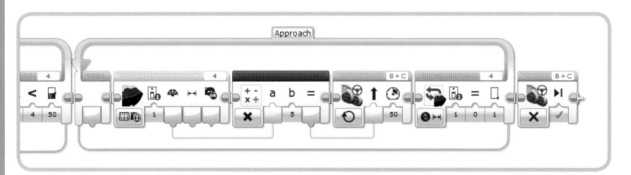

Figure 18-20: Step 2: Driving forward while adjusting the steering to the Beacon Heading value until the beacon is between the robot's claws

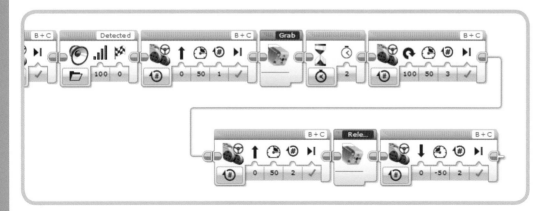

Figure 18-21: Step 3: Grabbing, lifting, and moving the IR bug

DISCOVERY #120:
KEEPING BUSY!

Difficulty: ▱ **Time:** ◷◷
Can you expand the *Autonomous* program so that the SNATCH3R repeatedly finds and moves the IR bug? Make the robot drive away from the object after it's released it, search for it again, and so on. If the SNATCH3R doesn't properly release the IR bug before driving away, program the robot to zigzag while moving backward to shake off the object.

HINT Use Random blocks to control the number of rotations the robot moves and turns after it has grabbed the object. This causes the robot to place the object in a random position each time.

DISCOVERY #121:
PATH FINDER!

Difficulty: ▱▱ **Time:** ◷◷
The Color Sensor in the base of the SNATCH3R enables the robot to see the color of the surface beneath it in order to follow lines. Can you make the robot follow the lines of a custom track, grab an object at the end of the line, and then return it to the start of the line?

TIP The test track you made in Chapter 7 (see Figure 7-4 on page 77) might not work very well because the SNATCH3R's treads can tear the soft paper apart. To solve this problem, glue the track to a tough piece of cardboard or create your own track on a sheet of plywood with black tape or a marker. Alternatively, you can use the Mission Pad and use the colored landmarks to design your own mission.

DISCOVERY #122:
PROXIMITY FINDER!

Difficulty: ▱▱▱ **Time:** ◷◷
Can you make the SNATCH3R autonomously find objects other than the infrared beacon, such as an empty water bottle? Use the infrared proximity measurement to detect the object closest to the robot. Then make the robot drive toward the object and grab it.

HINT Begin by creating and testing a modified version of the Search My Block. What operation mode should the Infrared Sensor use? What should be the initial value of the *Lowest* variable?

DESIGN DISCOVERY #29:
EXCAVATOR!

Building: ✹✹✹ **Programming:** ▱
Can you build a robotic excavator? Remove the robotic arm from the SNATCH3R so that only the base remains (see page 279). Use the Medium Motor to control the arm and the digger.

19

LAVA R3X: the humanoid that walks and talks

So far in this book, you've built vehicle robots, animal robots, and machines, but perhaps one of the coolest robot projects you can make with your EV3 set is a humanoid robot that walks on two legs. In this chapter, you'll build and program *LAVA R3X*, shown in Figure 19-1. LAVA R3X walks on two legs controlled by Large Motors, and it can move its head and arms using the Medium Motor.

Once you've made the robot walk, you'll be challenged to expand the program to make the robot interactive and lifelike by using the techniques you've learned throughout this book.

LAVA R3X is able to walk by continuously shifting its weight to one foot while moving the other foot forward. A mechanism in each leg turns the continuous forward motion of the motor into an alternating forward and backward motion of the foot and an alternating left and right tilting motion of the ankle (see Figure 19-2).

Figure 19-2: As the motor in each leg makes one complete rotation, the foot moves back and forth and the ankle tilts to the left and to the right.

Figure 19-1: LAVA R3X walks around on two legs while moving its head and arms, and it greets you when you shake its hand.

For the robot to walk in a stable manner, the mechanisms of both legs must be in the exact opposite position, and the motors should turn at the same speed. When these requirements are met, one foot is tilted so that it doesn't touch the ground while it moves forward, while the other foot is tilted so that it carries the robot's weight while it pushes backward, thereby propelling the robot forward.

As the motor in each leg makes one rotation, the mechanism moves left, right, forward, and backward, after which it's back in its initial orientation, having completed one step.

You'll use the Touch Sensor between the robot's legs to place the mechanisms in opposite positions, just as you did for ANTY's motors in Chapter 13, and you'll create a speed controller to ensure that both motors turn at the same speed.

building the legs

First, you'll build the robot's legs and create My Blocks to make the robot walk and turn. Next, you'll build the robot's upper body and create a program to make the robot interact with its surroundings using sensors. Select the pieces you'll need for the robot, shown in Figure 19-3, and then follow the instructions on the next pages.

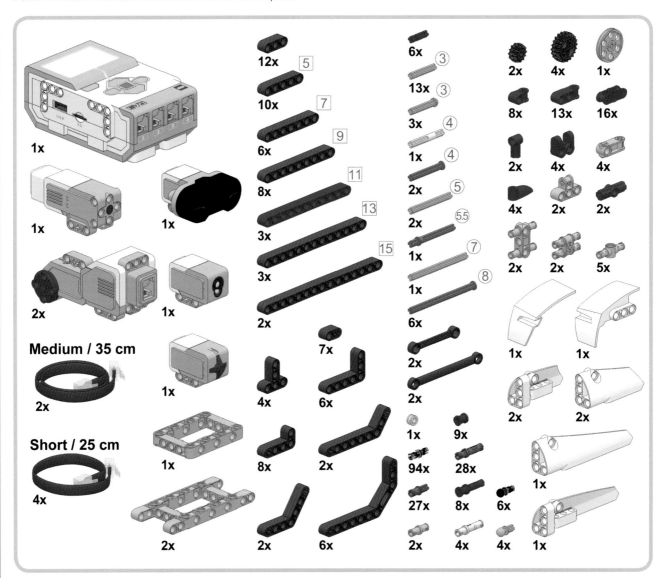

Figure 19-3: The pieces needed to build LAVA R3X. You should have a number of leftover pieces after you've built the legs; you'll use them to build the robot's head and arms later.

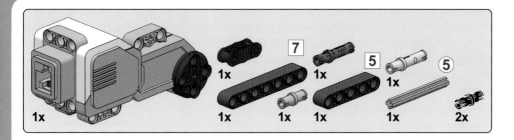

1

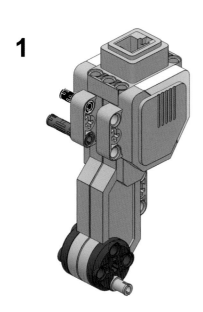

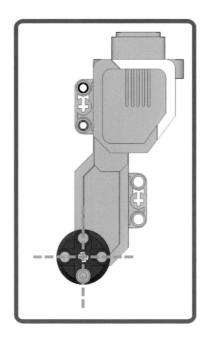

2

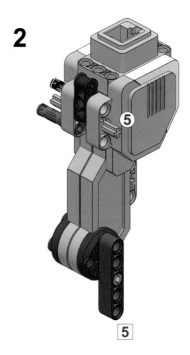

5

3

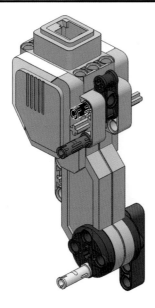

4

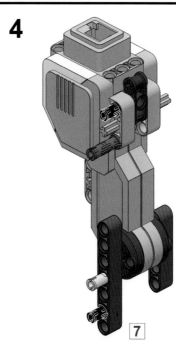

7

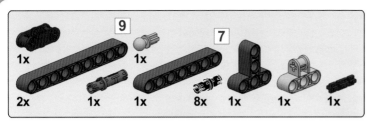

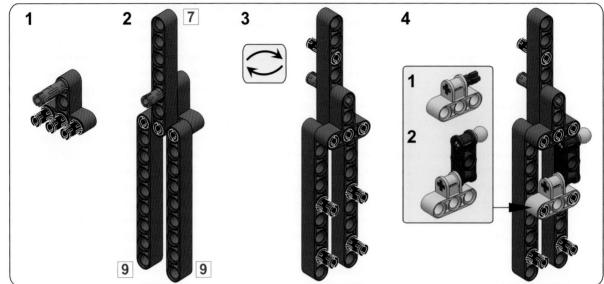

5

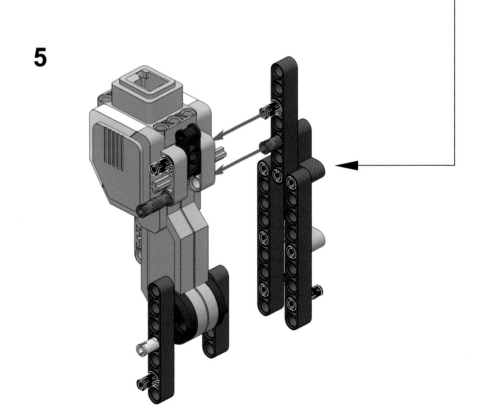

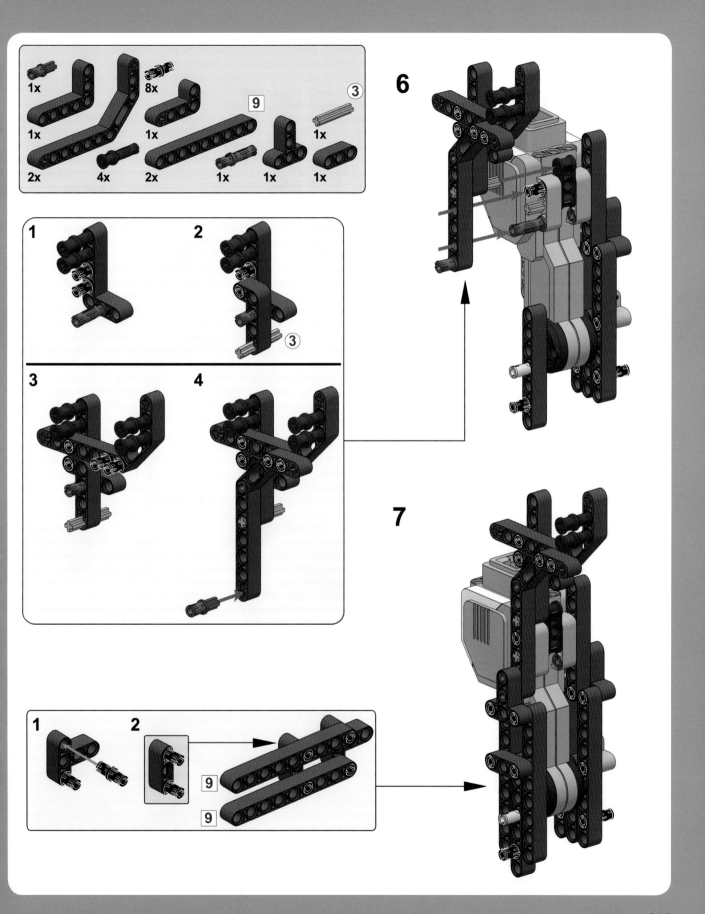

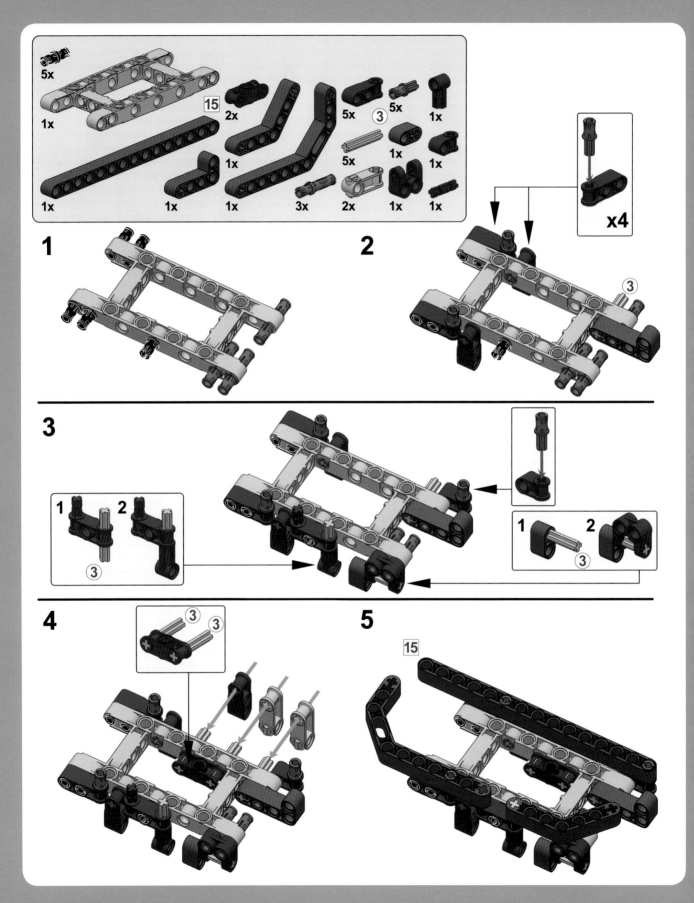

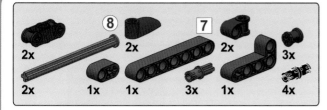

6

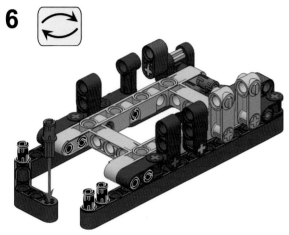

7

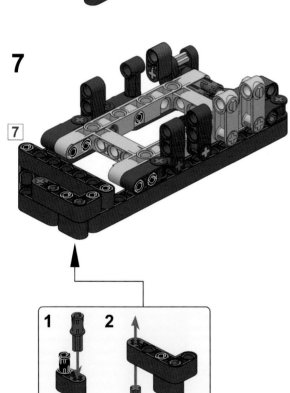

8

9

10

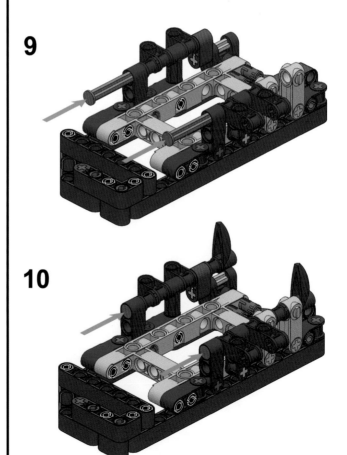

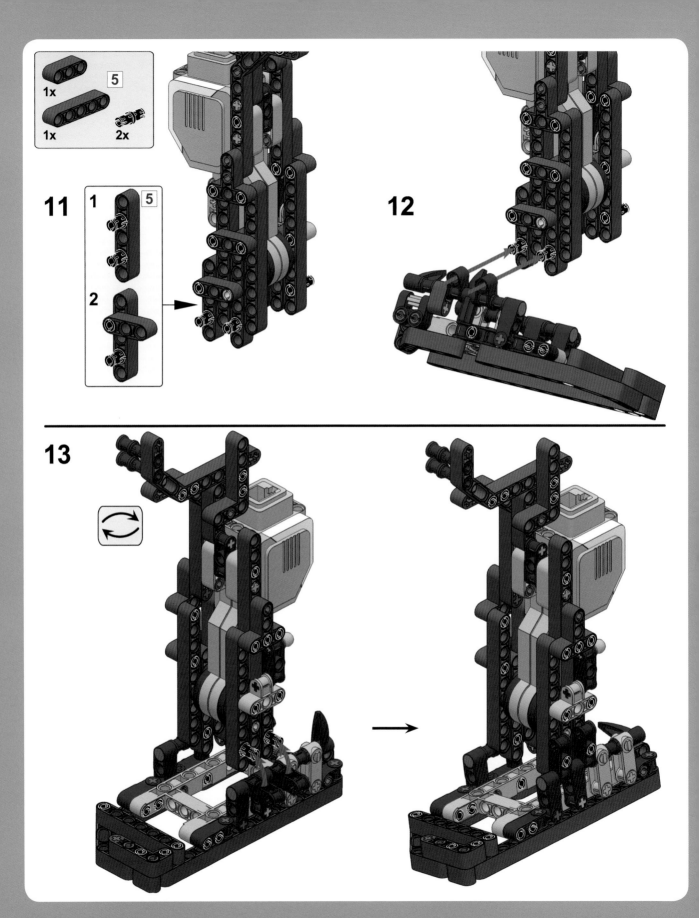

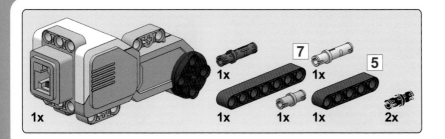

1

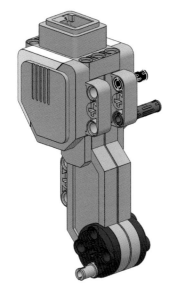

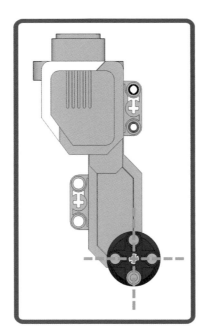

2

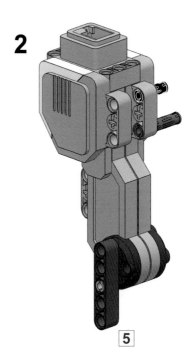

5

3

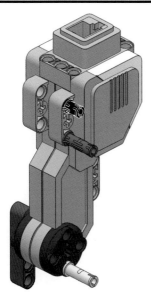

4

7

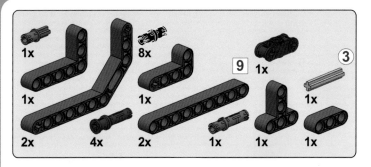

1x 8x

9 1x 3

1x 1x

2x 4x 2x 1x 1x 1x

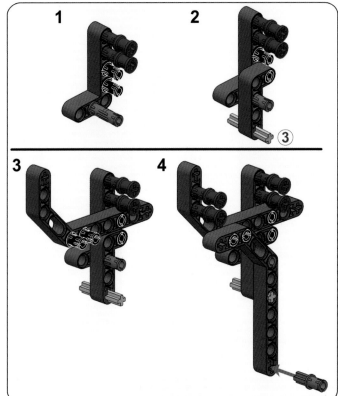

1

2

3

3

4

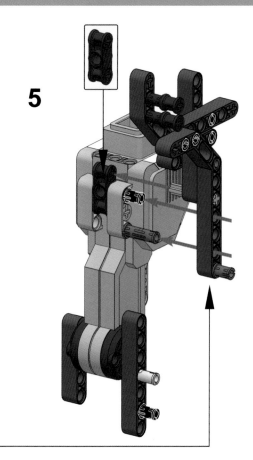

5

6

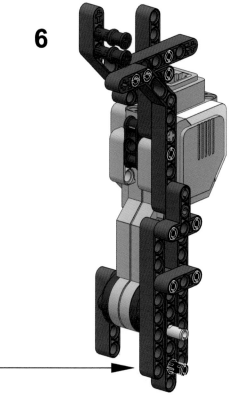

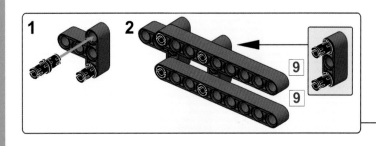

1

2

9

9

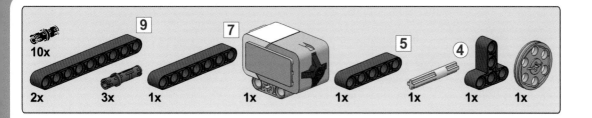

1

7

2

3

9
9

4

5

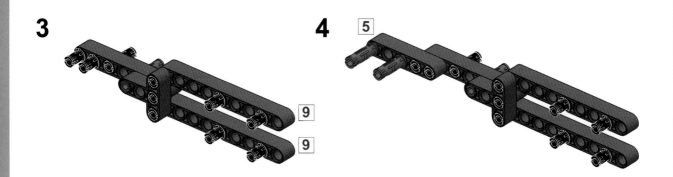

1 **2** **3**

4

5

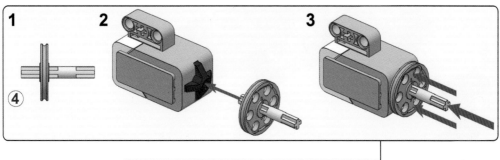

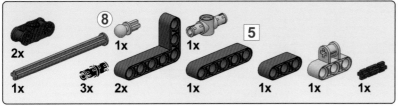

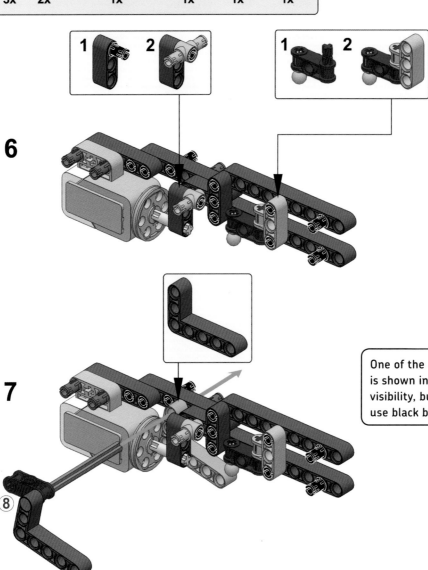

One of the angled beams is shown in blue for better visibility, but you should simply use black beams in this step.

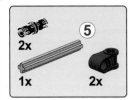

9

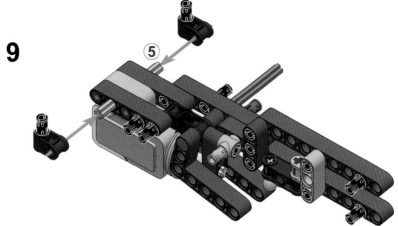

10

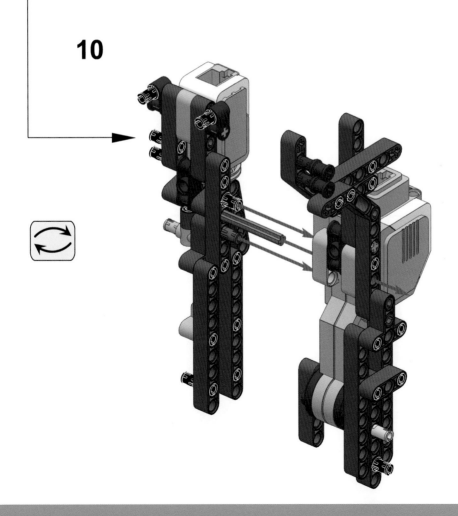

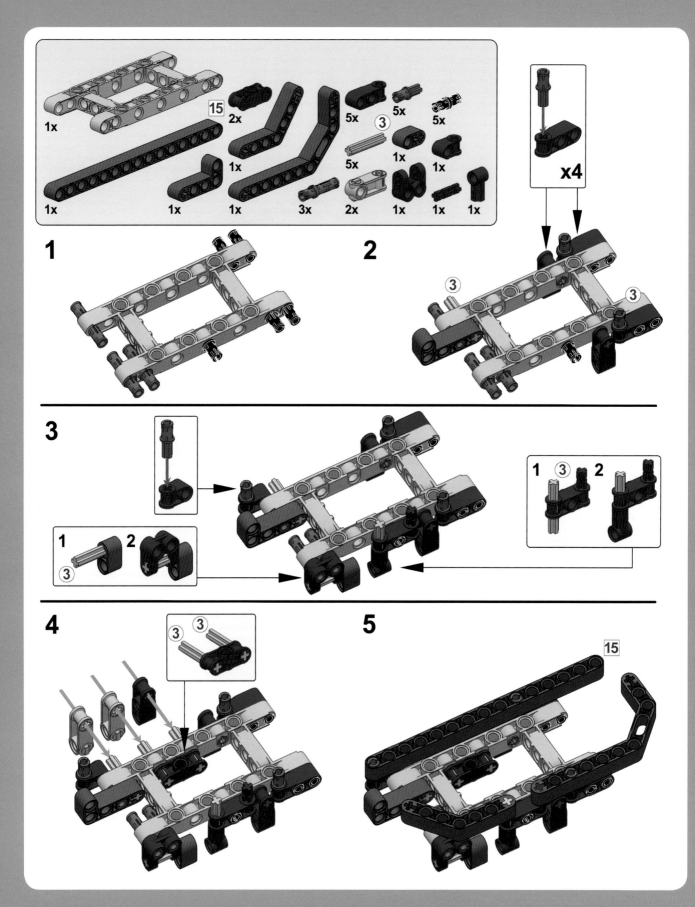

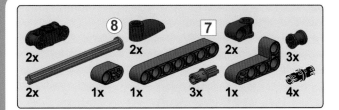

6

7

8

9

10

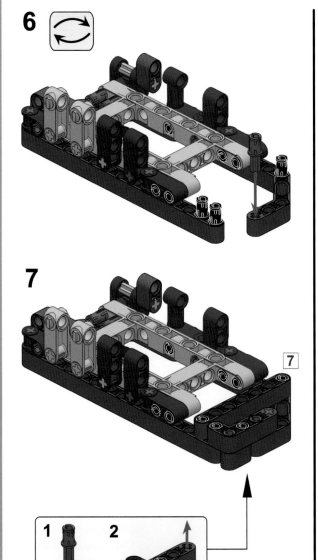

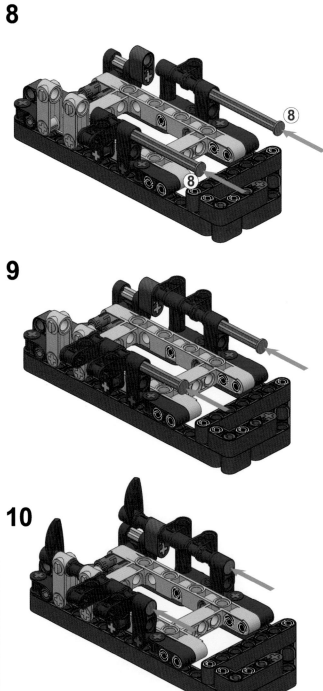

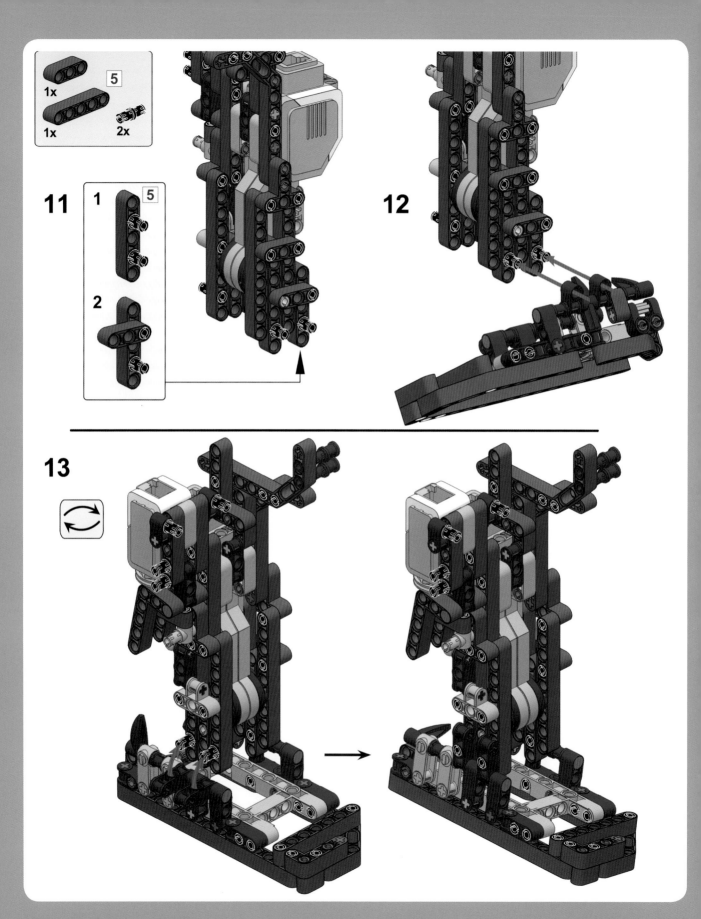

1

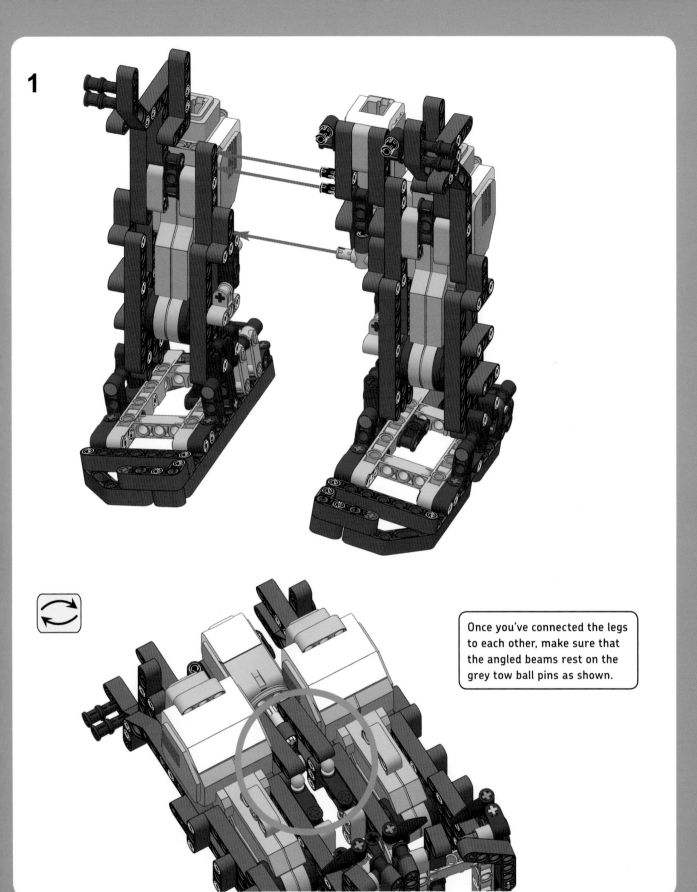

Once you've connected the legs to each other, make sure that the angled beams rest on the grey tow ball pins as shown.

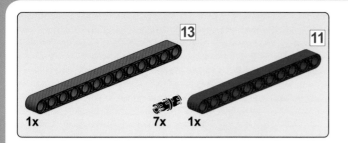

2 ↻

3 ↻

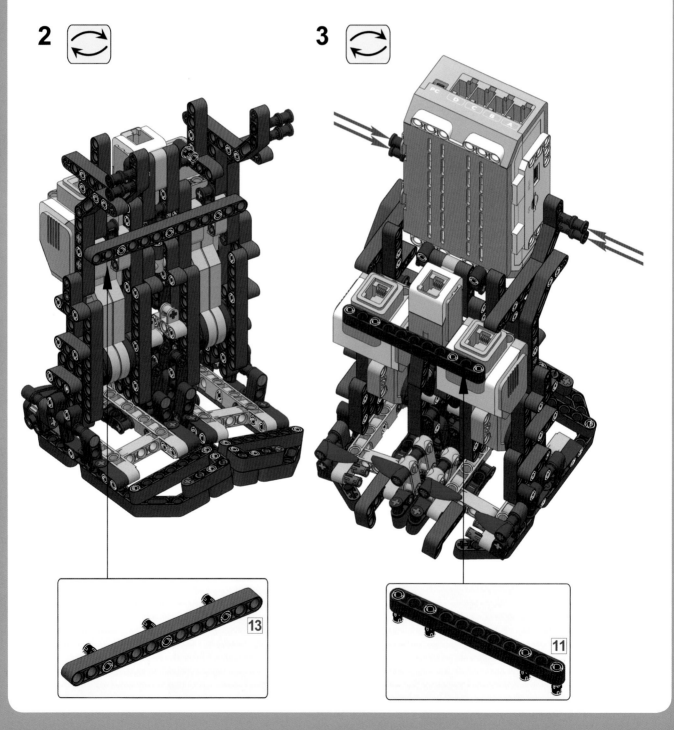

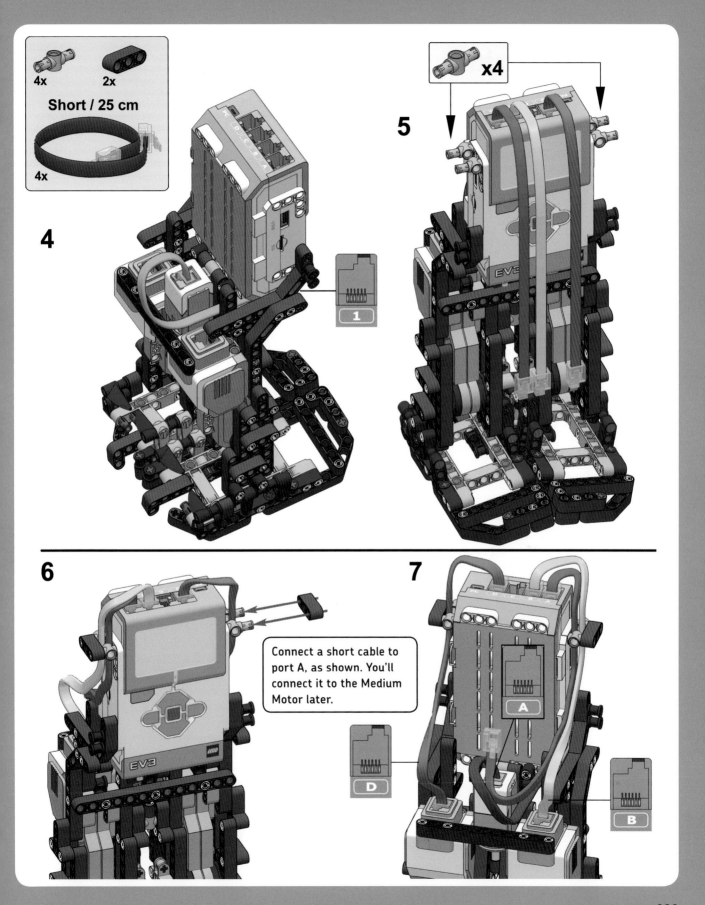

4

5

x4

1

6

7

Connect a short cable to port A, as shown. You'll connect it to the Medium Motor later.

D

A

B

Short / 25 cm

4x   2x

4x

making the robot walk

You'll now create several My Blocks to place the robot's legs in opposite positions, to make the robot walk forward, and to make the robot turn to the left. You'll also make a small program to test each block.

Testing these blocks without the robot's top heavy upper body makes troubleshooting easier because the robot is less likely to fall over. Once the robot walks in a stable manner using the My Blocks, you'll be ready to complete the design.

my block #1: reset

Each time a motor makes one complete rotation, the mechanism around this motor presses the Touch Sensor once (see Figure 19-4). LAVA R3X can use this sensor information to place the left and right leg in opposite positions.

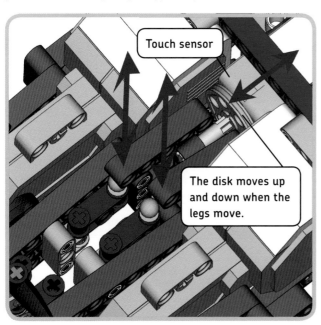

Figure 19-4: As each of the motors makes a rotation, it pushes a black angled beam against a grey disk, which in turn presses the Touch Sensor. The tan axle ensures that the grey disk remains aligned properly.

Because the robot can't detect which of the two leg mechanisms is pressing the Touch Sensor at a given moment, it should first position the legs in such a way that neither mechanism presses the sensor. To accomplish this, the robot moves the motors forward until the Touch Sensor is released, and then it stops the motors. Sometimes when the motors stop, a small amount of play in the mechanism causes the Touch Sensor to become pressed again. To make sure that the sensor stays in the released position, the robot pauses for 0.1 seconds. Then, it checks the state of the Touch Sensor again with a Loop block. If it's still released after this short pause, the loop ends; if the sensor has become pressed in the meantime, the loop runs again.

Once the robot knows that neither mechanism is pressing the Touch Sensor, it continues the reset procedure. First, it rotates the left motor (motor D) forward until the Touch Sensor is bumped, and then it rotates the motor 90 degrees farther *forward*. Next, the robot rotates the right motor (motor B) forward until the Touch Sensor is bumped, and then it rotates the motor 90 degrees *backward* so that the leg mechanisms are now 180 degrees apart, ready to begin walking. When the legs are in place, both Rotation Sensors are reset to 0. This means that while walking, the mechanisms are in opposite positions as long as the Rotation Sensor values of both motors are equal. Resetting the sensors will also make it easy to return to this starting position later on.

Create the *Reset* My Block that performs these actions, as shown in Figure 19-5. You'll use it at the start of each program for LAVA R3X.

my block #2: return

Throughout the program, LAVA R3X will not only walk straight forward but also turn to the left. After the robot turns, the leg mechanisms may no longer be in opposite positions, so you'll have to return the legs to these respective positions before the robot can continue to walk. But running the Reset block after each turn is impractical because it takes a long time to run. Fortunately, you can achieve the same effect by rotating each motor back to its 0 position.

To make a motor return to its starting position, the robot should measure its current position and rotate backward by the measured amount. For example, if a motor has turned 25 degrees forward, it has to turn 25 degrees backward. If the motor has made more than one rotation, it has to rotate backward only by the number of degrees that exceeds a whole number of rotations (see Figure 19-6). For example, if the sensor measures 450 degrees, the motor should rotate backward by only 90 degrees so that the sensor eventually measures 360 degrees. In this motor position, the foot has the same orientation as if the sensor measured 0 degrees, so the result is the same.

You can calculate the amount by which the position exceeds a whole number of rotations with the *modulo operator*, %. The modulo operator gives the remainder after one number is divided by another. For example, 7 divided by 3 gives 2 with a remainder of 1. That is, 7 % 3 = 1. The modulo operator is available in the Advanced mode of the Math block.

You can calculate the required number of degrees by taking the remainder of a division by 360. For example, 450 % 360 = 90.

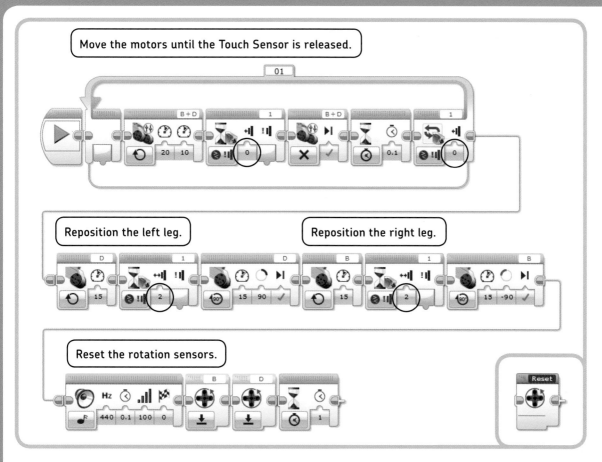

Figure 19-5: The configuration of the blocks in the Reset My Block (left) and the completed My Block (bottom right)

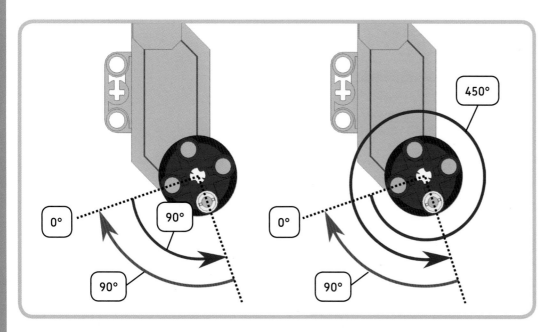

Figure 19-6: When a motor turns forward (blue arrow), it can return to the 0 position (0°) by measuring the current position (90°) and turning backward (green arrow) by the same amount. If the motor has made more than one rotation, it doesn't have to rotate all the way backward, but only the number of degrees that exceeds a whole number of rotations, as shown on the right (450° – 360° = 90°).

In other words, full rotations are removed from the number so that a smaller number of degrees remains (a number less than 360 degrees). You'll use this remainder to return the mechanism to its starting orientation.

Create the *Return* My Block that makes both mechanisms return to their starting orientation, as shown in Figure 19-7.

NOTE For division (÷), the Math block uses the slash symbol (/). For multiplication (×), it uses the asterisk (*). The modulo operator is represented by the percent sign (%). Instead of typing each symbol, you can select it from the list that appears when you enter the equation.

my block #3: onsync

To make the robot walk after the leg mechanisms have been placed in opposite positions, the robot must turn both motors forward at 20% speed (34 rpm). A Move Steering block in On mode might seem like a good candidate for this task, but slight speed deviations would eventually cause one motor to fall behind the other one so that the leg mechanisms would no longer be in opposite positions. Therefore, you'll have to make your own substitute that keeps the motors *synchronized*. That

is, you'll have to create a block that turns both motors at 20% speed on average while ensuring that the Rotation Sensor value of motor D is (almost) equal to the Rotation Sensor value of motor B.

To accomplish this, you'll make a motor run a little faster than 20% speed if it falls behind the other one, and you'll make the motor that's ahead turn a little slower than 20%. The farther the motor positions are apart, the larger the speed adjustment you'll make. For example, if motor B measures 790 degrees while motor D measures 750 degrees, you'll make motor D turn at 22% speed and motor B at 18% speed so that motor D can catch up with motor B.

To calculate the speed adjustment (2%, in this case), you first determine the difference between the two motor positions by subtracting the position of motor D from motor B (790 – 750 = 40 degrees, in this example). As in the Return My Block, you apply a modulo 360 operation to this number. That's because if a motor falls behind for more than one rotation, it is much easier for the motor to catch up by turning only the amount that exceeds a whole number of rotations (you get the same result with less effort). Place and configure the blocks that do this, as shown in Figure 19-8.

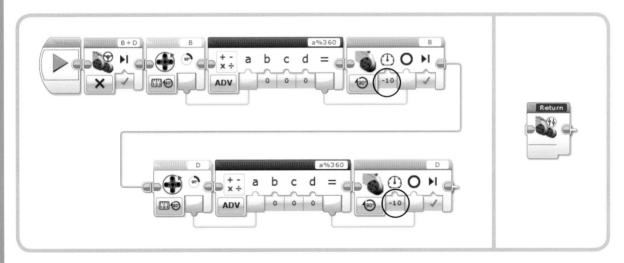

Figure 19-7: The configuration of the blocks in the Return My Block (left) and the completed My Block (right)

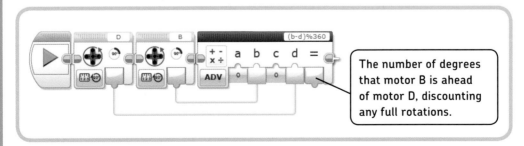

> The number of degrees that motor B is ahead of motor D, discounting any full rotations.

Figure 19-8: Step 1: The first blocks in the OnSync My Block calculate the difference between the two motor positions

You now have the number of degrees by which motor B is ahead of motor D. Sometimes this difference is greater than 180 degrees or less than –180 degrees. For example, motor B might be 220 degrees ahead of D. In these cases, you'll subtract 360 degrees (220 – 360 = –140, in this example) to find a more efficient rotation. In other words, you can now say that motor B is 140 degrees *behind* motor D. Because the motors rotate in a circle, the meaning is exactly the same, but the required speed adjustment will be less, and the movement will be less erratic.

Similarly, if the difference is less than –180 degrees, you'll add 360 degrees to the difference between the motor positions. Use Compare blocks to apply these steps in the program (see Figure 19-9). The output of the first Compare block is *true* (1) if the difference is greater than 180 degrees and *false* (0) if it's not. For *true*, the Math block subtracts 1 × 360 = 360 from the difference; for *false*, it subtracts 0 × 360 = 0 from the difference,

therefore leaving it unchanged. The second set of Compare and Math blocks works the same way, except it adds 360 degrees if the difference is less than –180 degrees.

Now we have the difference in degrees between motor B and motor D, and the number is adjusted to be the shortest distance between the two motor positions. For a difference of 40 degrees, we want a speed adjustment of 2% for each motor, so we divide the difference by 20 before adding it to motor D's speed and subtracting it from motor B's speed (see Figure 19-10). Turn the blocks on the Canvas into a My Block called *OnSync*, as shown in Figure 19-11.

The OnSync My Block may look complicated, but its function is straightforward. Just remember that when you place it in a Loop block, it functions like a Move Steering block in On mode, but it ensures that both motors stay synchronized by adjusting the motor speeds when one gets ahead of the other. This function is required for LAVA R3X to walk properly.

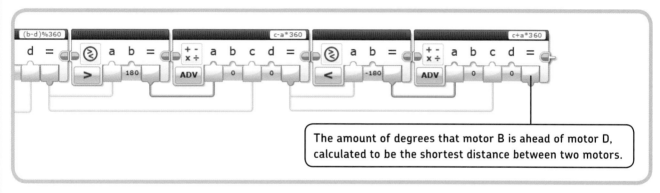

The amount of degrees that motor B is ahead of motor D, calculated to be the shortest distance between two motors.

Figure 19-9: Step 2: These blocks further process the difference between the two motors to find the shortest distance between the two motors. This distance is found by subtracting 360 degrees from the difference if it is greater than 180 degrees and adding 360 degrees to the difference if it is less than –180 degrees.

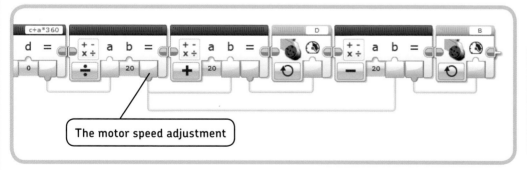

The motor speed adjustment

Figure 19-10: Step 3: To get the speed adjustment, you divide the difference between the motor positions by 20. Then, you add the adjustment to motor D's speed to make it turn faster and subtract it from motor B's speed to make it turn slower. (The block also works if motor D is ahead of B: The difference becomes negative, and motor B will turn faster to catch up with motor D.)

Figure 19-11: Step 4: Turn the blocks into a My Block called OnSync.

my block #4: left

LAVA R3X can turn left by rotating the right motor backward while keeping the left foot in a fixed position. The fixed position (120 degrees behind the starting orientation) is chosen such that the left foot just touches the ground each time the motor on the right makes one rotation. Each time the left foot touches the ground, the robot drags itself to the left by a small amount. The amount the robot turns varies depending on the type of floor the robot is on, but rotating the right motor backward for 10 rotations makes the robot turn left by roughly 90 degrees.

To ensure that the My Block works regardless of the current position of the motors, place a Return block at the start and end of the My Block. Create the *Left* My Block, as shown in Figure 19-12.

taking the first steps

The *WalkTest* program uses the My Blocks to make LAVA R3X repeatedly walk forward for 15 seconds and turn to the left (see Figure 19-13).

Place the robot on a flat, smooth surface, like a wooden floor, and run the program. The Reset My Block at the start places the legs in opposite positions. A beep played by a Sound block in the Reset My Block indicates that the reset procedure is complete, so the robot is ready to begin walking. The inner Loop block with the OnSync My Block makes the robot walk forward for 15 seconds, and the outer loop repeats the walking forward and turning behavior.

NOTE If the robot doesn't seem to walk properly, visit *http://ev3.robotsquare.com/* to see a video of how the robot should work and to download the ready-made program so you can compare it to your own.

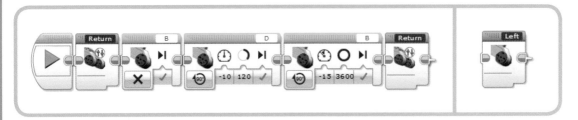

Figure 19-12: The configuration of the blocks in the Left My Block (left) and the completed My Block (right)

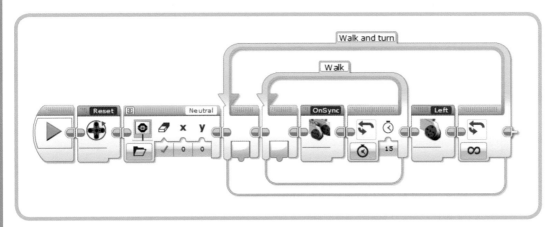

Figure 19-13: The WalkTest program

DISCOVERY #123:
WALK MY BLOCK!

Difficulty: ⬜ **Time:** 🕐🕐

Can you create a My Block that makes the robot walk forward for a given number of seconds? Create a My Block called *Walk* with one Numeric input called *Seconds*, as shown in Figure 19-14.

HINT Turn the inner Loop block of Figure 19-13 into a My Block, and use the Numeric input to control how long the Loop block repeats.

Figure 19-14: Walk My Block

DISCOVERY #124:
REVERSE!

Difficulty: ⬜⬜ **Time:** 🕐

Can you create a modified version of the OnSync My Block that allows LAVA R3X to walk backward? Make the motors move at –20% speed, instead of 20% speed, while keeping the motors synchronized.

HINT All you need to do is create a copy of the OnSync My Block (call it *OnRev*) and change two values. Which values determine the average speed of the motors?

DISCOVERY #125:
RIGHT TURN!

Difficulty: ⬜⬜ **Time:** 🕐🕐

Create a My Block called *Turn* with one Logic input value called *Direction*, as shown in Figure 19-15. Make the robot turn to the left if you choose *true*; make it turn right if you choose *false*. Substitute the Turn My Block for the Left My Block in the *WalkTest* program to test it.

Figure 19-15: Turn My Block

building the head and arms

Now build the robot's head and arms and attach them to the robot using the instructions on the following pages. When you're ready, verify that the moving elements of the arm mechanism don't interfere with the cables on top of the EV3 brick. To test this, manually rotate the axle connected to the Medium Motor and rearrange the cables if necessary.

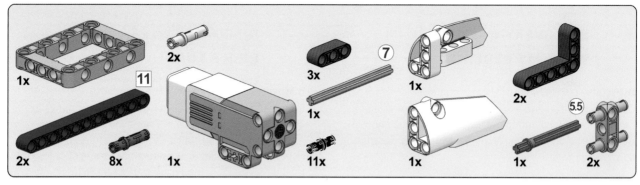

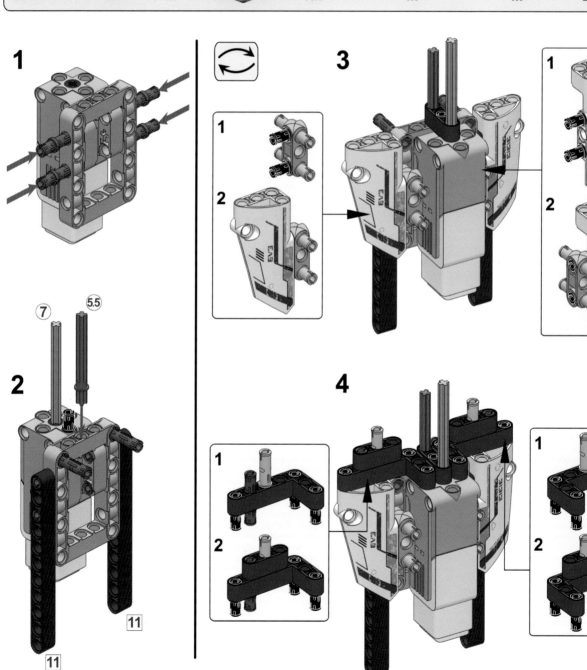

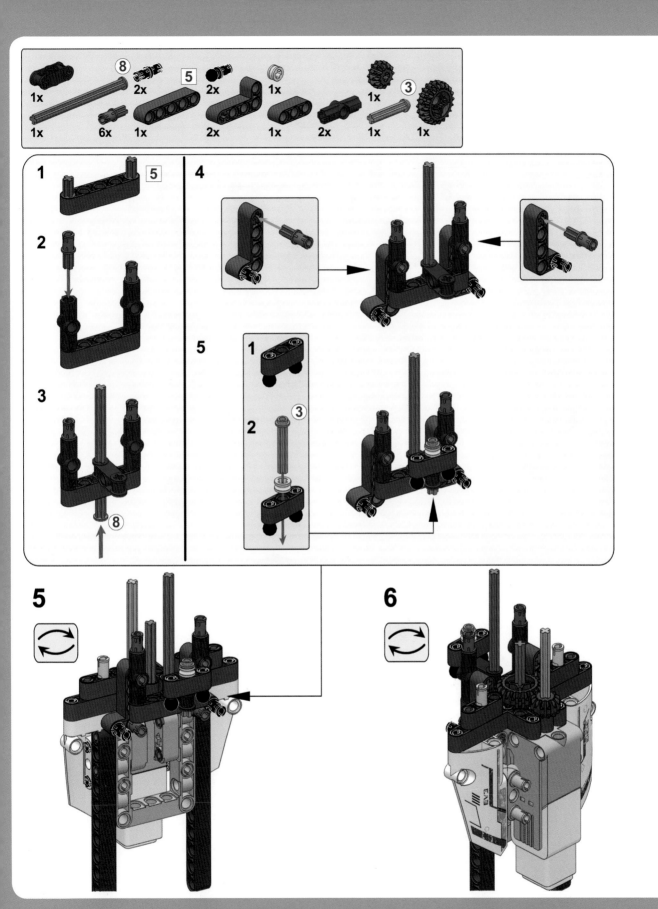

7

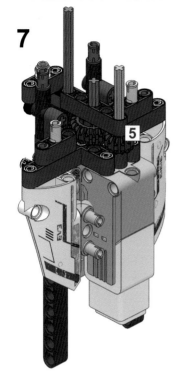

8

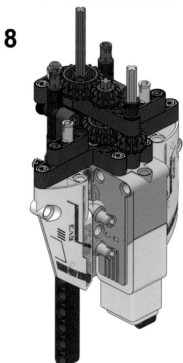

9

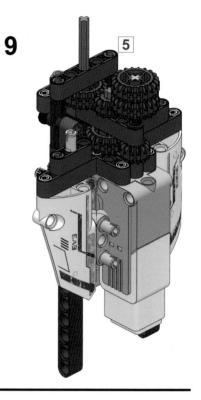

10

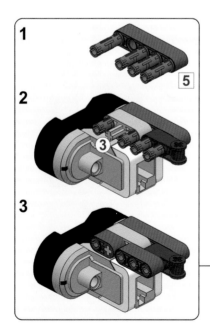

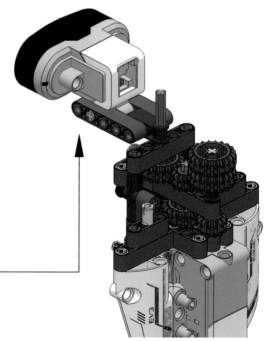

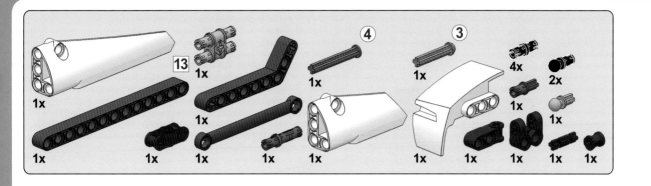

1

2

3

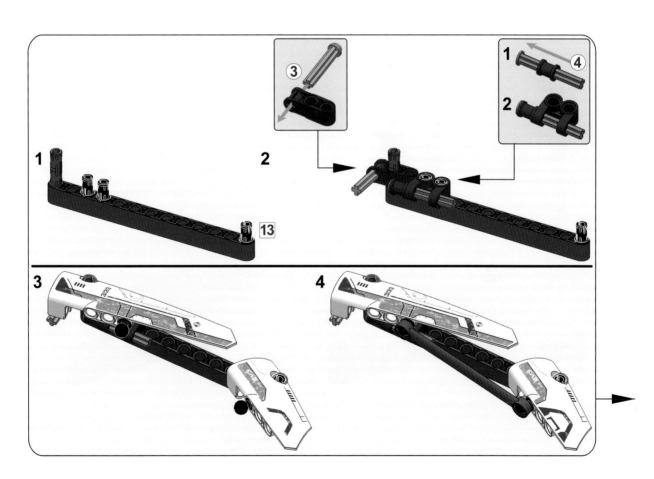

4

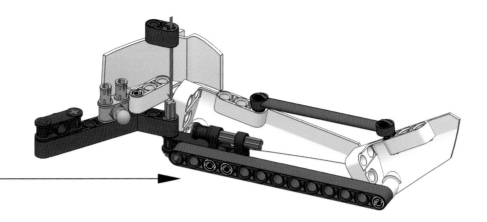

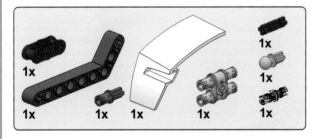

1

2

3

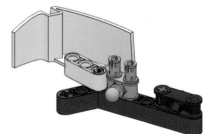

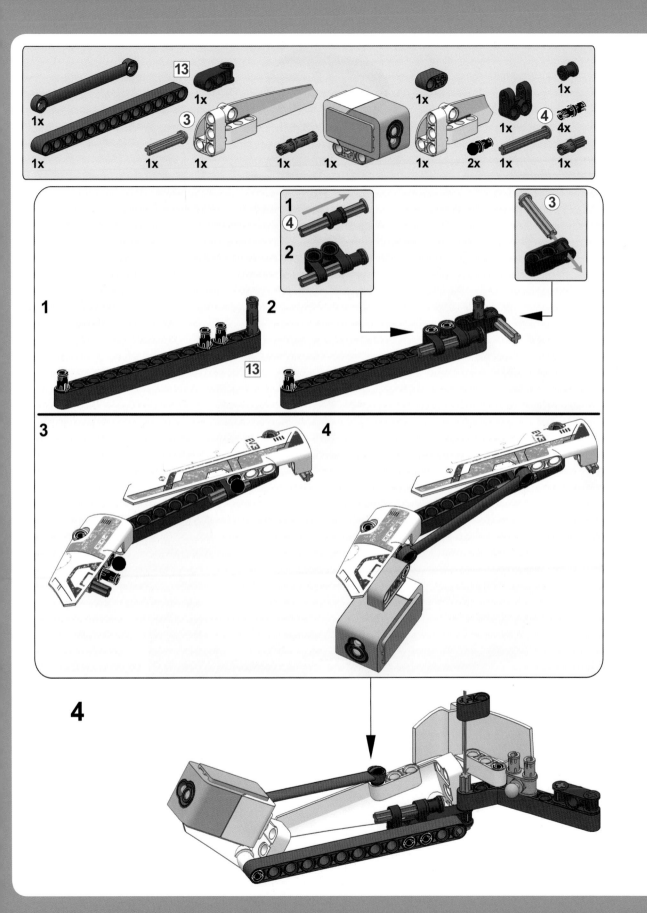

5

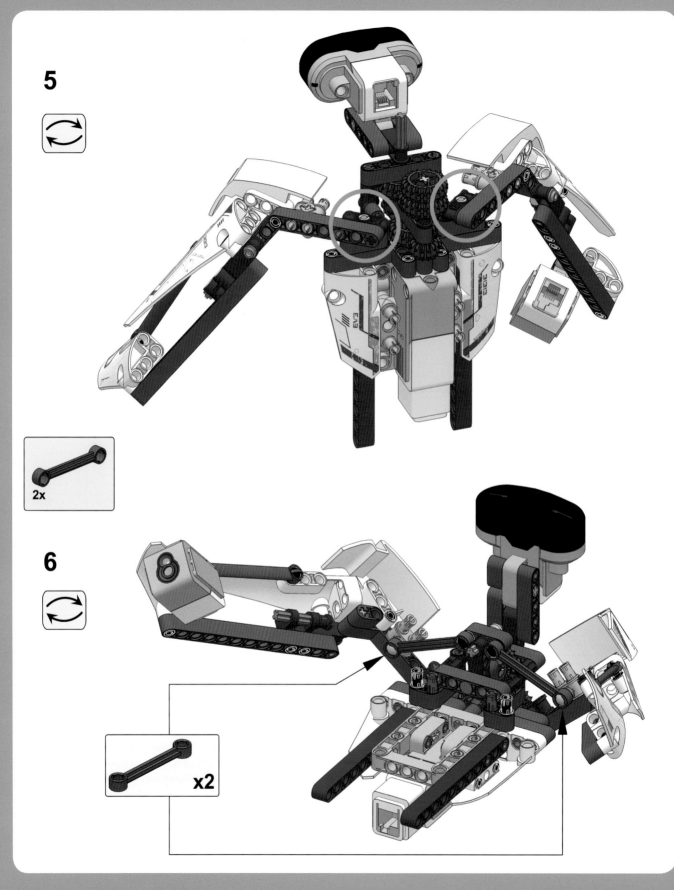

2x

6

x2

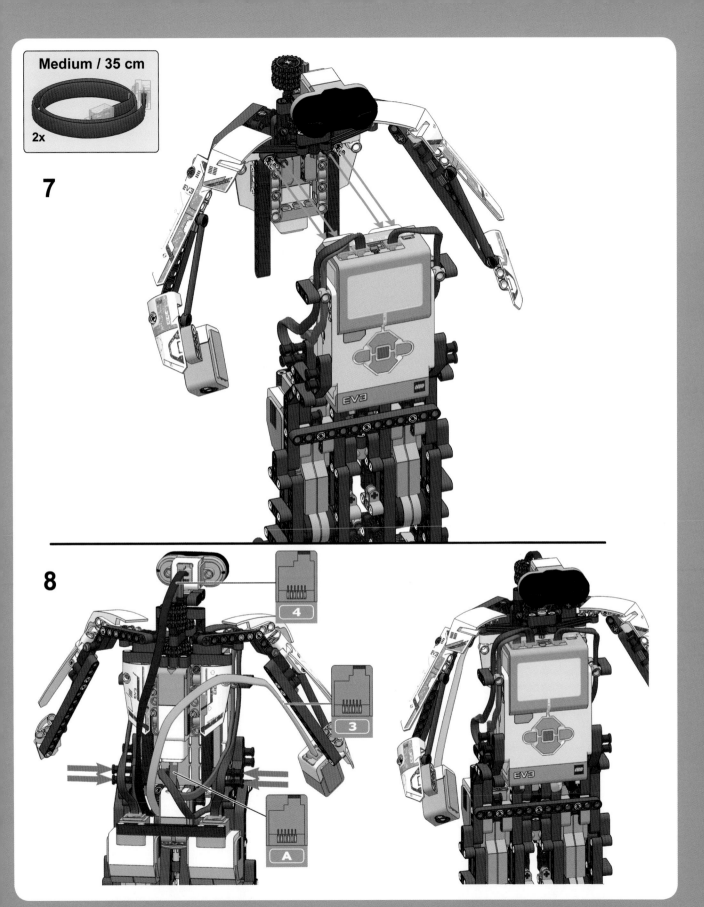

Medium / 35 cm

2x

7

8

4

3

A

controlling the head and arms

Having finished building the robot, you're ready to create a program that makes the robot walk, move its head and arms, and respond to its environment.

You control the movement of the arms and the head with the Medium Motor: Turning the motor forward makes the head and the arms move to the right; turning the motor backward makes them move to the left.

my block #5: head

To make it easier to control the movement of the head and arms, you'll create a My Block and place it parallel to the main walking program.

The *Head* My Block first places the Medium Motor in a known position by rotating the head all the way to the right. Then, it continuously makes the head move left and right, as shown in Figure 19-16. This movement allows the Infrared Sensor to see obstacles to its left and right in addition to obstacles ahead of it.

avoiding obstacles and responding to handshakes

Now that you've made the My Blocks, it's easy to create programs that make the robot walk and respond to sensors. For example, you can change the inner Loop block of the *WalkTest* program to make the robot walk forward until the Infrared Sensor sees an obstacle instead of walking forward for 15 seconds.

The final program will make the robot walk around while avoiding obstacles and will make it respond to handshakes. If you shake the robot's right hand, the robot will stop walking, say "Hello, good morning," and then continue walking. If the robot sees an obstacle, it will say "Detected" and then turn to the left.

resetting the legs and making the head move

The program begins by placing the legs in opposite positions with the Reset My Block. Then, it runs a loop that makes the robot walk and respond to sensors.

As the robot walks, it moves its head left and right using the Head My Block attached to its own Start block (see Figure 19-17). This configuration makes it possible to control the head and the legs independently: You can change the behavior of either sequence of blocks without worrying about interfering with the other sequence. You can change the behavior of the head by modifying the Head My Block, and you can change the walking behavior by changing the blocks in the Loop block.

Create a new program called *ObstacleAvoid*, and add the blocks shown in Figure 19-17.

walking until one of two sensors is triggered

Now you'll add blocks to the main loop that make the robot walk forward until it sees an obstacle or detects a handshake (see Figure 19-18).

The robot walks forward by moving the Large Motors forward with the OnSync My Block. The block is placed in a loop so that the motor speed continuously adjusts to keep the motors synchronized. The loop runs until the Infrared Sensor

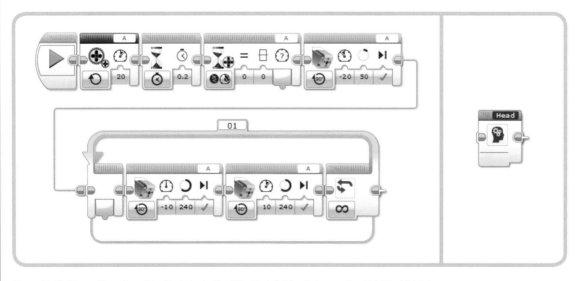

Figure 19-16: The configuration of the blocks in the Head My Block (left) and the completed My Block (right)

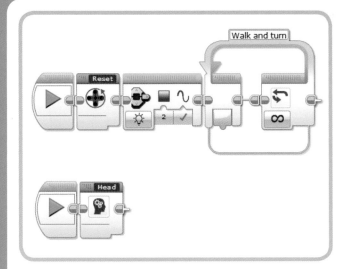

Figure 19-17: Step 1: These blocks place the legs in opposite positions and make the robot's head move. Note that the Loop block inside the Head My Block runs indefinitely so that the head keeps moving left and right.

responding to the triggered sensor

The robot is now ready to respond to the sensor that was triggered. If the Infrared Sensor was triggered, you'll make the robot say "Detected" and turn to the left. If the Color Sensor was triggered, you'll make the robot temporarily stop walking and say "Hello, good morning."

After the loop completes, you can determine which sensor caused the loop to end by looking at the *output value of the Infrared Sensor block*: It's *true* if the Infrared Sensor was triggered, and when it's *false* it means that the Color Sensor was triggered. (Because the loop has ended somehow and the Infrared Sensor wasn't triggered, you know that the Color Sensor must have been triggered.)

Use a Switch block to decide which blocks to run based on this value, as shown in Figure 19-19.

Even though it's unlikely, the loop may also have ended because both sensors were triggered at the same time so that the outputs of both Sensor blocks are *true*. Because the output of the Infrared Sensor is *true*, the program simply runs the blocks at the top of the switch, which are the same blocks that would run if only the Infrared Sensor was triggered. In other words, the program ignores the Color Sensor in this case.

Add the blocks to the switch that make the robot say "Detected" and turn to the left for *true*, and that make the robot say "Hello, good morning" for *false* (see Figure 19-20).

Now run the program and test it. If you use a USB cable to program the robot, you need to manually position the head straight forward to make room for the USB cable. To do this, turn the gears attached to the Medium Motor. Use the **Download** button to send the program to your robot, unplug the USB cable, and then start the program manually using the EV3 buttons.

The robot should now walk around autonomously and greet you if you shake its hand.

is triggered (it detects an obstacle), until the Color Sensor is triggered (it detects a handshake), or until both sensors are triggered simultaneously.

The Color Sensor block is able to detect a handshake by comparing the Reflected Light Intensity to a threshold. If the sensor value is greater than 10%, the robot sees your hand and the output is *true*. If the value is 10% or less, the sensor doesn't detect your hand and the output is *false*. Similarly, the Infrared Sensor block is configured to output *true* if a proximity measurement less than 50% is detected; otherwise, it outputs *false*.

A Logic Operations block compares both logic values. The output is *true* if at least one input value is *true*, causing the loop to end.

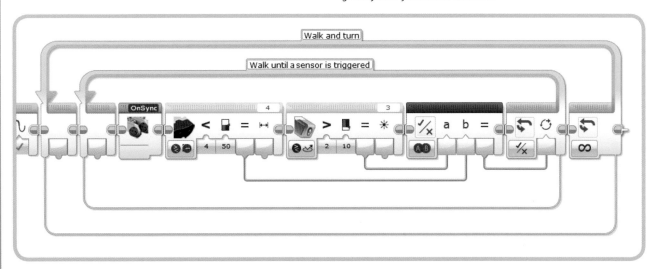

Figure 19-18: Step 2: The inner loop makes the robot walk forward until the robot detects an obstacle or a handshake.

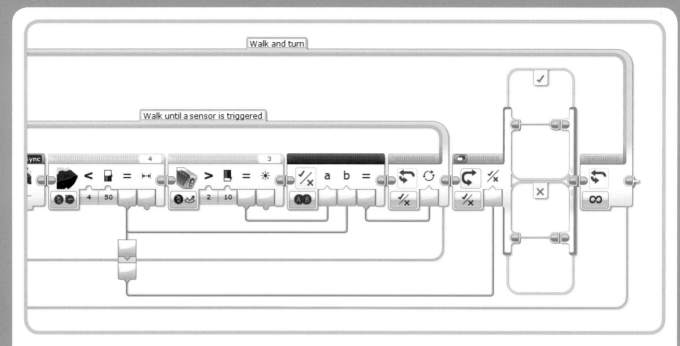

Figure 19-19: Step 3: The Switch block determines which sensor triggered the loop to end. If the Infrared Sensor was triggered, the blocks at the top of the switch will run (true); if it wasn't, the blocks at the bottom of the switch will run (false).

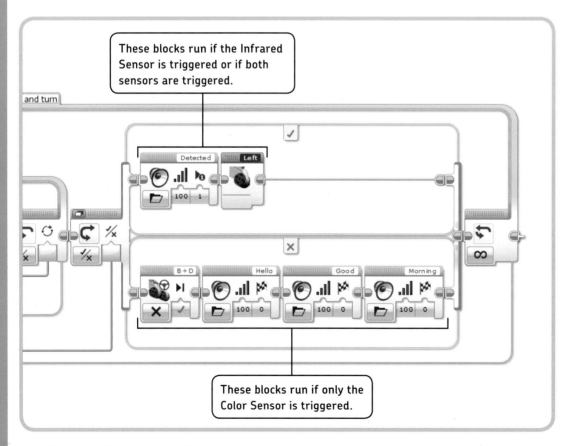

These blocks run if the Infrared Sensor is triggered or if both sensors are triggered.

These blocks run if only the Color Sensor is triggered.

Figure 19-20: Step 4: The robot turns left if it detects an obstacle (true) and stops and greets you if it detects a handshake (false).

further exploration

You've reached the end of this book. Congratulations! I hope you've enjoyed learning the ins and outs of the LEGO MINDSTORMS EV3 robotics set as well as building and programming the robot projects presented in this book. You're now ready to start creating robots on your own and sharing your ideas with the world. Whether your robots drive, grab, walk, or talk, the possibilities are endless with LEGO MINDSTORMS EV3!

But before you close this book, try the following Discoveries to expand the program for LAVA R3X to make it more interactive. When you're ready, be sure to check out building and programming instructions for a robot that sorts LEGO bricks by color and size (as in Figure 19-21) on the book's companion website, *http://ev3.robotsquare.com/*.

DISCOVERY #126:
DANCING ROBOTS!

Difficulty: ⬜ **Time:** ⏱⏱

LAVA R3X is able to walk by placing the leg mechanisms in *opposite* positions and running both motors forward. What happens if you place the legs in the *same* position and then turn both motors at 10% speed for five rotations but in opposite directions? (You don't have to worry about motor synchronization in this Discovery.)

HINT Once you've placed the legs in opposite positions with the Reset My Block, you can easily place the legs in the same position by rotating one motor forward. How many degrees should you rotate it?

Figure 19-21: Feel like building another robot? The BRICK SORT3R sorts LEGO bricks by color (red, yellow, green, and blue) and size (2×2 and 2×4). You can find building and programming instructions at the book's companion website.

DISCOVERY #127: WHAT'S THE DIFFERENCE?

Difficulty: ⬚⬚ **Time:** 🕐

To get a better understanding of the OnSync My Block, display the difference between the two motor positions on the screen. To accomplish this, place a Display block in the OnSync My Block that displays the value of the last Math block of Figure 19-9 on page 333. Then place the modified OnSync My Block in a loop configured to run forever. What happens to the difference if you slow one motor down manually by holding the leg still? How does the other motor try to compensate for the difference? (Don't try this for more than a few seconds!)

DISCOVERY #128: ROBOT COACH!

Difficulty: ⬚⬚ **Time:** 🕐🕐

Can you make the robot detect how long you work at your desk? Program LAVA R3X to display the amount of time you've been working at your desk, and make it advise you to take a break after one hour. If it detects that you're not following the advice, make it shake its head and play sounds to get your attention.

DISCOVERY #129: ROBOT FOLLOWER!

Difficulty: ⬚⬚⬚ **Time:** 🕐🕐

Can you program the robot look at you as it walks by making its head turn in the direction of the infrared beacon? Make the speed of the Medium Motor proportional to the Beacon Heading value using a technique similar to the one you used for the SNATCH3R in Figure 18-20 on page 308. Note that the robot's head cannot make a full turn. How do you limit the motor's movement in your program so that it doesn't try to turn farther than it actually can?

DISCOVERY #130: SYNCHRONIZED PACE!

Difficulty: ⬚⬚⬚ **Time:** 🕐🕐

In the *ObstacleAvoid* program, LAVA R3X's arms move left and right, independent of the pace at which the robot walks. Can you synchronize both movements so that it walks in a more elegant and realistic way?

DISCOVERY #131: REMOTE WALK!

Difficulty: ⬚⬚⬚ **Time:** 🕐🕐🕐

Can you make a program that allows you to control LAVA R3X remotely so that it can walk in any direction? You can use the techniques you learned in Chapter 8, but you'll need to use the techniques you learned in this chapter to ensure that the motors in the robot's legs remain synchronized. How do you make the robot begin to walk or turn when you press a button on the remote and stop when you release it?

DISCOVERY #132: TAMAGOTCHI!

Difficulty: ⬚⬚⬚⬚ **Time:** ○○○○

Can you turn LAVA R3X into a lifelike robot with different moods and behaviors? Use the infrared remote to command the robot to walk, talk, eat, and sleep. Create Numeric variables to keep track of the robot's health by monitoring its hunger level, energy level, and happiness.

Make the energy level decrease with each step the robot takes, and make it increase when you command the robot to sleep. Similarly, make the hunger level increase while the robot is walking, and make it decrease when you feed the robot. Make the robot's happiness slowly decrease over time, and make it increase each time you shake the robot's hand.

If the hunger level, energy level, or happiness reaches a certain critical limit, the robot should ignore new commands and make its own decisions. For example, if the robot is too tired (its energy level is below 10%), it should fall asleep for a while to restore its energy level. Or, if it's sad, you can make it say "No!" and cry each time you try to send new commands.

Display the energy level, hunger level, and happiness on the EV3 screen so you can diagnose the robot's health and mood. Experiment with different types of behavior, and add sounds and light effects to make the robot show emotions and appear more lifelike. For example, display smiling eyes on the EV3 screen to show that the robot is happy, and have it play snoring sounds while the robot sleeps.

You can use the flow diagram shown in Figure 19-22 as a guide when designing your program, but feel free to come up with your own ideas.

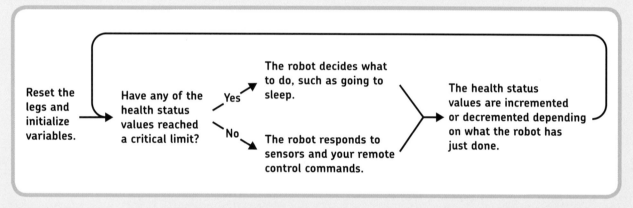

Figure 19-22: One possible implementation of Discovery #132. Note that this is a very basic overview to help you get started; each step listed here may consist of many programming blocks of your choice, including many more Switch and Loop blocks.

DESIGN DISCOVERY #30: BIPED ROBOT!

Building: ✻✻ **Programming:** ⬚⬚

Can you create an animal robot that walks on two legs? Remove the robot's upper body and the EV3 brick so that only the legs remain. Now you can create any type of robot that walks on two legs. Can you build an ostrich or perhaps a dinosaur? You can use the Medium Motor to control the robot's head, tail, or even its claws. Use the My Blocks you made in this chapter to control the legs of your robot.

troubleshooting programs, the EV3 brick, and wireless connections

When building and programming the robots in this book, you may occasionally run into problems transferring your programs to the EV3, and this appendix aims to help you find solutions to such problems. You'll also learn how to manage wireless connections to the EV3 brick, how to reset it, and how to update its firmware.

trouble-shooting compilation errors

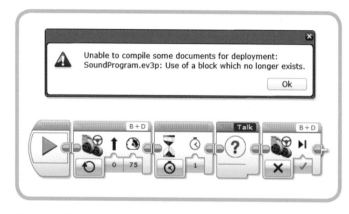

Figure A-1: The program cannot be compiled because the Talk My Block is missing.

When you download your program to the EV3 brick, the EV3 software turns your *source code* (the programming blocks you see on the screen) into a file with more compact code representing the actions that the EV3 brick will perform. This process is called *compilation*. Compilation sometimes fails, and if it does, you'll see an error message such as the one shown in Figure A-1.

missing my blocks

Compilation will fail if a program tries to run a block that no longer exists. If this happens, a question mark will be displayed on the missing block (see Figure A-1). Let's say you have a My Block called *Talk* and a program called *SoundProgram* that uses the Talk My Block. The project fails to compile if the Talk My Block is missing—for example, if you've deleted the Talk My Block from the Project Properties page.

Note that the error message doesn't tell you which My Block is missing (Talk); rather, it tells you which program (*SoundProgram*) contains references to the missing My Block.

To solve the problem, you can create a new My Block with the same name (Talk) or copy it from another project into your current project, using the instructions in Figure 5-13 on page 55. Or, if you just want to continue without the missing blocks, delete all blocks with a question mark on them.

errors in programming blocks

Compilation can also fail if your program contains instructions that the software doesn't understand. For example, compilation will fail if you enter an unknown symbol in the Equation setting of the Math block, as shown in Figure A-2.

The software doesn't tell you which block causes the compilation to fail, but you can track down the error by selecting a few blocks and trying to run them using the Run Selected button. If a selection runs, it contains no compilation errors; if it fails to run, the selection contains an error. As you systematically rule out which sections work and which ones don't, you should eventually find the block that contains an error. Then, you can resolve the error if you know what caused it or delete the block and replace it with a new one.

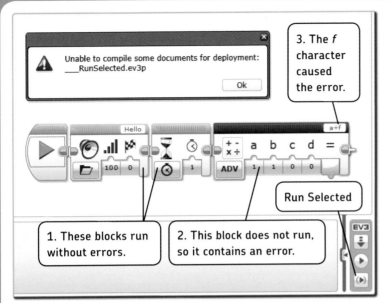

Figure A-2: Tracking down compilation errors in a program by running selections of blocks. The Equation setting on the Math block can work only with a, b, c, or d, so the f symbol causes the compilation to fail.

*Figure A-3: The Numeric variable called TestVar is unavailable because it isn't yet defined. To define it, click **Add Variable** and choose TestVar as the variable's name.*

missing variable definitions

When you copy a Variable block from one project and paste it into another, the *definition* is not always copied along with it. This means that even though the variable's name appears on the block that you copied, it's not available for use in other Variable blocks (see Figure A-3). To solve this problem, define a new variable of the same type with the same name.

The same thing can happen when you import a program or My Block containing Variable blocks into your project. You can see an overview of all variables on the Variables tab of the Project Properties page (see Figure 16-3 on page 246). These variables should be available for use in all programs throughout your project.

trouble-shooting a running program

The previous section can help you solve certain technical problems, but what if you have a program that runs, just not in the way you expect it to? This outcome can have many causes, and in many cases, there may simply be a user error. I've programmed robots for many years, and I still often make

mistakes. For example, sometimes I forget to connect a data wire, and then my robot won't work.

On the other hand, making mistakes can help you discover new techniques and solutions that you might not have found by following the steps perfectly. Troubleshooting might not seem useful or fun while you're at it, but it's an essential element of robotics. And, once you make your robot work by solving the problem on your own, programming is even more rewarding.

You may find the following tips useful for troubleshooting and preventing errors in your own programs:

* *Add comments to your program.* Comments don't affect the way your program works, but they can help you remember what each part of your program is supposed to do when you look at it later. You can place comments in your program using the Comment tool, or you can use the *Comment block*, as shown in Figure A-4. Use the Comment *tool* if you want your comment to stay in the same place on the *Canvas*; use the Comment *block* if you want the comment to stay in the same place in your *program*. For example, if you place another programming block before the Move Steering block, the Comment block automatically moves to the right along with the rest of the blocks, whereas the normal comment stays in the same place. To further document your project, you can add text and pictures to the Content Editor (see Figure 3-19 on page 33).

* *Choose descriptive names for programs, My Blocks, and variables.* For example, if you have two variables to count how many times the Left and Right buttons have been pressed, respectively, it's better to call them *CountLeft* and *CountRight*

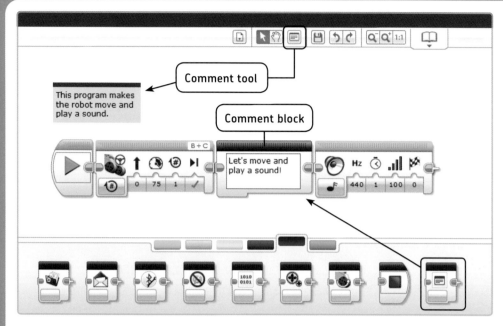

Figure A-4: You can add comments to your program using a Comment block or the Comment tool. The Comment block isn't available in version 1.0 of the EV3 software. To use it, install version 1.1 or higher using the instructions in Chapter 1. If you're not sure which version you have, click **Help ▸ About LEGO MINDSTORMS EV3**.

instead of *Count1* and *Count2*. Then you're less likely to pick the wrong one later on.

* *Use sounds, the display, and the brick status light to signal the program's progress.* For example, place a Sound block after a block that waits for a sensor to be triggered, or make the status light red while a certain My Block is running. This helps you see which part of the program is running. If the program fails, you'll know roughly where to look for the problem.

* *Display values on the screen.* For example, if you use a Switch block to make a decision based on a sensor value, it can be helpful to display that value so you can see what value triggered the *true* or *false* part of the switch. For example, an unexpected value, such as 0, might indicate that the sensor isn't connected.

* *Use the EV3 software to see which block is currently running.* This can help you see where a program gets stuck (see Figure A-5). For example, if the program stalls on a Large

Motor block, something might be preventing the motor from reaching its target—perhaps it's blocked.

* *Test changes to your program often.* If your program works fine at first but stops working when you add a new block, the last block you added may be the cause of the problem. When you make several big changes to your program, be sure to test your program after each change.

* *Test your program under different conditions.* If your robot works once, there's no guarantee it'll work the next time around if the conditions are different. A line-following program may work fine in one room, but it might not correctly detect the line in another room that has a lot of external lights, such as a robot competition table.

NOTE If one of the programs in this book doesn't work as described, you can visit the book's companion website (*http://ev3.robotsquare.com/*) to download the ready-made programs and compare them to your own.

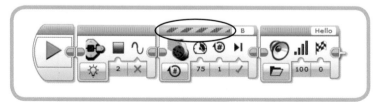

Figure A-5: The moving stripes indicate that the Large Motor block is currently running. You'll see the stripes only if you use the Download and Run button; you won't see them if you start the program using the EV3 buttons.

trouble-shooting the EV3

This section will show how you can diagnose several aspects of the EV3 brick, such as the battery level, free memory, and the USB connection between the EV3 and the computer, and it'll illustrate how to reset or update the EV3.

using the hardware page

You can find information about the EV3 brick and devices connected to it on the Hardware Page, as shown in Figure A-6. The Brick Information tab shows the name, the battery level, and the firmware version of the EV3 brick that's currently connected. To change the name of the EV3, type a new name in the indicated field and press the ENTER key. The personalized name should also appear at the top of the EV3 brick's display. This name helps to distinguish EV3 bricks if you have more than one. The Port View tab displays the sensor value of each motor or sensor connected to the EV3 (see Figure 6-5 on page 66).

managing connections

The Available Bricks tab lists the personalized name of each EV3 brick detected by the EV3 software. Depending on how your EV3 is configured, you can choose to connect the EV3 via USB, Bluetooth, or Wi-Fi by clicking the checkbox in the appropriate column. For example, the EV3 in Figure A-7 is called EXPLOR3R, and it's connected to the computer via USB.

Use the **Refresh** button to update the list. Use the **Disconnect** button to cancel a current connection so you can connect to another EV3 brick.

managing the EV3's memory

You can manage the EV3's files from your computer by opening the *Memory Browser* from the Brick Information tab, as shown in Figure A-8. You'll see all of the programs currently on the EV3, with one folder for each project.

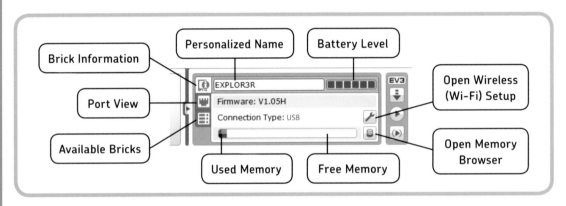

Figure A-6: The Brick Information tab of the Hardware Page

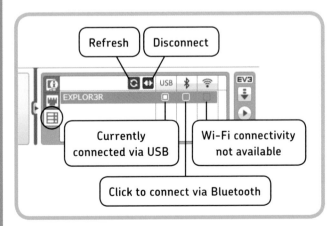

Figure A-7: You use the Available Bricks tab to manage connections to the EV3 brick. If a checkbox is greyed out, the connection type isn't currently available.

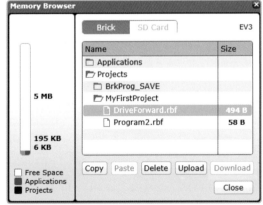

Figure A-8: Open the Memory Browser from the Hardware Page (see Figure A-6) or click **Tools ▶ Memory Browser**.

Compiled programs are very small (several KB) compared to the amount of free space (more than 4MB) on the EV3, so you shouldn't run out of space quickly. To avoid clutter on the File Navigation tab on the EV3 brick, however, you can remove unused projects by selecting them and clicking **Delete**.

You can send a file *from the EV3 brick to a computer* by selecting it and clicking **Upload**. You can send a file *from the computer to the EV3 brick* by selecting the destination folder on the EV3 and clicking **Download**. Note that transferring compiled programs back to the computer doesn't reveal the program's source code, which is what you normally program and edit using the EV3 software. You cannot "decompile" an EV3 program to continue working on it, so you should always save the source code for later use. On Brick programs are an exception to this rule, as you'll see in Appendix B.

solving problems with the USB connection

When you connect the EV3 brick to the computer with the USB cable, the EV3 software should detect the EV3 automatically, as indicated by the red EV3 symbol (EV3) on the Hardware Page. If the symbol stays grey (EV3), try following these directions to resolve the issue:

1. Make sure the EV3 brick is turned on.

2. Make sure the EV3 brick is connected to the computer with the USB connection labeled *PC* (see Figure 2-5 on page 20). The other end of the cable should be plugged into any of the computer's USB ports.

3. If you're sure everything is connected properly, try unplugging the USB cable and then plugging it back in, or try using another USB port on the computer.

4. Close the EV3 software and restart it, or reboot the computer.

5. If that doesn't help, unplug the USB cable, turn off the EV3 brick, turn it back on, wait for it to fully start, and then plug the cable back in.

You may experience problems when trying to connect an EV3 to a public computer, such as a computer in a classroom. If so, ask the system administrator to log in, launch the EV3 software, and check whether it connects to the EV3. When you're done, you should be able to connect to the EV3 using your own account.

restarting the EV3 brick

If an EV3 program stalls and you can't abort it with the Back button, you can restart the EV3 brick by pressing the Back button and Center button simultaneously until the brick status light turns off (see Figure A-9). The EV3 brick should restart

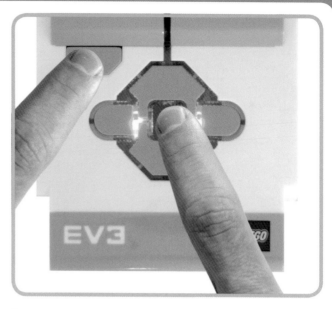

Figure A-9: To restart the EV3, hold the Back and Center buttons until the brick status light turns off.

when you release the buttons. Note that you will lose all programs and settings added to it since you last turned it on.

If your EV3 brick fails to start or if you see the brick status light blink red only briefly, replace the batteries with fresh ones and try updating the firmware, as discussed in the next section.

updating the EV3 firmware

If the software prompts you to update the EV3's firmware, you can do so by going to **Tools ▶ Firmware Update**. Connect the EV3 to the computer with the USB cable, choose the latest firmware version from the list, and click **Download**. The EV3 brick should automatically go into Update mode and display "Updating.." on the EV3 screen. When the two progress bars on the computer screen complete after a few minutes, the EV3 should restart automatically, and the process will be complete.

If the software fails to put the EV3 brick in Update mode, you can do this manually by pressing the Back, Center, and Right buttons simultaneously until the brick status light turns off, as shown in Figure A-10. Then release only the Back button. As soon as you see "Updating.." on the EV3 screen, you can release the other buttons, too. Now that the EV3 is in Update mode, plug in the USB connection again and retry updating the firmware.

NOTE Updating the firmware will remove all programs and files from the EV3 brick. If you enter Update mode by accident, you can restart the EV3 to continue normal operations.

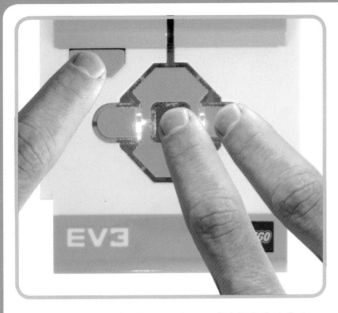

Figure A-10: To put the EV3 in Update mode manually, hold the Back, Center, and Right buttons until the brick status light turns off. Then, release only the Back button. Once the screen reads "Updating..", release the other buttons.

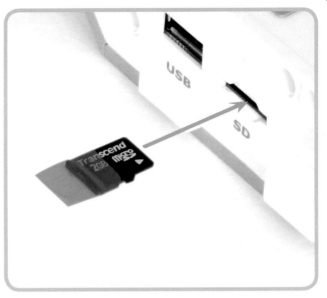

Figure A-11: You can use a microSD card to avoid data loss. Be sure to insert the card with the metallic contacts facing down. Also add some tape around the edge of the card to make it easier to remove from the EV3 brick later.

avoiding data loss using a microSD card

When you send a program to the EV3 brick, it's saved in temporary memory. The EV3 saves files and settings to permanent storage only when you turn off the EV3 brick. (This is why shutting down the EV3 takes a while.)

If you restart the EV3 without shutting down first, or if you remove the batteries while the EV3 is on, you'll lose all files and settings that changed since the EV3 brick was last turned on because the EV3 won't have had a chance to store them permanently. In fact, if you remove the batteries during shutdown (when the EV3 is busy saving files), you may lose older files, too.

While this can be frustrating, it's not usually a critical problem because you should have a copy of the source code on your computer. However, you can avoid this type of data loss by adding a microSD card to the EV3, as shown in Figure A-11. Projects will be saved to the card automatically each time you download a project to your robot; you shouldn't need to take any additional steps. The programs should stay on the card even if you restart the EV3 or update its firmware.

If you use a microSD card, you'll find your projects in the *SD_Card* folder on the File Navigation tab of the EV3 brick. Even large programs are just a few kilobytes in size, so a small microSD card should provide plenty of storage.

programming the EV3 wirelessly

Instead of using the supplied USB cable, you can connect your EV3 brick to the computer using either Bluetooth or Wi-Fi. Wireless transfers make programming a lot easier because you don't have to repeatedly connect and disconnect the USB cable each time you download a program.

Once you've set up the wireless connection, transfer programs to the EV3 using the Download and Run button as you would with the USB cable.

using Bluetooth to download programs to the EV3

The EV3 brick has built-in Bluetooth functionality that can be used for wireless programming, to communicate with another EV3 brick, or to connect with a smartphone or tablet for remote control. (Note that you can use only one of these features at a time.)

To use Bluetooth for wireless programming, you'll need either a computer with built-in Bluetooth functionality or a compatible Bluetooth dongle that plugs into the USB port on a computer (see Figure A-12).

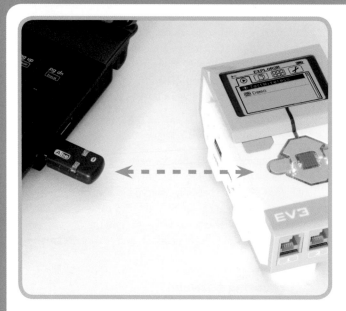

Figure A-12: One setup for wireless programming with Bluetooth

finding a Bluetooth dongle

There are many compatible Bluetooth dongles, many of which cost less than $10. Generally, it's not the dongle hardware but its drivers, in combination with the computer's operating system, that will determine whether the dongle is compatible or not. In many cases, you'll be able to simply plug the dongle into the computer, wait for the drivers to install automatically, launch the EV3 software, and follow the connection procedure in the next section. The drivers you'll need will depend on your operating system and your Bluetooth dongle. Visit the companion website (*http://ev3.robotsquare.com/*) for links to recommended Bluetooth dongles.

If you're experiencing problems with a computer's built-in Bluetooth, try disabling it and using an external Bluetooth dongle instead.

connecting to the EV3 with Bluetooth

Follow the next steps to set up a Bluetooth connection between the computer and the EV3 brick for the first time:

1. Plug a compatible Bluetooth dongle into a free USB port *on the computer*, or verify that built-in Bluetooth is enabled. Depending on the operating system, some drivers are automatically located and installed. Usually, it's *not* necessary to install the additional drivers that come with your dongle.

2. Turn on the EV3 brick and connect it to the computer with the USB cable.

3. Activate Bluetooth on the EV3 by going to the Settings tab and selecting **Bluetooth**. Then, check **Visibility** and **Bluetooth**, and uncheck **iPhone/iPad/iPod** with the Center button, as shown in Figure A-13.

Figure A-13: Go to the Settings tab on the EV3 brick, choose **Bluetooth**, and configure the settings as shown. (The iPhone/iPad/iPod setting should be checked for remote control with iOS devices only; it should be unchecked when using Bluetooth for wireless programming with a computer or when using it for remote control with an Android device.)

4. In the EV3 software on the computer, go to the Available Bricks tab on the Hardware Page and click **Refresh**, as shown in Figure A-7. The search process should take about 30 seconds. When ready, the list of EV3 devices is updated with the EV3s that are available for a connection.

5. In the list of EV3 bricks, there should be a checkbox for each available connection type. Create the Bluetooth connection by checking the box under the Bluetooth symbol (⁎). If the checkbox cannot be checked, click **Refresh** again. If that doesn't help, uncheck **Bluetooth** in the menu on the EV3 (see Figure A-13), check it again, and retry the connection procedure.

You can tell whether an EV3 has made a working Bluetooth connection by looking at the top left of the EV3 screen, which shows ⁎<> when connected and ⁎< when not connected to a computer. If successful, you can now unplug the USB cable and start downloading programs.

The next time you launch the software, you should need only to follow steps 4 and 5 to make the Bluetooth connection, and you shouldn't need the USB cable to configure the wireless connection.

If you didn't connect a USB cable in step 2 to make the first connection, the EV3 brick will ask you to confirm the connection and to choose a password to secure it when you reach step 5. Once you've set the password, the EV3 software will prompt you to enter the same password. In turn, the EV3 will ask you to confirm once more, and the connection should be ready. It's easiest if you stick with the default password (1234). If you use the USB cable to set up the Bluetooth connection, the software handles all of these security measures in the background.

using Wi-Fi to download programs to the EV3

You can add a Wi-Fi dongle to your EV3 brick so that it can connect to a wireless network, as shown in Figure A-14. When both the computer and the EV3 brick are connected to the same network, you can program your robot wirelessly. As of this writing, the EV3 brick supports only the *NETGEAR WNA1100 N150 Wi-Fi USB Adapter*. Visit *http://ev3.robotsquare .com/* for an updated list of compatible Wi-Fi dongles.

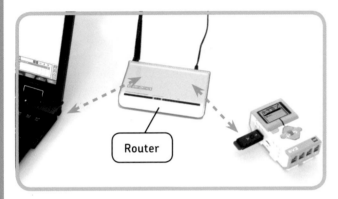

Router

Figure A-14: The setup for wireless programming with Wi-Fi

For the next steps, I'll assume that you already have a wireless network and that it's protected with a WPA2 password. I'll also assume that you know the network's name (SSID) and the password and that the computer is connected to this network. Follow these steps to establish the Wi-Fi connection:

1. Turn on the EV3 brick, and plug the compatible Wi-Fi dongle into the USB host port *on the EV3 brick* (see Figure A-14). Also, connect the EV3 brick to the computer using the USB cable (not shown in figure).

2. On the Brick Information tab of the Hardware Page, click **Open Wireless (Wi-Fi) Setup** (see Figure A-6). The EV3 brick should turn on Wi-Fi automatically and begin searching for wireless networks. When it's ready, choose your network from the list that appears on the computer screen and then click **Connect**. If you get an error message saying "No Wi-Fi Adapter found attached to the Brick," the EV3 brick didn't detect a (compatible) Wi-Fi dongle.

3. Enter the network's password on the dialog that appears, and click **Connect**. If successful, the symbol on the top left of the EV3 screen should change from 📶 (Wi-Fi is on) to 📶⇄ (Wi-Fi is connected). The EV3 brick is now connected to the router in your network but not yet to the computer.

4. Go to the Available Bricks tab on the Hardware Page, and click **Refresh**, as shown in Figure A-7.

5. There should be a checkbox for each available connection type. Create the Wi-Fi connection by checking the Wi-Fi checkbox (📶). If it becomes checked, the connection is successful and you can unplug the USB cable. If the checkbox cannot be checked, try disconnecting *the computer* from the network, reconnecting it, and clicking **Refresh** again.

choosing between Bluetooth and Wi-Fi

If you would like to program your robot wirelessly, I recommend that you use Bluetooth. First, you won't need a Wi-Fi network, nor will you need to configure its settings. Second, there's only one supported Wi-Fi dongle, and it takes up a lot of space in your robot. (By contrast, the EV3 already has Bluetooth built in.) Finally, Bluetooth requires just a few clicks to set up after the first time.

On the other hand, because the EV3 brick is actually a small Linux computer, Wi-Fi can be used to access more advanced features of the EV3 brick that can't be used with the standard EV3 software. However, unless you plan to learn how to use these features, you're probably better off using Bluetooth.

summary

I hope that this short appendix has helped you find a solution to your problem. Of course, only a few problems and solutions are listed here, and you may have other questions relating to one of the building or programming instructions in this book. Visit *http://ev3.robotsquare.com/* for links to other helpful resources, including forums where you can ask questions about LEGO MINDSTORMS EV3 in general.

B

creating on brick programs

Instead of using the EV3 software to create programs on a computer, you can create basic programs on the EV3 brick itself using the Brick Program application. This technique is useful for testing your robot if you're not near a computer. Sometimes it's sufficient to test your design with the IR Control app and to monitor sensors using Port View, but the Brick Program app can be used to create programs that involve both motors and sensors. For example, to test a mechanism with a Touch Sensor, such as the claws of the SNATCH3R, you can make a motor turn until the Touch Sensor becomes pressed.

This appendix will show you how to create *On Brick* programs as well as how to import them into the EV3 software so that you can continue working on them on a computer.

NOTE Creating On Brick programs is similar to creating EV3 programs on a computer. You enter commands differently, but the underlying principles are the same. For this appendix, I'll assume you've already mastered the

techniques covered in Chapters 1–6, and I'll show you how you can use the same techniques to create an On Brick program.

creating, saving, and running an on brick program

You can start working on a new On Brick program by selecting **Brick Program** on the Brick Apps tab of the EV3 brick, as shown in Figure B-1. On Brick programs always consist of one Loop block, and you can place a series of Action blocks and Wait blocks inside it.

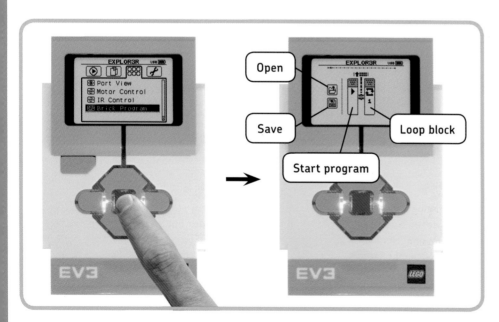

Figure B-1: Opening a new On Brick program

adding blocks to the loop

You navigate through a program using the Left and Right buttons, and the selected item is always positioned in the middle of the screen. To place a new block in the loop, navigate to an empty spot between two blocks and press the **Up** button (see Figure B-2). Next, select a block of your choice and place it in the loop with the **Center** button. Alternatively, cancel the selection with the **Back** button. For example, choose the Brick Status Light block to make the EV3 brick status light turn orange.

You can place up to 16 blocks in the loop.

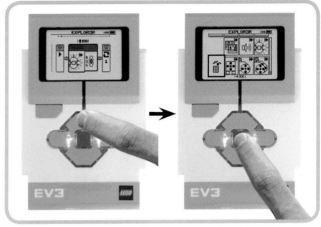

Figure B-3: Deleting a block from a program

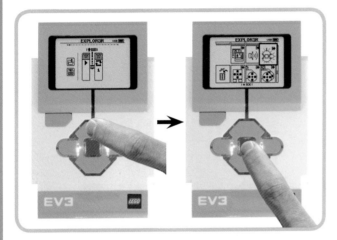

Figure B-2: Adding a block to a program

replacing a block

You can replace a block in your program with another one by selecting it (place it at the center of your screen with the **Left** and **Right** buttons) and pressing the **Up** button. Then, choose the new block with the **Center** button, or cancel the change with the **Back** button.

deleting a block

You can delete a block from your program by selecting it, pressing the **Up** button, and selecting the trash bin, as shown in Figure B-3.

configuring a block's setting

Each block has one configurable setting. To change it, select the block you want to change and press the **Center** button, as shown in Figure B-4. Then, use the **Up** and **Down** buttons to change the setting. Confirm with the **Center** button or cancel with the **Back** button.

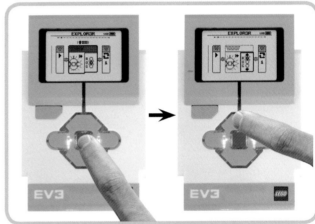

Figure B-4: Changing the setting on a block. In this example, you set the brick status light color to red (R) instead of orange (O).

running a program

Now create the *OnBrickStatus* program (see Figure B-5). Place two Brick Status Light blocks and two Wait blocks in the loop. Then, configure the program to change the light color to red, wait 2 seconds, change the light color to green, and wait 2 more seconds. Finally, change the setting of the Loop block so that it keeps repeating the blocks within it.

Now run the program by selecting the left part of the Loop block and pressing the **Center** button, as shown in Figure B-5. The brick status light should blink red for 2 seconds, blink green for 2 seconds, and continue alternating back and forth. You can abort the running program by pressing the Back button.

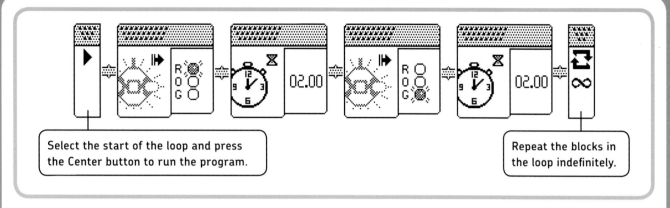

Select the start of the loop and press the Center button to run the program.

Repeat the blocks in the loop indefinitely.

Figure B-5: The OnBrickStatus *program changes the light color every 2 seconds. The full program is shown here, but you'll see only a subsection of it on the EV3 screen. Use the Left and Right buttons to verify that your program matches the figure.*

saving and opening a program

Select the **Save** symbol to save your current program, as shown in Figure B-1. On the screen, you'll see a keyboard that you can control using the EV3 buttons. Select backspace (⌫) to remove the current name, enter *OnBrickStatus* by choosing letters with the **Center** button, and then select the check mark.

To open and edit a previously saved On Brick program, select the open symbol and choose a program from the list that appears. If you just want to run your saved program, you can find it in the *BrkProg_SAVE* folder on the File Navigation tab of the EV3 brick.

If you close the Brick Program app using the Back button, you'll be prompted to save the program. Choose **X** to close without saving, or select the check mark to choose a name and save it.

NOTE You'll need to enter the name each time you save the program, so choose a short and simple name if you expect to modify your program often.

using the on brick programming blocks

Table B-1 shows which blocks are available for your On Brick programs, alongside the equivalent programming block in the EV3 software. You can choose from a selection of Action blocks or a Wait block in one of several modes. Use the page numbers in the table to find more information about each block or sensor mode.

The Brick Program app allows you to change just one setting on each block. The equivalent setting on the block in the EV3 software is marked in blue. Other settings, such as the sensor or motor ports, can't be changed. For example, if you want to use the Touch Sensor in an On Brick program, it must be plugged into input port 1.

The numeric settings on the Display and Sound blocks correspond to different image and sound files, respectively. You can find which file corresponds to each number in the *brick program app assets list* on page 55 of the EV3 user guide. (Click **User Guide** in the EV3 Software Lobby, as shown in Figure 3-2 on page 26.)

You can use these blocks to create programs similar to those you made throughout Chapters 4–6. For example, you can create a program that makes a Large Motor turn for 1 second each time you press the Touch Sensor. Attach the Touch Sensor to input port 1, attach a Large Motor to output port D, and create and run the *OnBrickTouch* program (see Figure B-6).

Note that this program makes the motor move for 1 second by switching the motor on, waiting 1 second, and then setting the motor speed to 0.

importing on brick programs

You can import a saved On Brick program into an EV3 software project by connecting the EV3 brick to the computer and clicking **Tools ▸ Import Brick Program**, as shown in Figure B-7. Choose the program on the dialog that appears, and click **Import**. The program should now appear on its own tab inside your current project.

table B-1: the blocks available in the brick program app and the equivalent blocks in the EV3 software

Action Blocks	Page	Wait Blocks (mode)	Page
Medium Motor	46	Time	49
Large Motor	46	Brick Buttons	98
Move Steering	35	Motor Rotation *Degrees*	99
Display	42	Color Sensor *Color*	78
Sound	40	Color Sensor *Reflected Light Intensity*	81
Brick Status Light	44	Infrared Sensor *Proximity*	90
		Infrared Sensor *Remote*	92
		Touch Sensor *State*	66
		Gyro Sensor* *Angle*	n/a
		Temperature Sensor* *Celsius*	n/a
		Ultrasonic Sensor* *Distance Centimeters*	n/a

* These sensors are not included in the EV3 home set (#31313). For more information about these sensors, visit *http://ev3.robotsquare.com/*.

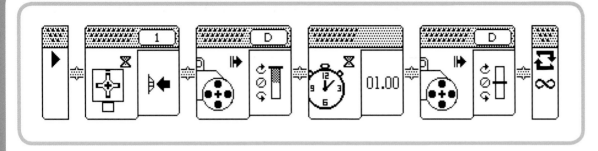

Figure B-6: The OnBrickTouch program makes the Large Motor on port D move for 1 second each time you press the Touch Sensor.

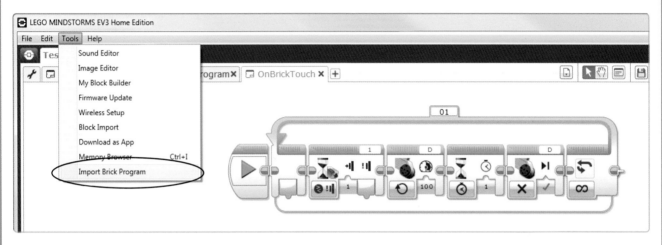

Figure B-7: Importing an On Brick program into the EV3 software

Once your program is imported into the EV3 software, you can edit it as you would any other EV3 program. As you can see, the blocks of the *OnBrickTouch* program have turned into the equivalent set of blocks in the EV3 computer software, according to Table B-1. When you run this program, it should behave just like the program you made on the EV3 brick.

Once imported into the EV3 software, a copy of the original On Brick program remains on the EV3 brick, but if you modify the program in the computer software, you can't turn it back into an On Brick program.

NOTE There is one exception to Table B-1: In an On Brick program, a Wait block in Motor Rotations – Degrees mode always resets the Rotation Sensor to 0 before it runs, but once the program is imported, it will no longer do this. You can reset the Rotation Sensor in your program if necessary with an additional block (see Figure 9-4 on page 99).

summary

You can create basic programs on the EV3 brick without a computer using the Brick Program app. Doing so can be a useful way to test a robot when you're away from a computer. On Brick programs always consist of a series of Action blocks and Wait blocks placed in one Loop block. Once you've imported an On Brick program into the EV3 software, you can edit and run it as you would a normal EV3 program.

index

The LEGO MINDSTORMS EV3 Discovery Book is set in Chevin. The book was printed and bound by Lifetouch Printing in Loves Park, Illinois. The paper is 70# MPC Silk.

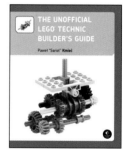